THE SMOKE OF LONDON

The Smoke of London uncovers the origins of urban air pollution, two centuries before the industrial revolution. By 1600, London was a fossil-fuelled city, its high-sulfur coal a basic necessity for the poor and a source of cheap energy for its growing manufacturing sector. The resulting smoke was found ugly and dangerous throughout the seventeenth and eighteenth centuries, leading to challenges in court, suppression by the crown, doctors' attempts to understand the nature of good air, increasing suburbanization, and changing representations of urban life in poetry and on the London stage. Neither a celebratory account of proto-environmentalism nor a declensionist narrative of degradation, *The Smoke of London* recovers the seriousness of pre-modern environmental concerns even as it explains their limits and failures. Ultimately, Londoners learned to live with their dirty air, an accommodation that re-frames the modern process of urbanization and industrial pollution, both in Britain and beyond.

WILLIAM M. CAVERT is Assistant Professor of History at the University of St. Thomas, and focuses on urban and environmental history in early modern Britain.

THE SMOKE OF LONDON

Energy and Environment in the Early Modern City

WILLIAM M. CAVERT
University of St. Thomas

CAMBRIDGE
UNIVERSITY PRESS

CAMBRIDGE
UNIVERSITY PRESS

University Printing House, Cambridge CB2 8BS, United Kingdom

One Liberty Plaza, 20th Floor, New York, NY 10006, USA

477 Williamstown Road, Port Melbourne, VIC 3207, Australia

4843/24, 2nd Floor, Ansari Road, Daryaganj, Delhi - 110002, India

79 Anson Road, #06-04/06, Singapore 079906

Cambridge University Press is part of the University of Cambridge.

It furthers the University's mission by disseminating knowledge in the pursuit of education, learning and research at the highest international levels of excellence.

www.cambridge.org
Information on this title: www.cambridge.org/9781107421318

First published 2016
First paperback edition 2017

A catalogue record for this publication is available from the British Library

Library of Congress Cataloging in Publication data
Names: Cavert, William M., author
Title: The smoke of London : energy and environment in the early modern city / William M. Cavert.
Description: Cambridge : Cambridge University Press, 2016. |
Series: Cambridge studies in early modern British history |
Includes bibliographical references and index.
Identifiers: LCCN 2016004339 | ISBN 9781107073005 (hardback)
Subjects: LCSH: Smoke – Environmental aspects – England – London – History. |
Air – Pollution – England – London – History. | Air – Pollution – Social aspects –
England – London – History. | Coal – Environmental aspects – United States.
Classification: LCC TD884.C37 2016 | DDC 363.738/70942120903–dc23
LC record available at http://lccn.loc.gov/2016004339

ISBN 978-1-107-07300-5 Hardback
ISBN 978-1-107-42131-8 Paperback

Contents

Figures

Acknowledgements

This book was written with the help of many friends, colleagues, and institutions, all of whom in their own ways were indispensible. Thanks go first to Ethan Shagan for believing in this project from the beginning and for repeatedly telling me what I needed to hear. The book's finishing touches were applied after a homecoming to my native mid-western soil, first as a guest at Grinnell College in Iowa and then as a new member of the University of St. Thomas in St. Paul, Minnesota. My home while writing the text was Clare College, Cambridge, which I thank most gratefully for welcoming me during the three years I spent there as a Junior Research Fellow. I can hardly imagine a more perfect place to research and write, a green and pleasant college in the shadow of the University Library. If this book succeeds in speaking to a range of historians, much of the credit must go to colleagues and mentors at Cambridge who have taught me so much. Thanking them all would require three years worth of daily diary entries, but those who cannot go unmentioned include Amy Erickson, John Guy, Mark Hailwood, Lawrence Klein, Craig Muldrew, John Robertson, Brodie Waddell, Alex Walsham, and Phil Withington. John Morrill has been especially important, and it is an honour to join the long list of historians who have been indebted to his generosity.

Beyond Cambridge my other intellectual home in England, both pre- and post-doctoral, has been the University of London's Institute of Historical Research, where I learned a great deal from the speakers, convenors, and audiences of the Tudor and Stuart Seminar, the British History in the Seventeenth Century Seminar, the Economic and Social History of the Early Modern World Seminar, and the Medieval and Tudor London Seminar. At least as important has been the invisible college of the after-hours seminars, where I have benefitted particularly from learned conversation with Alex Barber, Peter Lake, David Magliocco, Noah Millstone, Jason Peacey, Tim Reinke-Williams, Laura Stewart, and Elliot Vernon. Simon Healy has been exceptionally generous in sharing his formidable research and passing on archival references. I also thank John Adamson, Ian Archer, David Scott Gehring, Paul Griffiths, Robert Harkins, Mark Jenner, Fredrik Albritton Jonsson, Stuart Minson, Leona Skelton, Scott Sowerby, Koji Yamamoto, and Nuala Zahedieh for illuminating conversations and helpful references.

Before this was a book it was a thesis, and my thanks therefore go to everyone who helped me as a graduate student. At Northwestern Ken Alder, Joel Mokyr, and Edward Muir supervised my dissertation. They, along with Joe Barton, Tim Breen, Regina Grafe, Bill Heyck, Jeff Masten, Sara Maza, and Dylan Penningroth were generous mentors. Among my cohort of graduate students, Genevieve Carlton, Jason Johnson, Michael Martoccio, Lonnie Robbins, Meghan Roberts, and Strother Roberts offered much helpful criticism and encouragement, as well as abundant support and camaraderie. Karl Gunther and Sara Ross shared sage wisdom. They were joined by many other NU friends, including Charlotte, Dave, Nat, Stef, Rick, Josh, Celeste, Erik, Abram, Liz, Anna, Dan, Andy, and Bettina.

Even before Northwestern, this book was first a glint in my eye when I was a student at Loyola University Chicago. Something about living in my studio apartment above that alley suggested to me that studying the history of urban pollution might be interesting, and Robert Bucholz patiently helped me move from vague question to research agenda. Also at Loyola Robert Bireley, Anthony Cardoza, and Barbara Rosenwein provided essential guidance and encouragement. Finally, I hope it is not stretching the conventions of scholarly acknowledgements too far to thank those teachers without whose inspiration and instruction none of the above adventures would have been possible. At Carleton these were, above all, Jackson Bryce, Victoria Morse, Bill North, Susannah Ottaway, Diethelm Prowe, and Carl Weiner. And I was only able to take and to pass their classes because of the tireless work of earlier teachers in the public school system of Hopkins, MN.

Earlier versions of many of this book's chapters have been presented at several conferences, and I thank all of the organizers and audience members at: The North American Conference on British Studies (Minneapolis, Denver, and Baltimore); Stereotyping in the Public Sphere (IHR); The Country and the City Conference (Renmin, Beijing); The Medieval and Tudor London Seminar (IHR); The British History in the Seventeenth Century Seminar (IHR); The History of Science, Technology and Medicine Seminar (Imperial College, London); The Early Modern Britain and Ireland Seminar (Cambridge); The Eighteenth-Century Seminar (Cambridge); The American Historical Association Conference (Chicago); The Anglo-American Conference (IHR); The European Studies Conference (Montreal); The Social History Society Conference (Glasgow); The Newberry Library Center for Renaissance Studies Graduate Conference; The World Congress on Environmental History (Copenhagen); The Environmental History

Seminar Working Group (Birmingham); The Urban History Working Group (Warwick); The Global History Workshop (Northwestern and Cambridge); and The Workshop on Environmental History (Northwestern).

Funding to conduct and present this research has been generously provided by many benefactors, the most important of which were multi-year fellowships from Northwestern and Clare College. Additionally I gratefully acknowledge the support of: a Humanities Research Grant (Cambridge); the Harold Perkin Prize for Outstanding Doctoral Dissertation, T.W. Heyck Prize for Research in British History, and grants from the Buffett Center for International Studies, Science in Human Cultures Program, The Graduate School, and History Department Travel Grants (all Northwestern); a Dissertation Completion Fellowship from the Mellon Foundation and American Council of Learned Societies; a Huntington Library Francis Bacon Foundation Short-Term Fellowship; the Martinus Nijhoff Award in Policy Implications of Science, Horowitz Foundation for Social Policy; and a Mellon Pre-dissertation Fellowship, Council for European Studies at Columbia University. Besides needing money, historians also need books and manuscripts. I would therefore like to thank the staff at all of the archives listed in the bibliography, as well as the libraries of Clare College, Grinnell College, the Institute of Historical Research, the Newberry Library, Northwestern University Library, the Cambridge University Library, and the libraries of Cambridge's history and English faculties. Many of the printed books cited in the bibliography were first identified using digital tools including EEBO, ECCO, The Making of the Modern World, Google Books, and British History Online.

Finally I would like to thank my family. My parents and grandparents for love, support, encouragement, and (in the case of my grandfather) meticulous copy-editing. Diana and Sam for both the joy and the fun. And most of all, Katie. She has been with this book since its beginning, has heard the first rehearsals of many of its arguments and conclusions, and has moved across the ocean more than once so that I could write it. For that, and much more for everything else, this book is for her.

Conventions

Quotations from primary sources have been modernized in spelling and punctuation. Tons are imperial units of 2240 pounds. Dates are taken to begin on 1 January, except in the case of publications appearing between 1 January and 25 March, which are rendered as: 1643/4. Place of publication is London unless noted otherwise. Dates are old style before 1752. All websites were accessed on 14 July 2015.

Prologue: The Smoke of London

Among the innumerable people who arrived in eighteenth-century London, for work or pleasure, as tourists or immigrants, were two men from different corners of the British Empire. They were both born early in the century during the reign of Queen Anne, and both enjoyed long and distinguished careers before their deaths many decades later. They were nearly contemporaries, both raised in the world of printing and bookselling, both coming to the capital as young men, both achieving fame through their pens and conversation. They differed strongly from each other in politics and religion, so that while they moved in similar circles they were not friends. Their only known meeting occurred, inevitably, in London, the capital of England and of the increasingly global British Empire. It was a city that played a central role in both of their careers, and yet when they described London in writing neither tried to pretend that it was a healthy, clean, or beautiful place.

On the contrary, for both men one of the capital's defining characteristics was its dirty, smoky air. The elder man, raised in the colonies, once contrasted life in the capital with the 'sweet air' of a friend's country home in Hampshire. 'I now breathe with reluctance the smoke of London,' he wrote upon returning from 'the agreeable retirement' of the country to the city's busyness and dirt. Urban smoke, he wrote elsewhere, was 'sulphury' and rendered 'thick-built towns and cities … half suffocated', a problem that he aimed to fix through new fireplace designs. The younger man, raised in the Midlands, also thought urban living was an inherently smoky experience. One of his many essays pronounced that both brilliant talk and urban beauty were similarly rare:

A transition from an author's books to his conversation, is too often like an entrance into a large city, after a distant prospect. Remotely, we see nothing but spires and temples, and turrets of palaces, and imagine it the residence of splendor, grandeur, and magnificence; but, when we have passed the gates, we find it perplexed with narrow passages, disgraced with despicable cottages, embarrassed with obstructions, and clouded with smoke.

This author only knew one 'large city', but he knew it well. It was a place so vibrant that he famously declared that 'a man that is tired of London is tired of life, for there is in London all that life can afford'. But it was a smoky life, and was not for everyone. Our author's own beloved wife, we are told, lived in airy Hampstead, four miles north of the City, 'while her husband was drudging in the smoke of London'. Moreover the recorder of this observation, the author's Scottish companion and biographer, himself worried about a wife whose health 'renders her quite unfit to live in the smoke of London'.[1]

For authors like Benjamin Franklin, Samuel Johnson, and James Boswell, 'the smoke of London' was a useful phrase because it assimilated the city's manifold possibilities into its physical environment. For these men the foggy haze that rarely lifted from London during the best weather, and which made it quite dark for much of the winter, was a defining feature of metropolitan life, something that set London apart. As a city London was neither uniquely big nor uniquely rich nor especially grand; Paris was comparably large and boasted better monumental architecture, Amsterdam was the centre of a similarly bustling commercial network, and Madrid and Istanbul were also capitals of great empires. But London's urban environment was seen to be clearly distinct, a fact which was tolerable for Johnson and Franklin, but which for some foreign travellers was unambiguously a bad thing.

For them, London was defaced. Its buildings, wrote one German visitor, were 'blackened with the unmerciful smoke of coal-fires'. Coal smoke, for him, was 'that bane of London'. Another German found the House of Lords 'tarnished with smoke', Westminster Abbey's coronation chairs 'wretched and smoke-blackened', and items in the Royal Society's museum 'ruined by dust and smoke, so that they look utterly black and wretched'.[2] When the Finnish-Swedish naturalist Pehr Kalm stayed in London in 1748 – a few

[1] Benjamin Franklin to Jonathan Shipley, 24 June 1771; *An Account of the New Invented Pennsylvanian Fire-Places* (Philadelphia, 1744), both accessed at www.franklinpapers.org. Samuel Johnson, *The Rambler* Issue 14, 5 May 1750 in *The Yale Edition of the Works of Samuel Johnson* (New Haven, 1968), III, 79–80; George Birkbeck Hill, ed., *Boswell's Life of Johnson* (Oxford, 1887), I, 275; III, 187; Boswell to William Johnson Temple, 20 July 1784, Nellie Pottle Hankins and John Strawhorn, eds., *The Correspondence of James Boswell with James Bruce and Andrew Gibb, Overseers of the Auchinleck Estate* (New Haven, 1998), 97. For Franklin and Johnson attending a meeting together in 1761, 'Franklin, Benjamin' in *The Samuel Johnson Encyclopedia*, Pat Rogers, ed. (Westport CT, 1996), 150–1; for their antagonistic political engagement, Neill R. Yoy, 'Politics and Culture: The Dr. Franklin – Dr. Johnson Connection, with an Analogue', *Prospects* 23 (1998), 59–105.
[2] Karl Ludwig and Freiherr von Pöllnitz, *The Memoirs of Charles-Lewis, Baron de Pollnitz* (1739), II, 431, 433; W. H. Quarrell and Margaret Mare, trans. and eds., *London in 1710 From the Travels of Zacharias Conrad von Uffenbach* (1934), 74, 92, 101.

months before visiting Franklin in Pennsylvania – he could neither see St Paul's dome from Greenwich nor the rest of the City from atop St Paul's. Both were obscured by 'the thick coal smoke, which on all sides hung over the town'. Kalm complained:

> However free I was from cough when I now and again went into London from the country, I got one always as soon as I had been there a day ... but as soon as I left London and had been two days out in the country I lost my cough. All who lived far out in the country, and were not accustomed to coal smoke, even native Englishmen, had the same tale.[3]

For Kalm, coal smoke made London into a place that was difficult either to admire or to enjoy.

Another European acquaintance of Franklin's, the *philosophe* Jean-Pierre Grosley, was even more appalled by London's atmosphere. 'This smoke', he wrote in 1765, 'rolling in a thick, heavy atmosphere, forms a cloud, which envelops London like a mantle; a cloud which the sun pervades but rarely; a cloud which, recoiling back upon itself, suffers the sun to break out only now and then'. It stifled inhabitants, Grosley claimed, covered citizens with sooty rain, and blackened buildings both inside and out. And it was getting worse: 'if the increase of London proceeds as far as it may, the inhabitants must at last bid adieu to all hopes of ever seeing the sun'.[4] A less self-consciously enlightened but no less observant visitor, Atajuk from Labrador, found London in 1772 to be 'too many houses, too much smoke, too many people'.[5]

Such impressions struck visitors especially forcefully, unused as they were to any city with a pervasive smoke cloud, but the English themselves also described London as a city – even as *the* city – of smoke. In 1748, a few weeks after Kalm developed his cough, the blue-stocking (and friend of Johnson) Elizabeth Carter wrote to a friend that a summer in Enfield, 10 miles north of London, offered 'a more eligible situation than the noise and dust and smoke of a crowded city'. By the end of the summer, however, plans had changed: 'I am not so happy as to be running wild in the nettle groves of Enfield, but am panting for breath in the smoke of London'.[6] For Carter,

---

[3] Pehr Kalm, *Kalm's Account of His Visit to England on His Way to America in 1748*, Joseph Lucas, trans. (1892), 42, 138.

[4] Pierre-Jean Grosley, trans. Thomas Nugent, *A Tour to London: or, New Observations on England, and Its Inhabitants* (1772), 44.

[5] George Cartwright, *A Journal of Transactions and Events, During a Residence of Nearly Sixteen Years on the Coast of Labrador* (1792), I, 269; discussed by Coll Thrush, 'The Iceberg and the Cathedral: Encounter, Entanglement, and Isuma in Inuit London', *Journal of British Studies* 53:1 (2014), 59–79.

[6] Elizabeth Carter to Susanna Highmore, 8 June 1748, in Gwen Hampshire, ed., *Elizabeth Carter, 1717–1806: An Edition of Some Unpublished Letters* (Cranbury, NJ, 2005); Carter to Catherine Talbot, 5 August 1748, in *A Series of Letters between Mrs. Elizabeth Carter and Miss Catherine Talbot from the Year 1741 to 1770 ...* Vol. I (1808), 287.

and for her fellow intellectuals like Johnson, Boswell, Franklin, Kalm, and Grosley, much of what made London special was expressed by its material environment; this centre of human commerce and industry sat beneath a visible atmosphere of its own creation. John Evelyn's famous denunciation of urban smoke in 1661's *Fumifugium* is therefore not the isolated voice in the wilderness it is sometimes taken to be. Many claims that Evelyn developed in detail were also voiced, sometimes very earnestly, sometimes in jest, by innumerable other Londoners who could not help but notice that their city's environment had changed.

The process by which smoky air came to be a fundamental part of the image and experience of urban life in London is the subject of this book. No other city in the world contended with a similar atmosphere during the early modern period because, quite simply, no other city burned nearly as much dirty coal as London. This coal was a mineral fuel, a sedimentary rock found in the region surrounding Newcastle in north-eastern England. It was bituminous, a middle-grade coal high in energy but one whose combustion released smoke that was thick, dark, and contained concentrations of pollutants not emitted by wood fuels. Because of the concurrence of London's especially large size, its unequalled consumption of energy, and the particularly dirty nature of this fuel, England's capital achieved a dynamic unique in the early modern world but familiar thereafter: urban expansion, economic growth, and rising energy consumption contributed to each other in a series of positive feedback loops, collectively leading to environmental degradation. For a modern economist, such dirtiness is a negative externality or disamenity. During the seventeenth and eighteenth centuries, as this cluster was confronted for the first time, observers sometimes called it 'nuisance', 'noisome', 'pollution', or simply 'the smoke of London'.

Pollution has a history, and there are many ways to tell it. Environmental historians, or indeed others interested in humanity's changing relationship with its natural or built environments, have usually fit pollution's past into one of three grand narratives. For some, humans have always been wrecking their environment, and continuities and parallels are therefore as important as change over time. Others stress that before the twin arrivals of modern medical science and modern material wealth almost everybody in the past was poor and dirty but no one minded very much. A third story asserts the novelty of pollution, pointing out that problems like smoky air did not exist in significant ways before 1800, when industry transformed humanity's capacity to extract natural resources and change natural environments.

Different as they are, there is something to be said for all of these positions. Humanity has indeed been changing its natural environment for a very long time: land-clearance, burning, and hunting are far older than human civilization, and ancient cities from Rome to Xian consumed resources and

created waste on large scales.⁷ It is also true that pre-modern people probably did tolerate more dirt than their richer descendants, blessed as we are
with abundant water, chemical cleaning agents, and biomedical informa-
tion.⁸ Finally, modern economic growth raised levels of pollution to unprecedented levels all over the world. The smoky and smoggy air threatening
public health in cities like Beijing, Delhi, Cairo, and Mexico City is among
the results of this growth, as are a host of other pollutants, from nuclear
radiation to mercury to hormone disruptors. These threats to the global environment and public health have no pre-modern equivalents; this is indeed
something new under the sun.⁹

Each of these grand narratives reveal important truths, therefore, but each
is also insufficient because none of them can explain change on the scale
that happened in Europe during the centuries from 1500 to 1800. If we
stress that environmental degradation is always with us, we miss crucial
distinctions and risk lumping together very different types of environmental
intervention and stress, as if river salinization in ancient Mesopotamia were
equivalent to the Three Gorges Dam. To define the period before industrialization primarily through its lack of modernity – whether we interpret
this as a curse, since most were then poor and dirty, or as a blessing, since
none yet suffered a toxic planet – would be to erase distinctions within the
pre-modern, pre-industrial world. The centuries before 1800 witnessed too
much change, especially in England, for this approach to be reasonable. Just
as the world of William Shakespeare and Francis Drake was not the world
of Jane Austen and James Watt, so London's urban environment was profoundly different in 1800 from what it had been in the sixteenth century. The
dramatic changes that occurred during these early modern centuries have
little place either in the grand narrative of continuity or in the Manichean
dichotomy which divides history into the pre-modern and the modern. Both
approaches deny the possibility of saying much of anything about the early
modern period. This book, however, suggests that the history of the environmental changes experienced during these centuries is worth telling.

This history should be told not only as an act of recovery – written about,
as historians often say, 'on its own terms' – but also because the story of
this particular piece of the pre-modern, pre-industrial world is indispensible if we want to understand the enormous changes that came later.
Eighteenth-century England was powered by fossil fuels, a fact that was

⁷ J. Donald Hughes, *An Environmental History of the World: Humankind's Changing Role in the Community of Life* (Abingdon, 2009), ch. 2–4.
⁸ But see Mark Jenner's critiques of this narrative in 'Follow Your Nose? Smell, Smelling, and their Histories', *American Historical Review* 116 (2011), 335–51.
⁹ J. R. McNeill, *Something New Under the Sun: An Environmental History of the Twentieth-Century World* (New York, 2001).

particularly true and particularly noticeable in London. It therefore prefig-
ured in many important respects our modern world. It has been argued that
the 'fossil fuel revolution' was an epochal pivot in human affairs, comparable only to the agricultural revolution. If, as Edmund Burke III has suggested, we might divide *all* of history into two just phases, the age of solar
energy and the age of fossil fuels, then the moment of transition from the
former to the latter takes a central place within the story of humanity and
the world it occupies.[10]

From this perspective, the experience of London from its transition to coal
in the years before 1600 until the beginnings of industrialization around
1800 is not like other pre-modern experiences. There was indeed dirty air
in ancient Rome and deforestation in the Mayan Empire, but it was only
in early modern England that environmental challenges were solved in a
way that led ultimately to a new global energy regime. If we take seriously
Burke's suggestion that energy might define historical periods, then we need
to understand how and why the people of England converted to a coal-fired
society during the centuries between 1600 and 1800. British historians were
once confident that there was something special about these years, the ges-
tational period – so it has often been argued – of the modern state or parliamentary sovereignty or the British Empire or social classes or capitalism.
All of these grand narratives have been challenged, but it may perhaps be
defensible to suggest that from the perspective of energy regimes and their
relationship to environmental pollution, early modern England was, or became, 'the first modern society'.

The Smoke of London describes this transformation as a social and cultural, as well as an economic and environmental, development. It argues that
London's conversion to coal as its primary fuel led to two inter-connected
processes. First, coal produced smoke that Londoners often found ugly, unhealthy, or undesirable, and this book therefore recovers the unappreciated
and largely unknown early modern concern for urban air pollution. Second,
despite this concern Londoners' coal consumption expanded throughout
this period and beyond because it became deeply embedded in conceptions
of social stability, economic prosperity, and state power. The rise of the
coal-fuelled economy was therefore, and was seen at the time to be, also the
rise of the smoky urban atmosphere, as ultimately early modern Londoners
decided that coal brought benefits that rendered its dirtiness acceptable.
What follows is therefore an environmental history told through the experiences, ideas, conflicts, and goals of city-dwellers and their governors. It

---

[10] Edmund Burke III, 'The Big Story: Human History, Energy Regimes, the Environment,'
in Edmund Burke III and Kenneth Pomeranz, eds. *The Environment and World History*
(Berkeley, 2009), 33–53.

describes the ways that an increasingly urban and increasingly capitalist society confronted the consequences of its new energy regime, the ways that concern for air pollution led not to environmentalism but to accommodation. Many found London's smoke to be appalling, many others thought it was acceptable, and others, like the agriculturist Arthur Young, were content to embrace paradox and find it to be both at the same time. Towards the end of the eighteenth century he wrote that 'the clouds of coal smoke that envelop London' tainted the air year-round, and he launched into rapture upon escaping into the country's 'freshness and sweetness of air, the quiet and stillness, the sunshine unclouded by smoke'. And yet he could also exclaim 'thank God for the coal fires of England'.[11] Young spoke for many who found coal fires to be both blessing and curse. The smoke of London, then, was a space of multiple meanings, a symbol of a new kind of urban life with all of its grandeur and grime.

[11] Arthur Young, *Travels During the Years 1787, 1788, and 1789* (Dublin, 1793), I, 128, 503; Matilda Betham-Edwards, ed., *The Autobiography of Arthur Young, With Selections from His Correspondence* (Cambridge, 2012), 352.

Part I

Transformations

– I –

The early modernity of London

I. INTRODUCTION: *THIS BLACK DESTROYER*

During the 1680s an anonymous author was ready to solve two of London's most intractable problems, but something stood in his way. His targets were the deplorable state of public health, 'those many new but nameless diseases', and the social plague of the unemployed poor, 'those useless, idle hands which daily molest our streets and constantly pray on the labour and industry of others dwelling within this great hive the City'.[1] Luckily, something could be done. The 'ill disposition and gross temperament of the air' around London, on which he blamed its general unhealthiness, was subject to improvement. Examples of ancient and modern cities from Rome to Amsterdam showed that through 'vast toil and industry' human art could improve on nature. Not that London was naturally deficient; rather its first founders had sited it on a well-ventilated promontory above a great river, allowing the city and its inhabitants' bodies the possibility of 'a clear and good air'. London's natural advantages, however, had become a curse, as 'the great concourse of people flocking hither in trade help and further this almost unavoidable calamity, in that they occasion so much dirt and soil'. The author therefore advocated a policy whereby the idle poor, especially children and the elderly, would cleanse the dirty metropolis. They would be paid (minimally) for their work or punished if they chose idleness, a carrot-and-stick that would avoid the need for parish relief. Their honest toil would also cleanse themselves so that 'those stinking and loathsome vapors flying from their filthy garments and nasty bodies' would no longer trouble London's resident gentry.[2] This enlightened project, the author insisted, was so practical that readers should dismiss the most powerful argument against it: that any plan to improve London's air was doomed to failure because of coal smoke.

---

[1] Anon., 'Orvietan: or A Counter=Poison Against the Infectious Ayr of London,' BL Sloane MS 621, f. 4, 2v.
[2] Quotations from *ibid.*, f. 5, 5v-6, 6v, 12.

3

The obstacle in the way of this plan was that London's coal smoke, according to an opinion 'for some ages past ... current among us', was 'the only cause of polluting its air'. The author rejected this vehemently, arguing that if smoke's 'corrosive particles' really had such 'noxious effects' then why did iron degrade so quickly in the city's open air but not inside citizens' chimneys? If smoke were 'the sole cause of the corruption of the city's air', why did citizens take country houses in suburbs like Highgate, Hamstead, Chelsea, and Hackney where they were subjected to 'smoky effluvia' whenever the wind blew from the capital? He mocked the idea that 'that fuliginous and arsenical vehicle', 'those sulphurous emissaries of death', 'this black destroyer', was in any way as important as 'current' opinion thought. But this opinion, the author conceded, was very widely held, and he therefore denounced it with the anti-democratic rhetoric so familiar in political disputes. The idea 'of the smoke's being the sole destructive cause of the air in London' was 'so popular an error', 'so fond and vulgar an opinion', 'so crude and undigested an opinion', and 'so unsteady a basis to build their faith upon'. But its vulgarity and instability did not mean that only the poor, the uneducated, or the stupid perceived smoke's danger. Rather 'the judgments of persons not of the meanest apprehensions' also held this error. The Whig 'faction' itself, our author insinuated, derived its 'jealousies and fear' from 'spirits and faculties so stained and polluted' by a contaminated air.[3] This pollution he knew to proceed from dirty streets, but the fractious multitude blamed dirty urban air.

The London described in this tract was a busy, commercial place, sadly unphilosophical but brimming with potential. Its inhabitants' frustrating insistence that coal smoke was the root of their unhealthy air was matched by other philosophical failures, including the subordination of health to the desire for riches and 'getting a great estate', as well as the 'accursed' political maxims which made them 'the enemy of all order and good government'.[4] Londoners, then, were greedy, ambitious, and politically active, all of which had helped make their city the 'great hive'. Their city, for the author of *Orvietan*, was full of contradictions: a great and glorious metropolis and yet badly needing to be cleansed, a wealthy and opulent emporium yet capable of much improvement, a royal city yet subject to a contentious public sphere and riotous popular politics, a hive of 'labour and industry' yet full of the nastiness of the idle poor. It was an unfinished city, a place of possibilities, capable of wondrous renewal or further degradation.

Historians trying to make sense of London in this period have perceived a similarly Janus-faced city. It was an urban community defined by ancient

---

[3] *Ibid.*, f. 34–5, 9-9v, 39.   [4] *Ibid.*, f. 38–9.

walls and ancient privileges, and yet was also a sprawling metropolis that
seemed increasingly formless and boundless. Its wards and parishes, livery
companies and voluntary societies, alehouses and coffee houses allowed
both community and surveillance, and yet its constant traffic, constant
movement, and constant growth made it the best place in England to hide
in a crowd or to reinvent oneself. Its economy was richer, more special-
ized, arguably more capitalist, arguably more innovative, and certainly more
important to the state, than anywhere else in Britain, and yet historians
agree that it was not yet quite 'industrial'. London, in other words, has been
described, in numerous different ways, as a, even as *the*, quintessential early
modern city.

II. *POLLUTING ITS AIR*: EARLY MODERN LONDON IN ENVIRONMENTAL HISTORY

Orvietan's application of the word 'pollute' to the relationship between coal
smoke and unhealthy urban air should not be possible during the 1680s.
According to environmental historians' current narratives, both the thing
and the word 'pollution' were new in the nineteenth century. The thing was
new because the industrial revolution dramatically transformed human
abilities to manipulate nature and thereby also brought about the spectre
of environments newly subjected to large-scale dirtiness and destruction.
Industrial pollution is therefore categorically different from anything possible in the traditional economies of the pre-modern period, however mucky
and unhygienic such poor communities may have been. Pollution, in this
framing, is the environmental cost of new kinds of industrial production and
it is therefore inherently modern.[5] Air pollution has often had a special place
in this narrative, as industrial cities in nineteenth-century Britain, Europe,
and America consumed vast amounts of coal and so belched dark clouds out
of the huge smokestacks that became symbols of the new industrial city.[6]

---

[5] E.g., Joel Tarr, *The Search for the Ultimate Sink: Urban Pollution in Historical Perspective*
(Akron, OH, 1996); Christoph Bernhardt and Geneviève Massard-Guilbaud, eds., *Le Démon
Moderne: La Pollution dans les Sociétés Urbaines et Industrielle d'Europe/The Modern
Demon: Pollution in Urban and Industrial European Societies* (Clermont-Ferrand, 2002);
Thomas Le Roux, *Le Laboratoire des Pollutions Industrielles. Paris 1770–1830* (Paris,
2011); Jean-Baptiste Fressoz, *L'Apocalypse Joyeuse: Une Histoire du Risque Technologique*
(Paris, 2012).

[6] Stephen Mosley, *The Chimney of the World: A History of Smoke Pollution in Victorian
and Edwardian Manchester* (Cambridge, 2001); David Stradling, *Smokestacks and
Progressives: Environmentalists, Engineers, and Air Quality in America, 1881–1951*
(Baltimore, 2002); Melanie Dupuis, ed., *Smoke and Mirrors: The Politics and Culture of Air
Pollution* (New York, 2004); Angela Gugliotta, '"Hell with the Lid Taken Off:" A Cultural
History of Pollution – Pittsburgh' (University of Notre Dame, PhD Dissertation, 2004);
Frank Uekötter, *The Age of Smoke: Environmental Policy in Germany and the United States,
1880–1970* (Pittsburgh, 2009).

Perceptions and representations of air pollution, and the broader development of environmentalism of which they are a part, are often similarly seen as inherently modern because they responded to these material changes. A few scholars have pushed the emergence of environmentalism back into the enlightenment, arguing that eighteenth-century attempts to control and manipulate nature had to confront problems posed by nature's limits. Improvers sought to know how soils, climates, rivers, and forests worked so as to maximize productivity, often in the service of the state.[7] Other historians, focusing on the modern period, have seen environmental politics as contingent developments, rooted in particular moments, local contexts, and historically specific methods of mobilizing support.[8] But despite these varying approaches to the development of environmental awareness, concern, and activism, there remains an often unspoken assumption among modern environmental historians that these phenomena can only be in response to modern industrial capitalism.

Studies of the cultural and political history of air pollution, working within this framework, have therefore argued that industrial smoke emissions offered a challenge to modern societies. This challenge elicited various and contested responses, a series of debates through which modern notions of pollution and modern practices of environmental politics emerged. Peter Thorsheim's study of the construction of air pollution in modern Britain begins its story in the middle decades of the nineteenth century because before that it was widely believed that 'coal smoke was beneficial to health'.[9] Adam Rome similarly argues that urban smoke was not considered a problem in the United States before the late nineteenth century. He finds that 'air pollution' did not take on its modern meaning of 'the gaseous, chemical, and metallic by-products of combustion and industrial processes' until as late as the 1930s.[10] Others have offered a slightly different periodization,

[7] Richard H. Grove, *Green Imperialism: Colonial Expansion, Tropical Island Edens, and the Origins of Environmentalism, 1600–1860* (Cambridge, 1995); David Blackbourn, *The Conquest of Nature: Water, Landscape, and the Making of Modern Germany* (New York, 2006); Paul Warde, 'The Environmental History of Pre-Industrial Agriculture in Europe,' in Paul Warde and Sverker Sörlin, eds. *Nature's End: History and the Environment* (Houndmills, 2009), 70–92; Warde, 'The Invention of Sustainability', *Modern Intellectual History* 8 (2011), 153–70; Fredrik Albritton Jonsson, *Enlightenment's Frontier: The Scottish Highlands and the Origins of Environmentalism* (New Haven, 2013).

[8] E.g., Harriet Ritvo, *The Dawn of Green: Manchester, Thirlmere, and Modern Environmentalism* (Chicago, 2009); Robert W. Righter, *The Battle over Hetch Hetchy: America's Most Controversial Dam and the Birth of Modern Environmentalism* (Oxford, 2005); Gregory A. Barton, *Empire Forestry and the Origins of Environmentalism* (Cambridge, 2002).

[9] Peter Thorsheim, *Inventing Pollution: Coal, Smoke, and Culture in Britain since 1800* (Athens, OH, 2006), 3. See also p. 17 for his claim that John Evelyn was a marginal and unrepresentative figure.

[10] Adam W. Rome, 'Coming to Terms with Pollution: The Language of Environmental Reform, 1865–1915', *Environmental History* 1 (1996): 6–28, quotation from p. 6. See also Mark

finding concern over air pollution in the early decades of the nineteenth century.[11] Rome, Thorsheim, and other scholars stressing the environmental great divergence of modern from pre-modern environments clearly have important stories to tell, nor is there any doubt that both environmental change and environmental awareness were different in the nineteenth and twentieth centuries than previously. Moreover, Thorsheim may be right that at the beginning of the nineteenth century British people in general and Londoners in particular were much more concerned with rotting biological waste than with coal smoke.[12] But this was a new departure, a result of eighteenth-century studies of airs and their relationship to biological processes, not the legacy of an immobile classical natural philosophy.[13] As *Orvietan's* depiction of smoke-obsessed citizens in the 1680s suggests, there is in fact a rich story of environmental change and environmental concern to be told about London before the industrial revolution.

While environmental history is sometimes seen as an almost intrinsically modern field, scholars have also developed an increasingly wide range of approaches to the human relationship with nature before the industrial revolution. Nature has been central to recent explorations of contacts between Europeans, Africans, Americans, and Asians in the early modern centuries, as the 'Columbian Exchange' and subsequent movements of peoples, goods, plants, and microbes transformed the world.[14] John Richards' monumental survey of the early modern world showed how very different societies and polities, across and beyond Eurasia, came to exploit nature in parallel ways.[15] Geoffrey Parker has argued that global early modern political history can be explained by the Little Ice Age, as bad weather and poor harvests

Whitehead, *State, Science and the Skies: Governmentalities of the British Atmosphere* (Oxford, 2012), which makes 1843 its point of departure.

[11] Lee Jackson, *Dirty Old London: The Victorian Fight Against Filth* (New Haven, 2014), ch. 9; Ayuka Kasuga, 'Views of Smoke in England, 1800–1830' (University of Nottingham, PhD Thesis, 2013).

[12] Thorsheim, *Inventing Pollution*, 10.

[13] See, for example, Simon Schaffer, 'Measuring Virtue: Eudiometry, Enlightenment, and Pneumatic Medicine,' in Roger French and Andrew Cunningham, eds. *The Medical Enlightenment of the Eighteenth Century* (Cambridge, 1990), 281–318. The persistent influence of classical environmental thought, as well as the diversity of this tradition, is masterfully surveyed in Clarence J. Glacken, *Traces on the Rhodian Shore: Nature and Culture in Western Thought from Ancient Times to the Eighteenth Century* (Berkeley, 1967).

[14] Alfred Crosby, *The Columbian Exchange: Biological and Cultural Consequences of 1492* (Westport, CT, 1972). Leading examples include William Cronon, *Changes in the Land: Indians, Colonists, and the Ecology of New England* (New York, 1983); Elinor G. K. Melville, *A Plague of Sheep: Environmental Consequences of the Conquest of Mexico* (Cambridge, 1994); J. R. McNeill, *Mosquito Empires: Ecology and War in the Greater Caribbean, 1620–1914* (Cambridge, 2010); James C. McCann, *Maize and Grace: Africa's Encounter with a New World Crop, 1500–2000* (Cambridge, MA, 2005).

[15] John F. Richards, *The Unending Frontier: An Environmental History of the Early Modern World* (Berkeley, 2005).

led to rebellion and revolution across Eurasia in the mid-seventeenth century.[16] From Egypt to Japan early modern historians have found crucial environmental aspects to older narratives of state building and imperial expansion.[17] Within Europe early modernists have examined the relationships between environmental management and state formation in Spain and Venice, while medieval environmental history is the subject of an excellent recent synthesis that reflects vigorous and multifaceted expansion.[18] A few have even described pre-industrial pollution problems and laws enacted to combat them, though without trying to show the extent of such concern or how it changed over time.[19]

Collectively, this work presents a picture of an early modern world in which natural and man-made environments changed frequently and in which people responded to such change in sophisticated and interesting ways. Environmental history is clearly not, therefore, an exclusively modern subject. While certain aspects of the modern concern with toxic pollutants indeed are specific to the recent past, broader problems of urban waste disposal and perceptions of cleanliness have much deeper and richer histories. Many studies of this draw on, or fade into, the history of medicine. In doing so, much pre-modern environmental historiography has differed from Mary Douglas, whose structuralist approach to pollution explicitly rejected the

[16] Geoffrey Parker, *Global Crisis: War, Climate Change, and Catastrophe in the Seventeenth Century* (New Haven, 2012). For a more nuanced exploration of the relationship between climate change and political instability, Sam White, *The Climate of Rebellion in the Early Modern Ottoman Empire* (Cambridge, 2011). For an interrogation of Parker's thesis in the European context, see Paul Warde, 'Global Crisis of Global Coincidence?' *Past and Present* 228 (2015), 287–301.

[17] Anglophone examples include Alain Mikhail, *Nature and Empire in Ottoman Egypt: An Environmental History* (Cambridge, 2011); Peter Perdue, *China Marches West: The Qing Conquest of Central Eurasia* (Cambridge, MA, 2005); Mark Elvin, *Retreat of the Elephants: An Environmental History of China* (New Haven, 2004); Conrad Totman, *The Green Archipelago: Forestry in Pre-Industrial Japan* (Berkeley, 1989).

[18] Karl Appuhn, *A Forest on the Sea: Environmental Expertise in Renaissance Venice* (Baltimore, 2009); John T. Wing, *Roots of Empire: Forests and State Power in Early Modern Spain, c.1500–1750* (Leiden, 2015); Richard C. Hoffmann, *An Environmental History of Medieval Europe* (Cambridge, 2014). See also Scott G. Bruce, ed., *Ecologies and Economies in Medieval and Early Modern Europe: Studies in Environmental History for Richard C. Hoffmann* (Leiden, 2010).

[19] For medieval and early modern England, William H. Te Brake, 'Air Pollution and Fuel Crises in Preindustrial London, 1250–1650', *Technology and Culture* 16 (1975), 337–59; Keith Thomas, *Man and the Natural World: Changing Attitudes in England 1500–1800* (New York, 1983), 243–54; Emily Cockayne, *Hubbub: Filth, Noise, and Stench in England 1660–1770* (New Haven, 2007); Peter Brimblecombe, *The Big Smoke: A History of Air Pollution in London Since Medieval Times* (1987). See also Michael Stolberg, *Ein Recht auf saubere Luft? Umweltkonflikte am Beginn des Industriezeitalters* (Erlangen, 1994), 18–23; Richard W. Unger, 'Energy Sources for the Dutch Golden Age: Peat, Wind, and Coal', *Research in Economic History* 9 (1984), 225; Conrad Totman, *Japan: An Environmental History* (2014), 174.

sufficiency of 'medical materialism', stressing instead the role of the body as an image of the community and its social order.[20] But, as Mark Jenner has most forcefully demonstrated, medieval and early modern European understandings of health are not explicable without reference to medicine.[21] Jenner has therefore led the way in showing how cultural analysis must incorporate medical thought and professional practice in order to understand how and why early modern English people cleaned their streets, buried their dead, emptied their cesspits, and assessed their smoky capital.[22] Similar studies of pre-modern conceptions of dirtiness often focus on public health, prevention of plague, and the importance of airs in the classical medical tradition. In so doing they have shown that popular and elite conceptions of healthy living and healthy spaces overlapped to a substantial degree, as medical doctrines related to washing, scouring, and cleansing influenced both individual and collective behaviour.[23]

---

[20] Douglas, *Purity and Danger: An Analysis of Concepts of Pollution and Taboo* (New York, 2006). Douglas is famous for the idea that pollution is 'matter out of place', but this phrase (borrowed from William James) is the beginning, not the conclusion of her analysis. She argued most fundamentally that pollution is that which threatens social relations and conceptions of order, and is ultimately concerned with 'bodily disintegration' (p. 213) and death. She suggested in chapter 2 that medical materialism *and* conceptions of order structure notions of pollution in both modern and 'primitive' cultures, but the discussion is almost entirely limited to the primitive. She argued that to understand modern pollution as deriving from an ordered system we would need first to 'abstract pathogenicity and hygiene from our notion of dirt'. (44) Rejoining the symbolic and the medical in an analysis of modern cultures seems not to be possible because modernity produces 'disjointed, separate areas of existence'. (50) Thus while 'matter out of place' is a memorable formulation, it is not at all clear that her ideas are easily compatible with the concern for medicine and science that has informed most early modern and modern historians' studies of environmental pollution.

[21] Cf. the economic determinism of Bas van Bavel and Oscar Gelderblom, 'The Economic Origins of Cleanliness in the Dutch Golden Age', *Past and Present* 205 (2009): 41–69, which critiques the cultural approach of Simon Schama, *The Embarrassment of Riches: An Interpretation of Dutch Culture in the Golden Age* (Berkeley, 1988), esp. ch. 6, i, 'Cleanliness and Godliness', 375–97.

[22] Mark S. R. Jenner, 'Early Modern Conceptions of Cleanliness and Dirt as Reflected in the Environmental Regulation of London, c. 1530–1700' (Oxford: D. Phil Thesis, 1992); '"Another *epocha*"? Hartlib, John Lanyon and the Improvement of London in the 1650s,' in Mark Greengrass, Michael Leslie, and Timothy Raylor, ed. *Samuel Hartlib and the Universal Reformation: Studies in Intellectual Communication*, (Cambridge, 1994), 343–56; 'The Politics of London Air: John Evelyn's *Fumifugium* and the Restoration', *The Historical Journal*, 38 (1995): 535–51; 'Death, Decomposition and Dechristianisation? Public Health and Church Burial in Eighteenth-Century England', *English Historical Review* 120 (2005), 615–32; 'Follow Your Nose? Smell, Smelling, and Their Histories', *American Historical Review* 116 (2011): 335–51; 'Polite and Excremental Labour: Selling Sanitary Services in London, 1650–1830', paper at the Cambridge Early Medicine Seminar, November 2013.

[23] Guy Geltner, 'Healthscaping a Medieval City: Lucca's Curia Viarum and the Future of Public Health History', *Urban History* 40:3 (2013), 395–415; Dolly Jørgensen, '"All Good Rule of the Citee": Sanitation and Civic Government in England, 1400–1600', *Journal of Urban History* 36.3 (2010), 300–15; Leona Skelton, *Sanitation in Urban Britain, 1560–1700* (2015); Keith Thomas, 'Cleanliness and Godliness in Early Modern England,' in Anthony Fletcher and Peter Roberts, eds. *Religion, Culture, and Society in Early Modern Britain: Essays in*

In the case of late-medieval England, both wood and coal smoke were among the nuisances frequently regulated in towns in the interests of public health and beauty.[24]

The environmental history of early modern Europe can be told, therefore, but it must differ in crucial ways from the modern stories that dominate the field. The concern with wilderness that has been so fundamental in American and some other literatures simply does not apply in places like Britain where dense populations had farmed and hunted for millennia.[25] Early modern perceptions of nature entailed inherently moralizing attempts to order individuals into societies, as Douglas's approach suggests. But they were also based in learned medical and natural philosophical traditions that sought to explain health and disease. Early modern conceptions of cleanliness and dirt, therefore, stand somewhere between what some historians have identified as the modern attitude, characterized by scientific and technical language, and the 'primitive' attitude described by Douglas, which has nothing to do with 'science' but rather is focused on symbolic systems and social orders. In the early modern period, recent work suggests, urban dirtiness offended both morally and medically.[26]

These ideas, finally and perhaps most importantly, are inseparable from the social as well as legal, political, and institutional contexts in which they were expressed. The story of attitudes towards coal smoke in early modern London is therefore also the story of when such attitudes were voiced, to what purpose, and through what mediating structures and genres. As Emily Cockayne has argued, daily life in early modern English cities required the negotiation of endless annoyances and nuisances, any of which could threaten the crucial bonds of neighbourliness and community.[27] If Douglas was right that pollution is always, at least in part, about social order, then it makes sense that perceptions of pollution would be complex, variable, and contested in a city where social relations and social identities were particularly subject to negotiation and re-invention. If pollution is a certain kind of matter out of place, then understanding the meaning of urban coal

Honour of Patrick Collinson (Cambridge, 1994), 56–83; Sandra Cavallo and Tessa Storey, *Healthy Living in Late Renaissance Italy* (Oxford, 2013).

[24] Carole Rawcliffe, *Urban Bodies: Communal Health in Late Medieval English Towns and Cities* (Woodbridge, 2013), 163–9.

[25] E.g., Roderick Nash, *Wilderness and the American Mind* (New Haven, 1967). The classic critique of this focus is William Cronon, 'The Trouble with Wilderness: or, Getting Back to the Wrong Nature,' in William Cronon, ed. *Uncommon Ground: Rethinking the Human Place in Nature* (New York, 1996), 69–90.

[26] Relationships between the religious or moral implications of the word 'pollution' and other terms to assess material dirtiness in early modern France is explored in Patrick Fournier, 'De la souillure a la pollution, un essai d'interpretation des origines de l'idee de pollution,' in Bernhardt and Massard-Guilbaud, eds. *Le Démon Moderne*, 33–56.

[27] Cockayne, *Hubbub*.

smoke demands close attention to the rapidly changing places of early modern London.

III. *THE GREAT HIVE*: LONDON IN EARLY MODERNITY

London is not a case study in how polluted early modern cities could be. Rather, it was a place that mattered so much precisely because it was so atypical. No other city in the early modern world so dominated its country the way that London dominated the urban landscape of England. Edo, the world's largest city for most of the seventeenth and eighteenth centuries, shared pre-eminence in Japan with Osaka and Kyoto. Other great capitals sat atop networks of comparably important urban centres: Beijing was joined by Nanjing and Shanghai; Agra by Shajahanabad/Delhi, Lahore, and Surat; Istanbul by Cairo and Aleppo. All of these capital cities, which held from 500,000 to 1,000,000 people at some point between 1500 and 1800, were complemented by other major urban centres of over 100,000.[28] Paris, always the dominant city in France, was joined by a series of regional centres like Lyon, Rouen, Marseilles, Bordeaux, Toulouse, Orléans, Lille, Nantes, and Rennes, each with 40,000–100,000 people in 1700.[29] Even in the Kingdom of Naples, where no urban centre came close to the capital in either size or importance, Lecce reached around 30,000 people in 1600, more than 1/10 the population of Naples itself.[30]

London, in stark contrast to all of these cases except Naples, stood entirely alone as the only great city in early modern England, almost twenty times larger than the second largest English city around 1700. In that year, London contained something over 500,000 people, while the next largest urban centres in England were Norwich with 30,000, Bristol with 21,000, and Newcastle, Exeter, and York each with 10–20,000 people.[31] There are many ways to express the magnitude of this gulf: metropolitan London contained within it several parishes that would have been England's second largest city; if the immigrants who arrived in London in the year 1700 had instead founded their own town it would have immediately ranked among the ten largest cities in England; the same can be said of the number of

---

[28] Population figures taken from William T. Rowe, 'China: 1300–1900,' in Peter Clark, ed. *The Oxford Handbook of Cities in World History* (Oxford, 2013), 310–27; Ebru Boyar, 'The Ottoman City: 1500–1800,' in *ibid.*, 275–91; James McClain, 'Japan's Pre-Modern Urbanism,' in *ibid.*, 328–45; Tapan Raychaudhuri and Irfan Habib, eds., *The Cambridge Economic History of India. Vol. 1: c. 1200-c. 1750* (Cambridge, 1982), 171.

[29] Philip Benedict, *Cities and Social Change in Early Modern France* (1989), 24.

[30] Brigitte Marin, 'Town and Country in the Kingdom of Naples,' in S. R. Epstein, ed. *Town and Country in Europe, 1300–1800* (Cambridge, 2004), 319–21.

[31] Paul Slack, 'Great and Good Towns 1540–1700,' in Peter Clark, ed. *The Cambridge Urban History of Britain. Volume II 1540–1840* (Cambridge, 2000), 352.

babies christened in 1700 (14,600), or of the number of Londoners over age 80, or its number of naturally left-handed girls; and so on.[32] This division of urban England into one huge metropolis and a series of large provincial towns meant that within England there was no general category of the urban. Instead there was London, and there was everywhere else.

For some historians this has meant that London was absolutely central to England's transition from a medieval to a modern society. Roy Porter's many writings presented this position vividly, arguing that eighteenth century fashion, polite culture, and enlightenment itself were primarily metropolitan in origin and orientation.[33] Jürgen Habermas's *Structural Transformation of the Public Sphere* described a 'model case of British development' in which the key institutions *c.* 1700 were emphatically metropolitan, especially the coffee house and the periodical press.[34] Since the English translation of Habermas in 1989, publics and news media have become crucial to historical debates about the relationships between social and political change during the seventeenth and early-eighteenth centuries.[35] London has been central in much of this work, though often historians have been reluctant to frame their accounts as merely metropolitan. Instead, studies of coffee houses or parliamentary lobbying, for example, have stressed the national spread of news culture and political engagement rather than the outsized importance of London to their stories.[36] Despite that, studies of news circulation and public opinion have continued to show the centrality of London to these processes. Provincial readers consumed books, newspapers, pamphlets, and manuscript newsletters produced in the metropolis so as to understand and perhaps even pull the capital's levers of power.[37] In narratives describing an increasingly self-aware public, an

---

[32] Christening total from Thomas Birch, ed., *A Collection of the Yearly Bills of Mortality, from 1657 to 1758 Inclusive* (1759). Edmund Halley found that 4 per cent of Breslau births lived to 80, but a proportion of only 1.7 per cent would have sufficed to give over 10,000 octogenarian Londoners in 1700. Andrea Rusnock, *Vital Accounts: Quantifying Health and Population in Eighteenth-Century England and France* (Cambridge, 2002), 186.

[33] E.g., Roy Porter, *Enlightenment: Britain and the Creation of the Modern World* (2000); *English Society in the Eighteenth Century* (1991); *London: A Social History* (Cambridge, MA, 2001), ch. 5–7.

[34] Jürgen Habermas, *The Structural Transformation of the Public Sphere: An Inquiry into a Category of Bourgeois Society*, Thomas Burger trans. (Cambridge, MA, 1989), 57–68.

[35] E.g., Steve Pincus, '"Coffee Politicians Does Create": Coffeehouses and Restoration Political Culture', *Journal of Modern History* 67:4 (December 1995), 807–34; Brian Cowan, *The Social Life of Coffee: The Emergence of the British Coffeehouse* (New Haven, 2005); Peter Lake and Steven Pincus, eds., *The Politics of the Public Sphere in Early Modern England* (Manchester, 2007); Brian Cowan, 'Geoffrey Holmes and the Public Sphere: Augustan Historiography from the Post-Namierite to the Post-Habermasian', *Parliamentary History* 28:1 (February 2009), 166–78, which points out that British historians' interest in news culture long pre-dates Habermas's translation into English.

[36] E.g., Cowan, *Social Life*, esp. 154–84; Pincus, 'Coffee', 811.

[37] Jason Peacey, *Print and Public Politics in the English Revolution* (Cambridge, 2013); Chris Kyle, *Theater of State: Parliament and Political Culture in Early Stuart Britain* (Stanford, 2012); Alex Barber, '"It Is Not Easy What to Say of Our Condition, Much Less to Write

ideologically divided political nation, or a growing fiscal-military state, crucial innovations in early modern politics have been found to have centred on or developed in London.[38]

The same cannot quite be said, somewhat surprisingly, for recent explorations of social and economic history. E. A. Wrigley argued in a classic article of 1967 that London's gravitational pull transformed much of rural, as well as urban, England. The vast demands of its markets motivated producers to specialize, thereby facilitating improved transport, communications, and a generally more integrated economy and society.[39] Such integration and commercialization, Wrigley argued, allowed for rising domestic consumption and therefore production, leading ultimately to the industrial revolution. More recently, however, London's place in narratives of economic change has diminished. Wrightson, while not ignoring London's primacy, focused on the dynamism of regional manufacturing cities during the century after 1650.[40] Joel Mokyr suggests that if Wrigley's thesis may have some purchase before 1750, during industrialisation proper the real story is London's marginality compared to the booming and innovative regions of the north.[41] Steven Pincus's account of urban growth, commercialization, and industrial production in the later-seventeenth century notes London's dominance but gives more space to the growth of much smaller centres, from expanding towns like Newcastle and Sheffield to tiny resorts like Buxton and Harrogate.[42] Other historians have also questioned to what extent provincial centres were really merely aping metropolitan fashions, arguing instead for distinctive local impulses behind the 'urban Renaissance' of the long eighteenth century.[43] In such work Wrigley's

---

It": The Continued Importance of Scribal News in the Early 18th Century', *Parliamentary History* 32:2 (2013), 293–316.

[38] For London's importance to state finance and bureaucracy, see for example John Brewer, *The Sinews of Power: War, Money and the English State, 1688–1783* (Cambridge, 1988); D. W. Jones, *War and Economy in the Age of William III and Marlborough* (Oxford, 1988); D'Maris Coffman, *Excise Taxation and the Origins of Public Debt* (Basingstoke, 2013); Peter Temin and Hans-Joachim Voth, *Prometheus Shackled: Goldsmith Banks and England's Financial Revolution after 1700* (Oxford, 2012). For London's centrality to popular politics in the age of party division see Tim Harris, *The Politics of the London Crowd in the Reign of Charles II* (Cambridge, 1987), as well as his subsequent monographs; Mark Knights, *Politics and Opinion in Crisis, 1678–81* (Cambridge, 1994); Gary De Krey, *London and the Restoration, 1659–1683* (Cambridge, 2005); *A Fractured Society: The Politics of London in the First Age of Party, 1688–1715* (Oxford, 1985).

[39] E. A. Wrigley, 'A Simple Model of London's Importance in Changing English Society and Economy 1650–1750', *Past and Present* 37 (1967), 44–70.

[40] Keith Wrightson, *Earthly Necessities: Economic Lives in Early Modern Britain* (New Haven, 2000), 235–6.

[41] Joel Mokyr, *The Enlightened Economy: An Economic History of Britain 1700–1850* (New Haven, 2009), 302.

[42] Steve Pincus *1688: The First Modern Revolution* (New Haven, 2009), 59–68.

[43] Peter Borsay, *The English Urban Renaissance: Culture and Society in the Provincial Town, 1660–1760* (Oxford, 1989); revised by Rosemary Sweet, *The Writing of Urban Histories in Eighteenth-Century England* (Oxford, 1997); Kathleen Wilson, *The Sense of the People: Politics, Culture, and Imperialism in England, 1715–1785* (Cambridge, 1995);

vision of a modernizing economy in which all roads lead to London is difficult
to discern.

Yet studies of social life in the capital itself have frequently, perhaps even
increasingly, argued for the distinctive and dynamic aspects of the urban
experience. Historians of women's experience and gender relations, for
example, have found London an ideal laboratory in which to examine what
happens when the established networks and restraints of town and village
life are absent. About 70–80 per cent of early modern London's young
women were new arrivals in the capital and so had to make a living, form a
family, make friends, secure a good reputation, get along with neighbours,
and generally transform themselves from provincial girls into city women.[44]
The metropolis was uniquely endowed with venues to enjoy a performance,
share a drink, mix with one's social superiors or inferiors, behold the latest
fashions, and taste foods imported from across the globe or grown to perfection in local gardens.[45] Theatre was not merely seen in London, it was also,
sometimes explicitly and very often implicitly, about the city (or the 'town'
of the West End) itself.[46] Science, including both learned natural philosophy
and the technical arts, flourished there as nowhere else in Britain, allowing
for an increasingly symbiotic relationship between theory and practice.[47]
A series of fairs provided fun and temptation for both rich and poor, and
illegal or socially marginal communities like sex workers, homosexuals,
and criminal networks abounded across the metropolis in ways unthink-
able elsewhere.[48] London was by far England's most cosmopolitan place,
home to thousands of Protestant immigrants from Western Europe and to

Jon Stobbart, *Sugar and Spice: Grocers and Groceries in Provincial England 1650–1830* (Oxford, 2012).

[44] Eleanor Hubbard, *City Women: Money, Sex, and the Social Order in Early Modern London* (Oxford, 2012); Tim Reinke-Williams, *Women, Work and Sociability in Early Modern London* (Basingstoke, 2014), and the earlier studies cited by them.

[45] John Brewer, *The Pleasures of the Imagination: English Culture in the Eighteenth Century* (New York, 1997), 28–55 *et passim*; Hannah Grieg, *The Beau Monde: Fashionable Society in Georgian London* (Oxford, 2013). For the cutting edge horticulture of subur-ban London's market gardens, Malcolm Thick, *The Neat House Gardens: Early Market Gardening Around London* (Totnes, 1998).

[46] E.g., Jean E. Howard, *Theater of a City: The Spaces of London Comedy, 1598–1642* (Philadelphia, 2007); Karen Newman, *Cultural Capitals: Early Modern London and Paris* (Princeton, 2007); Ian Munro, *The Figure of the Crowd in Early Modern London: The City and Its Double* (Basingstoke, 2005); Lawrence Manley, ed., *The Cambridge Companion to the Literature of London* (Cambridge, 2011), ch. 2–6.

[47] Deborah Harkness, *The Jewel House* (New Haven, 2007); Larry Stewart, *The Rise of Public Science: Rhetoric, Technology, and Natural Philosophy in Newtonian Britain, 1660–1750* (Cambridge, 1992).

[48] Anne Wohlcke, *The 'Perpetual Fair': Gender, Disorder, and Urban Amusement in Eighteenth-Century London* (Manchester, 2014); Jerry White, *London in the Eighteenth Century. A Great and Monstrous Thing* (2013), ch. 8–10; Faramerz Dabhoiwala, *The Origins of Sex: A History of the First Sexual Revolution* (Oxford, 2012).

Scots and Irish, Jews and Turks, Africans both slave and free, Americans and Asians.[49] It was the hub of England's rapidly expanding postal system, whose merchants had investments, employees, contacts, friends, and even relatives all over the world, from the Caribbean and North America to the Levant and Madras and beyond.[50] Its neighbourhoods contained central offices of government and royal palaces, its status as a capital erasing distinctions between local and national politics.[51] Several of the 'classic signifiers of modernity', from courts, bureaucracies, stock exchanges, and banks to hospitals, prisons, pleasure gardens, and boulevards, were located within and so fundamentally shaped the uniquely urban experience of early modern London.[52] Britain's capital, in such histories, was more heavily governed and surveilled, more intricately networked, and more abundantly and variously supplied than anywhere else, certainly in England, possibly in Europe, and arguably in the world. It was therefore a creative destroyer of manners and ideas, practices and policies, a great workshop always producing a world that was visibly different.

IV. CONCLUSION: THE SOCIAL HISTORY OF URBAN POLLUTION

In such ways London was inescapably central to the political, economic, cultural, and social life of its nation during the early modern period. This centrality produced distinctively urban experiences that might reasonably be described as modern. London's precocious modernity – or, more neutrally, its status as the capital of so much that seemed new – was embodied in its physical environment. Its endlessly sprawling streets, its seas of houses, its unequalled ports and markets, the busy traffic in its great river, and its worsening difficulties with sanitation, sewage, and water supply all marked urban spaces as sites of exuberant but also unmanageable growth. The smoky air derived from heavy coal burning was another such embodiment,

---

[49] Lien Bich Luu, *Immigrants and the Industries of London, 1500–1700* (Aldershot, 2005); Jacob Selwood, *Diversity and Difference in Early Modern London* (Farnham, 2010); White, *London*, ch. 4.

[50] Lindsay O'Neill, *The Opened Letter: Networking in the Early Modern British World* (Philadelphia, 2014); Nuala Zahedieh, *The Capital and the Colonies: London and the Atlantic Colonies* (Cambridge, 2010); David Hancock, *Citizens of the World: London Merchants and the Integration of the British Atlantic Community, 1735–1785* (Cambridge, 1995); Miles Ogborn, *Global Lives: Britain and the World 1550–1800* (Cambridge, 2008).

[51] Julia Merritt, *The Social World of Early Modern Westminster: Abbey, Court and Community, 1525–1640* (Manchester, 2005); *Westminster 1640–60: A Royal City in a Time of Revolution* (Manchester, 2013).

[52] The phrase is from Robert Batchelor, *London: The Selden Map and the Making of a Global City, 1549–1689* (Chicago, 2014), 22, but the list also draws on Miles Ogborn, *Spaces of Modernity: London's Geographies, 1680–1780* (1998); Paul Griffiths, *Lost Londons: Crime, Change, and Control in the Capital City 1550–1640* (Cambridge, 2008).

a material result of specifically metropolitan consumption that served to mark out the city even as it obscured it.

To ask, as an environmental historian might do, what such smoke meant to contemporaries, is therefore also to ask, with early modern social historians, about the meanings of consumption and space in a capital city. Answering these questions is the goal of what follows. Consuming coal, which as the next chapter shows Londoners did on a scale unmatched in the early modern world, was freighted with social, economic, and political implications. London's coal smoke, which Chapter 3 shows to have been significantly harmful to its inhabitants' health, was found objectionable in a variety of contexts. Chapters 4 and 5 show how particular visions of royal space led to attempts to restrict heavy coal burning in London's most politically symbolic places. Chapter 6 explores the many ways in which the period's learned natural philosophers and physicians assessed urban air as they applied new methods and theories to, among other topics, the healthiness of England's capital. The following section describes how burning coal became an essential aspect of London life, a vital component, so it seemed to contemporaries, of social stability, commercial progress, and state power. London's status as both capital and metropolis is crucial here, as problems with its market quickly became political concerns for royal officers and Members of Parliament. Coal smoke was thus thought by many to be ugly and/or unhealthy, even as there was increasing agreement that London could not do without it. The final section examines how these contradictory impulses led to real but limited changes in behaviour. John Evelyn's great 1661 denunciation of London smoke, *Fumifugium*, failed to reform the city but did influence the king to check his own exposure to pollution. As poets, playwrights, and essayists increasingly described urban smoke as intrinsic to urban life, ways to avoid or escape the city proliferated. Instead of either embracing or eliminating urban smoke, therefore, Londoners accommodated themselves to it. In so doing, the epilogue suggests, they provided a language and a set of behaviours that helped to frame later responses to urban growth during the industrial revolution. By the arrival of the dark Satanic mills of the nineteenth century, England's capital had over two centuries of experience wrestling with the consequences of being a fossil-fuelled city.

—2—

Fires: London's turn to coal, 1575–1775

Plants died, generation after generation, for millions of years. They died in swamps where they could not decay, instead sinking and accumulating into layers of peat that were eventually squeezed and pressed into harder and drier layers of coal. Three-hundred million years later, about 1,100 years before the present, the ground above those long-dead plants was within the area claimed by the first kings of England. Because of these two developments, one geological and one historical, we can talk about England's coal deposits. People living above and on these deposits, most easily accessible along the valleys of the Tyne and Wear Rivers near to the city of Newcastle, had been using them for unknown centuries when they started to ship some southwards during the thirteenth century. When England's population was halved by the Black Death of the fourteenth century that trade declined until population recovered during the sixteenth century. The growing city of London, where so many English people came during the long reign of Queen Elizabeth, soon used more of this coal than anywhere else. The consequences of this use are the subject of this book.

Coal, London's people found, was a remarkable resource – we could equally call it a gift – but a gift that also brought heavy costs. Many of these costs fell on those who earned their livings transforming the coal inside the ground into something that could be burned in a fireplace or furnace. Coal mining offered wages to people in areas with cold climates, poor soils and no large cities, but it must have been nearly intolerable work. Hewers descended into the dark earth – increasingly far downwards as mines were progressively deepened – crouching and lying down as they hacked the coal away. Others, often their wives or children, then carried or pulled it in baskets upwards towards the light of day, returning again and again for more.[1]

---

[1] Mining, including the broad range of tasks and skills hidden by that general term, is treated by John Hatcher, *The History of the British Coal Industry. Volume 1 Before 1700: Toward the Age of Coal* (Oxford, 1993), ch. 6, 9–11; Michael W. Flinn, *The History of the British Coal Industry. Volume 2 1700-1830: The Industrial Revolution* (Oxford, 1984), ch. 10–12; J. U.

Under the best of conditions this must have been very hard work indeed, and conditions must have been very often beset by difficulties like bumpy or wet surfaces. Sometimes, and it is impossible to know how often, roofs collapsed or gases caught fire. We know of a few such mine disasters because outsiders or visitors described them, but most must have been unrecorded.

Whether because such accidents and such work were rarely observed and publicized, or because consumers preferred not to think about the labour upon which their consumption depended, or because such hardship was nearly ubiquitous in the pre-modern world and therefore unremarkable, Londoners who burned mineral coal seem to have spared very few thoughts for miners. Nor were they very interested in the sweat of the people who led wagons from the pithead to the riverside, who shovelled endless amounts of coal from river craft into coasting vessels, or who braved the North Sea and sometimes the guns of enemy ships to transport all of this coal into the Thames. For Londoners, such work was overlooked and not discussed. The rest of this book, following early modern Londoners, also largely ignores these workers. This allows it to focus on the metropolitan experience without distraction, and is defensible on the grounds that it follows the lead of its subjects, early modern people who rarely considered production and consumption together. But it should be remembered throughout that this story of coal consumption and its consequences is missing something, that there can be no consumption without production, no using without making.

The other great cost that coal burning entailed was the smoke and pollutants it released into the air. These, this book will show, were far from invisible to contemporaries. But before the consequences of coal comes its consumption. When, why, and to what extent did people in London use coal during the sixteenth through eighteenth centuries? If coal was a gift as well as a curse to the people who used it, what boons did it bring?

II. FUEL SCARCITY AND THE PHOTOSYNTHETIC CONSTRAINT

The most ambitious history of the British coal industry – indeed one of the most ambitious works ever written on the development of capitalism in early modern Britain – was John U. Nef's 1932 *The Rise of the British Coal Industry*. For Nef, the rise of British coal was at once the rise of capitalism and of the modern nexus of industry and resource extraction. This was part of the reason why he so confidently argued for an 'early industrial revolution' in England around 1600, some two centuries before conventional

Nef, *The Rise of the British Coal Industry*, 2 vols. (1932), I, 411–29, and II, 135–97; David Levine and Keith Wrightson, *The Making of an Industrial Society: Whickham, 1560–1765* (Oxford, 1991).

periodization would place it. For Nef, this revolution entailed a multifaceted set of transformations to property rights, state policy, and the role of urban capitalists. These were all prompted by a material problem, the pervasive timber scarcity 'which we may describe as a national crisis without laying ourselves open to a charge of exaggeration'.[2] Coal, for Nef, was a national solution to a national problem, a response to a resource bottleneck that allowed for the social, political, and economic modernization that, in his view, was the fundamental story of early modern English/British history.

Nef's confidence regarding his immunity from 'a charge of exaggeration' could hardly have been more misplaced, as a generation of economic historians attacked his work from several directions. The 'early industrial revolution' thesis, for example, was rejected by those who pointed out that shipyards were early modern England's largest industrial facilities and the factory-less textile industry by far its largest employer.[3] The 'timber crisis' thesis fared little better, as students of iron production and of the countryside agreed that industry was more likely to preserve than to destroy its vital wood sources.[4] The detailed local histories of the parts of Middlesex that lay only a few miles outside of London reinforce this revisionist case, as London's immediate hinterland retained substantial woods well into the early modern period. St John's Wood, for example, just three miles north of Westminster, only lost its woods during the 1660s and 1670s. Willesden, a further three miles to the north-west, similarly saw its woods diminish and disappear in a gradual and uneven way throughout the seventeenth and eighteenth centuries.[5] Far from being the treeless island that Nef described, early modern Britain in fact retained substantial woods long after the beginnings of the rise of coal.

Critics, then, have been unconvinced by the more immoderate aspects of Nef's argument, and it is clear that his description of a universal timber famine is not really defensible. In other ways, however, the connection

---

[2] Nef, *The Rise*, I, 161.

[3] D. C. Coleman, 'The Economy of Kent under the Later Stuarts' (Unpublished PhD Dissertation, University of London, 1951), ch. 8; D. C. Coleman, 'Naval Dockyards under the Later Stuarts', *Economic History Review* 2nd ser. 6:2 (1953), 134–55; one survey of the large literature on proto-industrialization, which often stresses the importance of textiles and its location in the countryside, is Sheilagh Ogilvie and Merkus German, *European Proto-Industrialization: An Introductory Handbook* (Cambridge, 1996).

[4] G. Hammersley, 'Crown Woods and Their Exploitation in the Sixteenth and Seventeenth Centuries', *Bulletin of the Institute for Historical Research* 30 (1957), 136–61; *idem.*, 'The Charcoal Iron Industry and Its Fuel, 1540–1750', *Economic History Review* 26 (1973), 593–613. For a summary of critiques, D. C. Coleman, *The Economy of England 1450–1750* (Oxford, 1977), 84–9, 167. Oliver Rackham, *History of the Countryside* (Dent, 1986), 90–2.

[5] T. F. T. Baker, *A History of Middlesex, Volume VII, Acton, Chiswick, Ealing, and Willesden Parishes* (Oxford, 1982), 221–2; *A History of Middlesex Vol. IX, Hampstead and Paddington Parishes* (Oxford, 1989), 122.

between decreasing wood supplies and the turn to coal remains persuasive for three main reasons. First, because many contributors to this debate have not done enough to stress that the existence of woods does not entail their accessibility. If critics have argued that royal forests contained trees rotting on the ground, this has been shown to result above all from lack of transport infrastructure.[6] City dwellers were not helped by trees in remote forests that could be reached only with great expense. Even for woods that were accessible, moreover, there were in many cases legal obstacles to bringing their products to market. The area around Acton and Ealing, now between Heathrow Airport and central London, retained woodlands well into the seventeenth century in part because tenants there had a customary right to take fuel and timber, the 'firebote, cartbote, ploughbote, and hedgebote' from common lands which were so often important contributors to the 'makeshift economy' of the poor.[7] Whether because of geography or customary law, there was much wood that remained beyond the reach of the market.

Second, of those woodlands that were able to supply the urban market there were strong forces pushing against the kind of sustainable management through which the medieval city seems to have been supplied.[8] Cutting down and selling trees, and then converting former woodlands to agricultural production could offer a quick supply of cash to landowners. This was done by some aristocratic landowners late in the sixteenth century, precisely when London's population was first exceeding its former late-medieval highpoint.[9] Such deforestation was likely to be condemned as improvident and short-sighted, but later in the seventeenth century improvers like John Houghton would extol the benefits of windfalls of capital brought in by clear cutting.[10] Felling trees, especially during a period of rising wood prices,

[6] Philip A. J. Pettit, *The Royal Forests of Northamptonshire: A Study in their Economy 1558–1714* (Gateshead, 1968), 5, 103, 127.

[7] Baker, *Middlesex Vol. IX*, 82. More broadly see Andy Wood, *The Memory of the People: Custom and Popular Senses of the Past in Early Modern England* (Cambridge, 2013), 158, 161, 179, 184, 193; Steve Hindle, *On the Parish? The Micro-Politics of Poor Relief in Rural England c. 1550–1750* (Oxford, 2004), 43–5.

[8] James Galloway, Derek Keene, and Margaret Murphy, 'Fuelling the City: Production and Distribution of Firewood in London's Region, 1290–1400', *Economic History Review* 49 (1996), 447–72.

[9] The Earl of Essex, for example, raised much-needed cash in the autumn of 1588 through selling woods, and the profligate young Earl of Northumberland raised some £20,000 to repay his mounting debts in the 1590s. London Metropolitan Archives [hereafter LMA] LMA COL/CA/1/1/, Corporation of London, Court of Aldermen Repertories [hereafter 'Reps'], Rep 22, f. 7v; G. B. Harrison, ed. *Advice to his Son: by Henry Percy Ninth Earl of Northumberland (1609)* (1930), 81–2.

[10] Opponents of deforestation include Arthur Standish, *The Commons Complaint* (1612); R. C. [Rooke Church], *An olde thrift nevvly reuiued* (1612). Cf. Houghton's celebration of a more commercialized fuel market discussed in Chapter 8.

was one of a few ways a landowner could gain much-needed cash, and converting cut lands to arable farming continued revenues, both of which worked against the practice of sustainable forestry.

Third, however, many trees really existed in early modern England, many urban people perceived, and with some justification, an imminent threat to their wood fuel supplies. Such concerns were most clearly voiced during the last three decades of the sixteenth century. Writing in about 1576 William Harrison lamented 'the great sales yearly made of wood whereby an infinite quantity hath been destroyed within these few years' which led directly to new types of fuel including sea coal, which 'even now' was in use in some London houses.[11] Brewers pointed out in 1575 and 1579 that using coal saved wood for 'the whole state' and 'the commonwealth's sake'.[12] By 1586, however, they no longer framed this as a recent conversion, but rather the universal practice of all manufacturers.[13] Perhaps most tellingly, London's governors themselves feared that their wood supplies were inadequate. They blamed shortfalls on the usual scapegoats, the same 'engrossers' and 'forestallers' of the market who were also blamed for grain scarcities.[14] Besides such blame they also undertook surveys of existing wood supplies during the 1570s and 1580s.[15] They found disturbing shortages, and in response quickly turned their attention to the coal available from Newcastle.[16] By the early seventeenth century London's adoption of coal was said to have already happened; it was the 'ordinary and necessary fuel', 'spent almost everywhere in every man's house'.[17] By 1601, the Privy Council wrote to the Lord Mayor and aldermen of London to state that the Queen took high coal prices seriously because she understood that London's 'greatest relief of fuel … is of coals brought by sea from Newcastle and other places of the North'.[18] Coal therefore, became London's fuel exactly when contemporaries believed that wood's availability had become threatened, and the

---

[11] Frederick J. Furnivall, ed. *Harrison's Description of England in Shakespere's Youth* (1877), 343.

[12] Guildhall Library, London [hereafter GL] MS 5445/5 (30 August 1575), 5445/6 (5 February 1579).

[13] GL MS 5445/7 (17 February 1586).

[14] This tradition is discussed in James Davis, *Medieval Market Morality: Life, Law, and Ethics in the English Marketplace* (Cambridge, 2012); John Bohstedt, *The Politics of Provisions: Food Riots, Moral Economy, and Market Transition in England, c. 1550–1850* (Farnham, 2010), 65–9.

[15] Rep. 18, f. 60-62v; Rep. 20, f. 16v; Rep. 21, f. 328; LMA COL/CC/1/1/, Corporation of London, Court of Common Council Journal [hereafter 'Jour.'], Jour. 20, f. 130–1; LMA COL/AD/1/23 Letter book Z, f. 6v, 40v-41, 341.

[16] Rep. 23, f. 388, 481.

[17] Elizabeth Read Foster, *Proceedings in Parliament 1610. Volume 2: House of Commons* (New Haven, 1966), 268; The National Archives, London [hereafter TNA], STAC 8/13/2.

[18] *Acts of the Privy Council of England: New Series*, 46 vols. (1890–1964), [hereafter *APC*], *APC 1601-4*, 67.

frequent governmental attention to wood supplies of the 1550s to 1570s diminished exactly when London switched to coal.[19] Nef may have been wrong, then, to see a 'national crisis' in wood supplies, but he was right to stress the connection between wood scarcity and the rise of coal, especially in England's only large urban market.

What this connection produced was a city that derived most of its energy from mineral sources by around 1600. Part III of this book describes how London's vast importance led to policies that intended to protect and expand its own coal consumption, and in so doing also advanced the coal industry throughout much of England more broadly. London was thus a crucial driver of the longer-term transition of the English economy into one that depended more and more on mineral fuel. Coal's steadily expanding reach inland from those areas accessible by water transport meant that even before the development of canal or rail networks, the great majority of England's energy needs during the eighteenth century were supplied by coal.

Economic historians have debated vigorously in recent years the significance of this shift, with E. A. Wrigley the leading advocate for coal's vital contribution to economic growth.[20] Coal, he argued, freed large amounts of land to produce food and animal pasture, even as it melted the lime that made this expanding acreage more fertile. Wrigley's term for the effects of this is the 'advanced organic economy'; 'advanced' in part because coal use allowed it to escape some of the natural limits that classical economists like Malthus had predicted, but 'organic' rather than 'mineral' because coal did not yet provide the motion that would drive the fully industrial economy of the nineteenth century. Wrigley's analysis of this is fundamental, but the term 'advanced organic economy' is perhaps an imprecise label for the bundle of changes he describes. Coal, strictly speaking, is no less organic

[19] For the practices and strategies of local government during this transition, see also William Cavert, 'Villains of the Fuel Trade' (forthcoming); Simon Healy, 'The Tyneside Lobby on the Thames: Politics and Economic Issues, c. 1580–1630', in Diana Newton and A. J. Pollard, eds. *Newcastle and Gateshead Before 1700* (Chichester, 2009).

[20] See most recently E. A. Wrigley, *Energy and the English Industrial Revolution* (Cambridge, 2010), summarizing many earlier publications. England and Wales derived more energy from coal than wood by c.1620, and by the 1770s the proportion was 10:1. Paul Warde, *Energy Consumption in England and Wales, 1560–2000* (Naples, 2007), 116, 118. Wrigley's work has been importantly developed by Kenneth Pomeranz, *The Great Divergence: China, Europe, and the Making of the Modern World Economy* (Princeton, 2000); Robert C. Allen, *The British Industrial Revolution in Global Perspective* (Cambridge, 2009); Prasannan Parthasarathi, *Why Europe Grew Rich and Asia Did Not: Global Economic Divergence, 1600–1850* (Cambridge, 2011); and Astrid Kander, Paolo Malanima and Paul Warde, *Power to the People: Energy in Europe over the Last Five Centuries* (Princeton, 2014). Those rejecting the importance of coal include Joel Mokyr, *The Enlightened Economy: An Economic History of Britain 1700–1850* (New Haven, 2009), esp. 267–72; and Deirdre McCloskey, *Bourgeois Dignity: Why Economics Can't Explain the Modern World* (Chicago, 2010), ch. 22.

than wood fuels. More importantly, the primacy of mineral over wood fuels already during the seventeenth century undermines the distinction between 'the mineral economy' of the nineteenth century and the 'organic' period preceding it. The fundamental importance of natural limits to growth and coal's role in transcending those are perhaps better captured by another phrase from Wrigley's work, though one which he has used less often and less prominently, 'the photosynthetic constraint'. This points more directly to his key distinction between energy flow and capital. The flow of energy from sun to plant to human consumer is renewable, but it is also constrained by climate and weather and, most importantly, it takes up land that might have been used for other productive purposes. The capital stocks of coal, while ultimately deriving from the same solar and biological sources, have accumulated in such quantities that they provide far greater energy supplies, at least until they are ultimately depleted. Capital stocks are not renewable but while they last they free up land for other uses, thus offering both larger amounts of energy and larger amounts of land (often described as *ghost acres*) to an economy. Insofar as we are still, in the twenty-first century, benefitting from (or, if you prefer, trapped by) the enormous energy stocks offered by fossil fuels, we are all living in a world made by the ability to transcend the photosynthetic constraint.

The unprecedented coal consumption that London attained by about 1600 played a key role in this more general rise of coal. London therefore holds a central place in a process that could be seen, with Nef, as a key contributor to the development of capitalism or, with Wrigley, as allowing for the economic benefits of escaping the photosynthetic constraint. Fuel seems an undramatic commodity, but the work of Nef and Wrigley show that the stakes of the story of fuel use in early modern London are in fact profound. London was at the very centre of England's transformation into a new kind of economy, one in which abundant energy made possible new kinds of work, life, and urban space.

III. COAL-BURNING LONDON

The contemporary perception that Londoners began using coal in significant quantities during the last decades of the sixteenth century is confirmed by the research of economic historians into tax records. Nef's claims regarding the rapid growth of London's coal imports, like his position on the wood crisis, attracted the scorn of subsequent scholars who noted mistakes or found his conclusions too thinly supported by evidence.[21] More judicious

---

[21] D. C. Coleman, *Industry in Tudor and Stuart England* (1975), 46–7; 'The Coal Industry: A Rejoinder', *Economic History Review* 30 (1977), 343–5.

recent work, however, has largely confirmed Nef's overall picture of urban coal imports growing dramatically during the late sixteenth century and at a steady rate thereafter.[22]

Measuring the volume of the coal trade is relatively easy because, in contrast to some high value goods, there is no reason to suspect substantial smuggling. Even during the eighteenth century when coal was heavily taxed such duties still only amounted to about 5 shillings per ton, rendering smuggling operations unduly difficult. The tax records generated for the coal trade into the Thames, then, seem to have produced reliable totals. These show a baseline of 10–15,000 tons during the middle years of the sixteenth century growing to over 27,000 tons by 1581 and then increasing very quickly to almost 50,000 tons by the late 1580s and over 68,000 tons in 1591–2. Within fourteen years imports then doubled again to 144,000 tons in 1605–6 and then nearly doubled yet again to over 283,000 tons in 1637–8. When Hatcher's records resume in the 1670s annual imports ranged from about 310–406,000 tons, exceeding 500,000 tons in some years by the late 1680s. There were occasionally dramatic fluctuations from year to year, but these are usually easily explained by insecurity at sea. The year ending at Michaelmas 1688, for example, exceeded 577,000 tons, but as the descent of the Dutch disrupted trade the following year's total reached only 57 per cent of this sum. Again in 1701 London imports reached 640,000 tons but the trade nearly collapsed at the declaration of war in May of 1702, declining by 78 per cent during the months of May and June.[23] Despite such short-term fluctuations, however, over the medium and long term the trend of the coal trade is consistently and clearly upward. From the beginnings of its expansion in the 1580s, London's coal imports grew steadily throughout the early modern period and beyond.

This growth tracks London's demographic expansion fairly closely. From about 1600 London consumed roughly one ton per person per year.[24] This trend suggests, and more detailed attention confirms, that it was the daily fires of London's inhabitants rather than the large furnaces of coal-fired industries that drove this growth. This finding would have surprised many early modern observers, and perhaps also a few modern historians, who have associated London coal burning and smoky air with industrial production.[25] In 1579, London's brewers were told by the Queen's officers how

---

[22] See the discussion of the historiography and the problems with sources in Hatcher, *History, Before 1700*, 483–5.

[23] Coal Duty Accounts 1701–15, Lambeth Palace Library MS 748, f. 1v–2v.

[24] Vanessa Harding, 'The Population of London, 1550–1700: A Review of the Published Evidence', *London Journal* 15 (1990), 111–28.

[25] For a more detailed treatment of industrial coal use, see William Cavert, 'Industrial Coal Consumption in Early Modern London', *Urban History*, [forthcoming].

she disliked their use of coal when she 'took her noble pleasure upon the water', and indeed at that time brewing probably did consume close to half of London's coal.[26] But within a few years, as coal became widely burned in households as well as furnaces, the brewers' share dwindled rapidly and probably never much exceeded 10 per cent of the entire metropolitan demand during the rest of the early modern period. No other industry came close to approaching this figure, and it is unlikely that manufacturing industries exceeded 25 per cent of London's coal demand at any time after the 1570s.

While a minority, the amount of coal required by London's industries was not negligible. In 1593, the Brewers Company surveyed their members' grain consumption in order to divide the expenses of lobbying parliament. The largest brewer was assessed for 200 quarters of malt per week, another for 150, another 140, and eight others at 100 quarters or more. If these large brewers used their coal no more efficiently than did the brewhouse in Westminster Abbey during these years, they would have burned from 11.8 to 23.5 tons of coal per week. Even if they achieved greater efficiency, as is likely, they still would have required hundreds of tons annually. In 1628 a brewer and MP, Joseph Bradshaw, complained that he was taxed on malt at a rate implying 720 quarters per month. Westminster Abbey would have required 85 tons of coal to brew so much malt, and even if the larger brewhouse only required something like 50 tons, the two tons of coal burned daily was soon enough to attract the angry attention of Bradshaw's Westminster neighbour, King Charles I.[27] Industrial coal burning was therefore limited in its total importance across the metropolis and yet also a locally important phenomenon which could quite reasonably have attracted the attention of neighbours.

The remaining three-quarters of London's fuel was burned not in large furnaces but in the small fires of London's inhabitants, rich and poor. It was quite true that, as the crown argued in a 1607 Star Chamber case, sea coal had become 'the ordinary and usual fuel ... almost everywhere in every man's house' in London and also throughout much of the rest of England.[28] Its use transcended class distinctions to a great extent, as coal was bought and burned by aristocrats, artisans, and almsmen. There were, however, three fundamental differences between how the rich and the poor used their fuel. The most basic and most important distinction is that the rich simply burned

[26] GL MS 5445/5, n.p., 6 February 1578/9. The Queen's 1586 displeasure with 'the taste and smoke of sea coals' has often been confused with this incident, but see the discussion in Chapter 4.
[27] Chapter 4 and William Cavert, 'The Environmental Policy of Charles I: Coal Smoke and the English Monarchy, 1624–40', *Journal of British Studies* 53:2 (2014), 310–33.
[28] TNA STAC 8/13/2.

more fuel and spent more time in well-heated spaces. The allowance provided to residents of almshouses, which has been taken as an indication of basic fuel needs, typically ranged from one-half to a full chaldron (.7–1.4 tons) per year, either per person or per married couple.[29] Middling households used more than this in absolute terms, but per capita measures for households of several inhabitants can hide as much as they reveal. Heat, of course, is not like food because it can warm all members of a home, or all inhabitants of a space, at the same time. And yet not all had equal access to it, as masters and mistresses, the respected elderly or the ill, may have enjoyed more frequent use of the best-heated and least-drafty parts of houses and workshops. The Bakers Company described, in 1620, typical household needs of 4 chaldrons (6.4 tons) for master, wife, children, apprentices, and servants. There is probably no per capita quantity that can adequately express the different experiences of warmth and cold within such a home.[30] Above such middling homes, the houses of the rich unsurprisingly burned much more fuel than the poor, both throughout their larger houses and within their private rooms. In 1623, the Earl of Middlesex bought half a ton of coal every week during the winter.[31] In the 1630s, the Earl of Rutland's London house normally consumed 30 chaldrons (42 tons) of coal plus additional supplies of wood and charcoal, and other elites exceeded even this huge quantity during the following century.[32] An eighteenth-century pamphleteer claimed, plausibly, that for one 'person of high quality in Bloomsbury Square' their usual annual consumption was 90 chaldrons, or 132 tons, easily 100 times more than many poor Londoners were able to burn.[33]

The second crucial difference between rich and poor fuel consumption was that while coal was generally *the* fuel of the poor, for the rich it complemented other varieties which each retained their niche. Rutland House in the 1630s needed, besides large amounts of sea coal, six tons of Scotch coal

---

[29] Ian Archer, *The Pursuit of Stability: Social Relations in Elizabethan London* (Cambridge, 1991), 192 on the Merchant Tailors almshouses, which provided 1 chaldron per capita, GL MS 34,010/7, f. 76v, 79v. The Brewers Company, however, provided 0.6 chaldrons for each almswoman, GL MS 5491/1.

[30] Sylvia Thrupp, *A Short History of the Worshipful Company of Bakers of London* (n.p. info, 1933), 17.

[31] Kent History and Library Centre [hereafter KHLS] U269/1 AP36.

[32] Royal Commission on Historical Manuscripts [hereafter HMC], *HMC Rutland* I, 499–500. See, for example, purchases of up to 28 tons of coal at a time in the late 1670s by Sir Richard Temple, Henry E. Huntington Library [hereafter HEHL] ST 152, f. 12, 46, 88v; and the expenditures by the Duke of Montagu and the Earl of Sunderland in the 1710s and the Duke of Marlborough in the 1750s. Second Duke of Montagu, stewards accounts, Bedfordshire and Luton Archives and Record Service [hereafter BLARS], X 800/7–12; Third Earl of Sunderland accounts, British Library [hereafter BL] Add MS 61,656, f. 121, 222; Third Duke of Marlborough accounts, BL Add MS 61,678, f. 19.

[33] *Frauds and Abuses of the Coal-Dealers Detected and Exposed: in a Letter to an Alderman of London* (3rd ed. 1747), 25.

(a higher grade mineral coal), twenty-six loads of Kentish faggots (bundled thick sticks), and 12,000 billets (common firewood).[34] The Earl of Middlesex had similar expenditures, and his few tons of coal were dwarfed by his consumption of wood fuels including billets, faggots, and charcoal.[35] The House of Lords spent over £700 annually on fuel during the Civil War, consuming huge amounts of wood and charcoal during a period when most Londoners were distressed by the high prices of coal.[36] Below the peerage, London's middling sorts or merchants also distinguished between occasions calling for cleaner-burning but more expensive wood fuels. Charcoal commonly heated the meetings of parish vestries and livery companies. Wood fuels were considered preferable for many purposes, but they also cost more and were therefore inaccessible to the poorest Londoners for whom winter warmth could be a rare luxury. Sea coal, therefore, despite being used by all, was particularly associated with the poor. London's rich and middling households burned a variety of fuels to suit a variety of spaces and occasions, but the poor could not afford such nice distinctions.

The final way in which the poor experienced London's landscape of fuel differently from the rich concerns their experience with the market. Aristocrats like the Earls of Rutland could leave purchases of mundane necessities like fuel entirely in the hands of stewards. Even if some knew details of market conditions, people like the Duke of Chandos who pinched pennies after losing much of his vast fortune in the South Sea Bubble, it was still an easy matter for them to monitor prices and then tell their servants when to take in a year's supply. Chandos issued such orders during the summer for the quantity of sea coal and Scotch coal, buying 100–200 tons to be distributed between his mansion at Cannons and his London townhouse. He described which wholesaler should provide them and which of his 'vaults' and storehouses should hold such huge piles of fuel.[37] Buying in the summer months in this way was normal for anyone with the cash or credit and space necessary to acquire and keep such a bulky commodity.

Many of the capital's poor, however, lacked all of these assets. Even a modest annual supply like 1 chaldron cost a little under 20 shillings (£1) during the late sixteenth century, rising to 30 and 40 shillings by the eighteenth century, meaning that acquiring it all at once would have entailed a substantial outlay for poor consumers.[38] Space, for many, was probably almost as scarce as cash or credit, as very many artisans lived in housing

---

[34] *HMC Rutland*, I, 500.   [35] KHLS U269/1 AP35–36.
[36] Parliamentary Archives [hereafter PA] HL/PO/JO/10/1/210, f. 13.
[37] Duke of Chandos, Household Audits and Orders, HEHL ST 24, Vol. 1, 90–3, 105, 215, Vol. 2, 114, 116, 182 for the summers of 1722, 1727, and 1731. Also ST 57, Vol. 44, 176 for 1739, Vol. 53, 17–8 for 1740.
[38] For details on price movements and seasonality see William Cavert, 'The Politics of Fuel Prices', forthcoming. For the importance of credit in daily and small-scale exchanges

that would feel unbearably cramped even compared to the overcrowded conditions of twenty-first century London.[39] Finding room for the 1.5 cubic metres taken up by each chaldron of coal would probably have been a significant inconvenience for such households.[40] For those sharing bedrooms or crowding families into small spaces it may well have been impossible. In such cases fuel was bought as needed and in small units, by the bushel or even less. John Houghton's price reports found that during the 1690s winter prices were 25 per cent above their summer averages, but in cases of war or severe winters differences were much more.[41] Deep freezes and, in particular, war were not small exceptions to normally prevailing conditions, as conflicts with the Dutch, French, or Spanish threatened coastal trade repeatedly throughout the seventeenth and eighteenth centuries.

Under those circumstances small dealers would have raised their prices both in response to the general market and to compensate themselves for having stored fuel since the previous summer. Very many, of course, must have taken advantage of their strong position relative to desperate customers and raised prices as high as possible. Even without such profiteering, buying fuel in winter by the half bushel was always a far worse deal than buying in summer by the chaldron. These market conditions were largely irrelevant for rich and middling purchasers, who would have been prudent enough to take advantage of the entirely predictable low prices of summer. Many of the poor would not have had that option. For them the period when fuel was most essential was also when it was most expensive. Altogether, then, coal was the fuel that London's poor could not do without but which they also found frustratingly inaccessible in sufficient quantities or at affordable prices.

IV. COMPARISONS: FUELLING EARLY MODERN CITIES

London's demand for fuel was entirely unique within Britain, where it stood out as the only major city and by far the largest consumer of coal (as well as nearly everything else). Comparisons with other large cities elsewhere are therefore especially helpful in order to see what was and was not unusual about London. All major capitals everywhere of course consumed large amounts of provisions. Cities were (and are) dependent upon supply networks for articles of consumption, and London's early modern peer cities

---

throughout the period see Craig Muldrew, *The Economy of Obligation: The Culture of Credit and Social Relations in Early Modern England* (New York, 1998).

[39] Peter Guillery, *The Small House in Eighteenth century London* (2004).

[40] An exchequer commission found in 1616 that the London chaldron consisted of 396 gallons, 1.5 cubic meters, equal to 1.07 cubic meters per ton. Hatcher, *History, Before 1700*, 568.

[41] John Houghton, *A Collection for Improvement of Husbandry and Trade* (1692–8), *passim*.

also required huge amounts of fuel. Paris in 1735 consumed 400,000 loads, amounting to 800,000 cubic metres, of wood – the approximate equal of London's sea coal in bulk, though providing only about one-quarter as much energy.[42] Some other cities consumed on comparable scales, though more research is needed to establish with any precision varying levels of energy consumption in the early modern world. Beijing, Edo, and Istanbul were all at least as large as London, and it is clear that they required very large amounts of fuel.

Despite these examples of other voracious urban markets, there are good reasons to believe that London's fuel consumption exceeded anywhere else in the early modern world. Warmer climates must surely have moderated the fuel needs of some of the world's largest cities, such as Agra, Delhi, and Cairo. Many cooler areas, and some warm places as well, shared the same ecological strains as Britain, as demographic expansion led to deforestation and fuel scarcity.[43] Conrad Totman's work has shown English readers how Japan, which included the world's largest city in Edo as well as other major centres, successfully responded to severe wood shortages through a state policy of conservation.[44] Japan also preferred fuel-efficient methods for cooking and home heating, including a variety of braziers to warm bodies directly rather than heating an entire room.[45] The English, by contrast, preferred the 'cheerful' but highly inefficient flame of an open fire. One would therefore expect Japanese cities to consume less fuel than London, and records from Osaka suggest that this was indeed the case. In 1714, its 375,000 inhabitants consumed about 128,000 tons of firewood plus charcoal valued at slightly over a quarter of this amount. Contemporary London was less than twice as large as Osaka, but its mineral coal alone provided about nine times as much energy as Osaka's wood.[46] Other early modern urban centres probably also used very large amounts of fuel, but still far less than London. More work needs to be done on fuel and energy in cities like Istanbul and especially Beijing, which did consume at least some coal during the early modern period.[47] Unless such work discovers that Beijing used coal

---

[42] Daniel Roche, *A History of Everyday Things: The Birth of Consumption in France* (Cambridge, 2000), 131. Energy density by volume figures from 'Typical calorific values of fuels' under 'facts and figures' at www.biomassenergycentre.org.uk, accessed 11 July 2015.

[43] John F. Richards, *The Unending Frontier: An Environmental History of the Early Modern World* (Berkeley, 2005), 53–5 (on the Netherlands), 144–7 (on China), 183–7 (on Japan), 409 (on Brazil), 421–2, 431–2 (on the Caribbean).

[44] Totman, *Green Archipelago*; *Early Modern Japan* (Berkeley, 1993) 225–9, 271.

[45] Louis G. Perez, *Daily Life in Early Modern Japan* (Westport, CT, 2002), 114–5; Richards, *Unending Frontier*, 179.

[46] London in 1714 consumed about 550,000 tons of coal, which provided about twice as much energy by weight as wood. For energy by weight see Vaclav Smil, *General Energetics: Energy in the Biosphere and Civilization* (New York, 1991), 323.

[47] Susan Naquin, *Peking Temples and City Life, 1400–1900* (Berkeley, 2000), 433; Parthasarathi, *Why Europe*, 158–9, 170–5, suggests that Beijing did consume some coal

on a hitherto unsuspected scale, there does not seem to have been any other early modern city that approached both London's size and its dependence on non-renewable fuel sources.

In Europe only Dutch cities, led by Amsterdam, obtained significant amounts of energy from non-renewable sources, primarily from peat. The precise amount, while undoubtedly large, has been debated by economic historians. If J. W. De Zeeuw's larger estimate is accepted, then London alone used almost as much energy as all of the Netherlands during the seventeenth century; if Richard Unger's downward revision is preferred, London far surpassed it.[48] The Dutch were Europe's second-leading coal consumers after the British, and yet during the early eighteenth century all of the Netherlands imported less than 10 per cent of the coal used by London alone.[49] The Dutch, like the English, used capital fuel stocks to escape the economic constraints imposed by the fuel shortages that were pervasive across the early modern world. But they did not do so to the extent of the English, whose only great metropolis employed more energy, both absolutely and in per capita terms, than Amsterdam or any other city during the seventeenth and eighteenth centuries. Thus, from the Netherlands to Japan both the fear and reality of deforestation and fuel scarcity were common across the early modern world.[50] There was a range of responses to this ecological pressure, but London's emphatic embrace of mineral fuels was unequalled. In a world of big cities facing scarce and expensive energy, coal set London apart.

V. CONCLUSION: A MEANINGFUL THING

The scale and nature of coal burning in early modern London have been described here in ways that will become crucial to the discussion that follows. Coal was a gift, endowment, and resource that allowed London, like much of the rest of England, to transcend the photosynthetic constraint. While contemporaries would not have used this term they were well aware

---

during the early modern period, and that a very significant Chinese mining industry existed by the arrival of European observers in the mid-nineteenth century, but there is very little in the secondary literature on the nature and scale of early modern urban fuel consumption. Pomeranz, *Great Divergence*, 63–4 suggests that Beijing's consumption was modest before the nineteenth century.

[48] Richard W. Unger, 'Energy Sources for the Dutch Golden Age: Peat, Wind, and Coal' *Research in Economic History* 9 (1984), 221–53. David Ormrod, *The Rise of Commercial Empires: England and the Netherlands in the Age of Mercantilism 1650–1770* (Cambridge, 2003), 247 accepts Unger's position.

[49] Ormrod, *Commercial Empires*, 252–4.

[50] The Americas lacked very large cities, but not industries that denuded forests. See Daviken Studnicki-Gizbert and David Schecter, 'The Environmental Dynamics of a Colonial Fuel-Rush: Silver Mining and Deforestation in New Spain, 1522–1810', *Environmental History* 15:1 (2010), 94–119.

of coal's economic importance, which as we will see was rehearsed in a series of policy debates across the period. London's was a fossil-fuelled economy. But the discussion of the predominance of domestic, rather than industrial, fuel use shows that it was in some ways also a fossil-fuelled society. In London's homes, parish churches, alehouses, coffee houses, and workshops, social relations were in part performed with and through access to fire and warmth, and this meant coal. Coal was central not only to work and production but also to health, comfort, sociability, and paternalism. For these reasons those who considered London's coal supplies were never thinking about a mere commodity, but rather about the webs of relationships and meanings that this mundane object made possible.[51] Coal is treated here, therefore, less as a factor for economic production than as one of the many *things* that became essential to the emergence of a certain kind of society and a certain kind of state. Things like coal were important because of the ways that they helped organize daily life, social relations, and political power. Economic historians have explored some of these considerations in their quest to understand British industrial development, but they were also unavoidable to contemporaries who wanted to understand, or even to protest against, London's degraded urban environment.

[51] For the growing literature on the social importance of objects, see Arjun Appadurai, ed. *The Social Life of Things: Commodities in Cultural Perspective* (Cambridge, 1986); Paula Findlen, ed. *Early Modern Things: Objects and Their Histories* (New York, 2012).

−3−

Airs: smoke and pollution, 1600–1775

I. *AN EXCELLENT AIR*: THE ENVIRONMENT OF EARLY MODERN LONDON

Authors of Renaissance urban panegyrics included good and healthy air within their catalogue of a city's blessings. Leonardo Bruni's 1403 *Laudatio Florenitae Urbis* celebrated the benefits of Florence's natural location as 'never vexed or threatened by pestilential climate, by fetid or impure air, by the humidity of water, or by autumnal fevers ... This happy situation cannot be praised enough'.[1] Bruni's English successors followed his lead and translated his praise from Florence to London. It was favoured by 'most cheerful air' according to Drayton's *Poly-Olbion* and was 'much blessed' as a place where 'air, land, sea, and all elements show favour every way' according to Camden's *Britain*.[2] Seventeenth-century compilers of facts in the natural historical tradition continued in this vein, asserting like Thomas Delaune in 1681 that 'with the wholesome gusts and fresh air of the country round about, it must needs therefore have an excellent air'. London, Delaune continued, 'is by experience found to be as healthy a city (considering its greatness and number of inhabitants, with the prodigious quantity of coals burnt yearly in it) as any in the known world'.[3] Among the advantages of the metropolis in the early eighteenth century, according to Guy Miège, was the benign influence of the 'fresh air' of the Thames from one side and suburban fields on the other.[4]

---

[1] Benjamin G. Kohl and Ronald G. Witt, *The Earthly Republic: Italian Humanists on Government and Society* (Manchester, 1978), 148.

[2] Michael Drayton, *The Second Part, or a Continuance of Poly-Olbion* (1622), 252; William Camden, *Britain, or A chorographicall description of the most flourishing kingdomes, England, Scotland, and Ireland, and the ilands adjoyning, out of the depth of antiquitie* (1637 ed.), 436–7. Also Peter Heylyn *Eroologia Anglorum. Or, An Help to English History* (1641), 321.

[3] Thomas Delaune, *The Present State of London* (1681), 4. See also *England's Remarques* (1682), 99, 105; Edward Chamberlayne, *Angliae Notitia: or the Present State of England* (1702 ed.), 336.

[4] Guy Miège, *The Present State of Great Britain and Ireland,* (1723 ed.), 104.

Unfortunately for its inhabitants, these encomia to London's climate were mistaken. Whatever natural advantages its situation afforded, the early modern city's environment was not pure and it was not healthy. Miège's praise was qualified by an acknowledgement that cities were subject to the effects of their own economies as well as their natural setting, and that in London this meant in large part an economy dependent on coal. As the next chapters discuss in detail there were a variety of ways in which early modern London's smoky air was considered unhealthy by contemporaries. Before turning to perceptions, however, this chapter examines what twenty-first century science and medicine have found regarding coal smoke, air pollution, and their effect on human bodies. This is important for the story that follows because perceptions of smoky air were not only cultural constructs; they also responded to and tried to make sense out of actual material realities. Some of these were evident to the senses, while others have only recently been understood through scientific research. But whether or not London's inhabitants realized it, burning coal really did change their air in ways that had real effects on their bodies.

II. COAL SMOKE AND HUMAN HEALTH

Smoke of almost any kind, in significant concentrations and over sustained periods, is detrimental to health. Archaeologists have attributed high rates of maxillary sinusitis (infection of the sinuses leaving detectible damage to bones), found in a range of medieval and early modern sites, to outdoor and indoor pollution.[5] Indoor air pollution, though until recently almost never given the kind of attention described either in this book or in other studies of modern environmental politics, nevertheless has been and remains a major problem. The World Health Organization has identified indoor air pollution as one of the leading threats to public health worldwide, causing an estimated 4.3 million premature deaths annually. The vast majority of these occur in poor countries where homes are heated by biofuels. The victims of indoor air pollution are therefore those who spend large amounts of time around home fires, overwhelmingly women doing cooking and the small children accompanying them. Exposure to such smoke leads to heightened rates of respiratory disease, lung cancer, heart disease, strokes, cataracts, and other problems.[6] Such conditions, now found almost exclusively

---

[5] Charlotte Roberts and Keith Manchester, *The Archaeology of Disease* (Ithaca, NY, 2007), 174–6; but Karen Bernofsky, 'Respiratory health in the past: a bioarchaeological study of chronic maxillary sinusitis and rib periostitis from the Iron Age to the Post Medieval Period in Southern England' (PhD Thesis, Durham University, 2010), is cautious regarding pollution's contribution to sinus and lung infection.

[6] http://www.who.int/indoorair/en; www.who.int/indoorair/en; www.epa.gov/iaq/cookstoves.

in the under-developed world, must have been very widespread throughout pre-modern societies. In early modern London, therefore, it seems extremely likely that wives, servants, and others who spent long hours over fires suffered a range of diseases. The documentary record, however, reveals no awareness of this at all.

Among the most important causes of such diseases for those exposed to smoke, whether from burning wood, dung, or mineral coal, are very fine particulates.[7] While common sense would suggest that thick smoke, full of ash, is the most unhealthy, recent medical research has focused on the dangers of these very fine micro-particles. PM_{10} are objects less than 10 micrometres in diameter, about one-fifth the width of a human hair, which the American Environmental Protection Agency defines as 'inhalable coarse particles'. 'Fine particles' are those below 2.5 micrometres, and 'ultrafine' are less than 0.1. Micro-particles are a complex category, defined by size rather than chemical make-up, and their precise relationship(s) with disease are therefore highly complicated and the subject of ongoing research. In general, however, recent findings suggest that the smallest particles are the most dangerous, as these enter most easily into the lungs and perhaps bloodstream. They have been shown to raise significantly the risks of lung and heart diseases. They also cause much of the haze that is characteristic of polluted areas. Fossil-fuel combustion causes fine particles ($PM_{2.5}$), so an area that burns sufficient coal will experience heightened levels of micro-particle pollution and accompanying health risks.

The other leading problem attendant on heavy coal combustion is sulfur dioxide (SO_2), which even in very short exposures can cause or exacerbate respiratory or heart diseases and further adds to urban haze.[8] Such haze filters sunlight and so reduces the solar energy available to plants and people. Vegetation is further damaged by SO_2's acidity. Under wet conditions, such as often prevail in London, SO_2 turns to sulfuric acid. When dry it becomes sulfate. These particles are among the micro-particles discussed above, causing inflammation of the lungs and also contributing to acid rain. SO_2's health impacts have been found most damaging to the most vulnerable, including asthmatics, children, and the elderly. Similar effects are cause by nitrogen dioxide (NO_2), which also results from coal combustion.[9] NO_2 inflames airways and exacerbates asthma, again acting most harmfully on the most vulnerable, and it too forms micro-particles that harm the lungs and heart.

---

[7] This discussion draws on Marquita K. Hill, *Understanding Environmental Pollution* (Cambridge, 2010), 130–4; and websites of the USA's Environmental Protection Agency available at www.epa.gov/airquality/particlepollution.

[8] This paragraph follows Hill, *Understanding*, 125–7; and www.epa.gov/airquality/sulfurdioxide/index.html.

[9] Hill, *Understanding*, 128–9; www.epa.gov/airquality/nitrogenoxides/health.html.

III. POLLUTANT LEVELS IN EARLY MODERN LONDON

Proceeding from the fact that pollutants have harmful properties to more specific descriptions of the actual state and effects of the air in London during previous centuries is difficult because there are no measurements of pollution levels before the modern period. It is simply not possible to say with certainty how polluted early modern London was. There is, however, data derived from a relatively simple model that provides useful insights into average conditions and the ways they changed over time.

Atmospheric chemists Peter Brimblecombe and Carlotta Grossi have estimated concentrations of pollutants, including SO_2, NO_2, and PM_{10}, in the London atmosphere beginning in 1125 and projected forward to 2090.[10] Their work suggests that levels of all three pollutants were dangerously high, with SO_2 the most strikingly elevated. From a base of 5–7 micrograms per cubic metre ($\mu g/m^3$) during the middle ages, SO_2 concentrations increased to 20 $\mu g/m^3$ in 1575, 40 in 1625, 120 in 1675, 260 in 1725, and 280 by 1775. Compare this with UK standards which declare 20 $\mu g/m^3$ as a target for annual mean concentrations, and establish 125 $\mu g/m^3$ as a 24-hour mean not to be exceeded more than three times per year.[11] The American EPA formerly used 78 $\mu g/m^3$ before moving to standards based on hourly rather than annual averages.[12] Current Chinese annual standards are 60 $\mu g/m^3$ for urban areas.[13] From 2002 to 2011, the City of London recorded annual average concentrations of just 2–5 $\mu g/m^3$.[14] Accordingly to Brimblecombe and Grossi's model, then, London would have exceeded modern air quality standards for SO_2 sometime in the middle decades of the seventeenth century, and during the eighteenth century would have been very far above them. Early modern London's air contained concentrations of SO_2 that were as much as seventy times its current level, and even exceeded extremely polluted contemporary cities like Beijing.[15] (See Figure 3.1)

Other pollutants were also high, but there are no parallels for the distance between SO_2 levels in the early modern and in contemporary developed cities. Bituminous coal was high in sulfur and early modern combustion

---

[10] Peter Brimblecombe and Carlotta M. Grossi, 'Millennium-long damage to building materials in London', *Science of The Total Environment* 407 (February 2009), 1354–61; which builds on Peter Brimblecombe, 'London Air Pollution, 1500–1900', *Atmospheric Environment* 11 (1977), 1157–62.

[11] http://uk-air.defra.gov.uk/assets/documents/National_air_quality_objectives.pdf.

[12] www.epa.gov/ttn/naaqs/standards/so2/s_so2_history.html

[13] Chinese standards are from Clean Air Asia, http://cleanairinitiative.org/portal/node/8163.

[14] City of London, *2013 Air Quality Progress Report*, 28, available at www.cityoflondon.gov.uk/business/environmental-health/environmental-protection/air-quality/Pages/air-quality-reports.aspx.

[15] Data for Beijing, Chongqing, and Delhi are from Clean Air Asia's CitiesACT database, http://citiesact.org/data/search/aq-data.

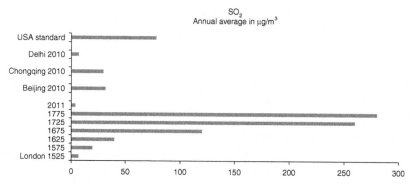

Figure 3.1. Sulfur dioxide annual concentrations: historical,
current, and standards.

was certainly often very inefficient in ways that modern technologies have mitigated. NO_2 is a somewhat different story, as it was both less elevated in early modern London and it remains more prominent today due to automobile combustion. Medieval London's air had almost no NO_2, but with the adoption of mineral coal its level rose to about 7 µg/m³ in 1625 and 40 µg/m³ by the mid-eighteenth century.[16] This latter mark is right at the UK's standard for annual mean concentrations, is very close to urban background levels measured in central London in 2011, and is somewhat less than levels recorded in Beijing, Chongqing, and Delhi.[17] (See Figure 3.2.)

Particle pollution was also probably high. Levels of PM_{10} in the medieval city averaged under 20 µg/m³ but reached 40 by 1625, 60 by 1675, and 130 by 1725.[18] The UK's standard is 40 µg/m³, a total that London far exceeded from the mid-seventeenth century onwards according to Brimblecombe's data.[19] These early modern totals substantially exceed the current background level, recorded as 28 in 2011, but are approached and even, in the extreme case of Delhi, exceeded by the contemporary world's dirtiest cities.[20] (See Figure 3.3.) Brimblecombe's study does not estimate $PM_{2.5}$, but this must also have been very high as these finer particles result from coal combustion.

Unless the data offered by Brimblecombe and Grossi are highly inaccurate, therefore, it is clear that the air of early modern London was full of

[16] Brimblecombe and Grossi 'Millennium-long damage', 1356.
[17] For London, *2013 Air*, 20. For UK standards, http://uk-air.defra.gov.uk/assets/documents/ National_air_quality_objectives.pdf.
[18] Brimblecombe and Grossi 'Millennium-long damage', 1356.
[19] http://uk-air.defra.gov.uk/assets/documents/National_air_quality_objectives.pdf.
[20] For London, *2013 Air*, 23.

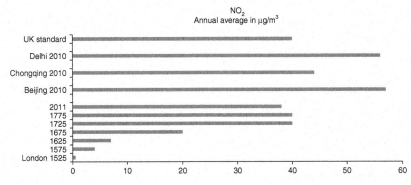

Figure 3.2. Nitrogen dioxide annual concentrations: historical, current, and standards.

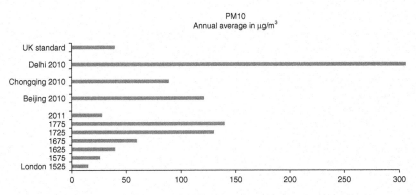

Figure 3.3. Particle pollution annual concentrations: historical, current, and standards.

pollutants that are known to cause hazy air and to harm human health. The air of England's capital during the early modern period, by these figures, was roughly as polluted as the very dirtiest cities in our contemporary world.

IV. LIVED EXPERIENCE AND LOCAL VARIATIONS

Annual averages such as these are both highly illuminating and yet also severely limited. They suggest that early modern air pollution was quite real because it was caused by concentrations of pollutants that are similar to those which are now measured, seen, and tasted by hundreds of millions of people daily. But as averages they hide variation in ways that recent regulatory standards find unacceptable. The harmful effects of even short periods of heightened pollution are not offset by contrary cleaner periods. Very short episodes can effect great damage. An extreme example of this was London's famous smog of December 1952. From an average background level of about 150 μg/m³, urban SO_2 levels spiked at about 1,500 μg/m³ and some smog metres recorded concentrations as much as fifty-four times their normal rates.[21] This episode, which lasted less than one week, killed over 7,000 people, a stark illustration of the power of the kinds of short-term variations that are obscured by annual averages.[22] Less extreme variations are quite normal in contemporary cities, as roadside readings for NO_2 are commonly far higher than urban background levels, meaning that drivers, cyclists, and pedestrians are regularly exposed to harmful levels as they move among vehicles.[23]

The data for early modern London discussed here must also hide such temporal and geographical variation. Brimblecombe's 'simple model' is obliged, because of a lack of precise data, to treat London as if it had been a circular city, burning uniform coal at a uniform rate throughout the year.[24] Of course, none of these assumptions is true. Geographical variation, for example, meant that during the late-seventeenth and eighteenth centuries the West End contained a less dense concentration of fires. Historians have long been aware that, as William Petty observed in 1662, westerly winds made those parts of the city much cleaner.[25] Less attention has been paid to the fact that many inhabitants of the West End would have been absent from London for significant portions of the summer and autumn, as they enjoyed their country houses and oversaw their estates. The fires burned by those that stayed behind would have been more likely to use higher-grade fuels, as the rich were willing and able to pay premiums to avoid excessively stinking or smoking fires. Many humbler inhabitants of the metropolis would have been away during summers as well, on board mercantile or military ships or working on farms. Therefore, in the late summer early modern London had fewer inhabitants while those that stayed had fewer occasions to require fires. On such days, and especially if there were westerly breezes, the air of the West End would probably have been quite clear.

[21] Brimblecombe and Grossi 'Millennium-long damage', 1356; Brimblecombe, *Big Smoke*, 168–9.
[22] Thorsheim, *Inventing*, 168–170 discusses why the often-cited figure of c.4,000 deaths is too low.
[23] *2013 Air*, 20–1 shows that London's annual average background level of NO_2 was 37 μg/m³, yet certain roadside sites were far higher, and often experienced hour-long averages of over 200 μg/m³.
[24] Brimblecombe, 'London Air', 1159.
[25] William Petty, 'A Treatise of Taxes and Contributions', in Charles Henry Hull, ed. *The Economic Writings of Sir William Petty, Together with the Observations Upon the Bills of Mortality, More Probably by Captain John Graunt* (Cambridge, 1899), 41.

During the winter, by contrast, coal consumption and resultant smoke were heavier than indicated in Brimblecombe's table. The city was fuller as elites returned for the social season, sailors were on shore, and fields offered less work than the capital. Its people were colder and its fires much more frequent. Wintertime air quality would therefore have certainly been significantly worse than an annual average. A temperature inversion, such as that which contributed to the smog of 1952, would have prevented breezes from dispersing pollutants and temporarily pushed concentration levels further upwards.

Perhaps most importantly for the discussion that follows, early modern air pollution had enormous geographical variation. While there was a general urban atmosphere and a collective urban cloud, the low height of many chimneys meant that there were many small smoke clouds that made tiny contributions to the collective metropolitan problem and yet were nevertheless enormously annoying to neighbours. In such very local contexts fuel quality might matter a great deal, as a manufacturer who burned lower grade coal (whether through his own frugality or his supplier's dishonesty) would have released substantially more sulfur and ash into his neighbourhood. If such a manufacturer practiced a trade that burned a large amount of fuel over a short time, such as sugar refining, glass making, or brewing, neighbours might well perceive coal smoke as a local, temporally specific, and particular problem, rather than a general aspect of urban life. In early modern London, as the following chapter shows, such neighbours were often very powerful people indeed.

Part II
Contestations

−4−

Royal spaces: palaces and brewhouses, 1575–1640

I. INTRODUCTION: 'WHERE THE KING LIVES'

Early in 1628, as a cold winter and a naval war conspired to drive London coal prices to unprecedented heights, the justices of the court of King's Bench debated where, when, and how coal smoke could become illegal.[1] A man in Gloucester, Jones, had sued his neighbour, Powell, for erecting a brewery next to his home and emitting excessive smoke that entered the house and ruined professional documents. Justices pointed out in Powell's defence that neither brewhouses nor coal were illegal, brewing being a necessary trade and coal 'the fuel of the realm'. But the plaintiff had suffered property damage so the key question was how to balance the competing interests of the two individuals and by extension how to reconcile the public need for both clean air and for economic production. One could, one justice argued, use a lawful trade in an unlawful way. To demonstrate this principle he turned to the relative clarity provided by the social geography of London. To be a butcher was clearly legal, but not to be a butcher on Cheapside, home of the kingdom's richest goldsmith bankers. Hat makers were prohibited near the front of St Paul's Cathedral but were acceptable near the prison for vagrants, as 'we all know that on the back side of Bridewell there is a great number of this trade'. The character of the immediate neighbourhood was therefore crucial. Another justice suggested that one who buys a house next to an existing brewhouse must accept any smoke he finds, but to build a brewery anew within an established neighbourhood was another matter. This was most clearly demonstrated by recent indictments against those erecting new brewhouses around Whitehall, just half a mile from 'where the King lives'. Thus, in their discussion of Jones's Case the justices chose

---

[1] For cold weather and the high fuel costs resulting from Dunkirkers raiding English shipping, Thomas Brook to Countess Rivers, 18 January 1628, Lyn Boothman and Sir Richard Hyde Park, eds. *Savage Fortune: An Aristocratic Family in the Early Seventeenth Century* (Woodbridge, 2006), 52–3; John Edwards to Sir Thomas Myddelton, 8 April 1628, Mary Frear Keeler, Maija Jansson Cole, and William B. Bidwell, eds. *Proceedings in Parliament 1628, Volume 6: Appendixes and Indexes* (New Haven, 1983), 206.

to work through the legality of urban coal smoke by examining the characters of London's various neighbourhoods, from Bridewell where hatmakers belonged to Whitehall where brewhouses did not.[2]

One of the remarkable features of the discussion in this case, a case which became an important precedent informing the doctrine of nuisance into the modern period, is the way justices grounded their examination of nuisance in the geography of London rather than the facts of the dispute itself. The surviving records do not discuss where exactly the brewhouse was nor the nature of its surroundings in Gloucester nor the status of the men involved, despite the fact that they were both among Gloucester's leading citizens. The plaintiff, John Jones, spent half a century as a notary and magistrate at the centre of the city's public affairs. He was a freeman of Gloucester from 1580, a sheriff by 1587, alderman from 1594, Mayor for the first time in 1597 and Member of Parliament in 1604. He served as registrar to the Bishop from 1581 until his death in 1630. When he died a funerary monument was erected on the cathedral's interior wall where Jones can still be seen today, surrounded by sculpted books, papers, pen, and ink – the implements of his clerical and magisterial professions. In August 1625, when Jones claimed the nuisance began, he was only a few weeks from election to serve his third term as Mayor. And when the case was discussed in the King's Bench in 1628 he was helping King Charles's government institute martial law in Gloucester.[3] His opponent, moreover, was also a leading citizen. James Powell was one of the city's few licensed brewers and held a brewery on Longsmith Street, just 16 feet away from Jones's house. The same summer that Powell was alleged to have begun brewing he sat with Jones on a commission to consider civic lands. He later served as a sheriff and, in 1639–40, as Mayor.[4] Both men were therefore substantial figures,

---

[2] The translated and collated discussions in two manuscript reports are provided by J. H. Baker and S. F. C. Milsom, *Sources of English Legal History: Private Law to 1750* (1986), 601–6; the discussion of Whitehall is in *Les Reports de Sir Gefrey Palmer, Chevalier & Baronet; Attorney General a Son Tres Excellent Majesty le Roy Charles Le Second* (1678), 536–9; the date of the complaint is given in *The Reports of that Reverend and Learned Judge, Sir Richard Hutton Knight; Sometimes on of the Judges of the Common Pleas* (1656), 135–6. No source provides the correct King's Bench manuscript reference, which is TNA KB27/1560 (3 Chas I Hil 2), r. 1115.

[3] Alan Davidson, 'John Jones', in Andrew Thrush and John P. Ferris, eds. *The House of Commons 1604–1629* (Cambridge, 2010), Vol. IV, 918–9; *HMC Beaufort* (1891), 482; *APC 1627–8*, 335. For Jones's role in civic political divisions, Peter Clark, '"The Ramoth-Gilead of the Good" Urban Change and Political Radicalism at Gloucester 1540–1640', in P. Clarke, A. Smith, and N. Tyacke, eds. *The English Commonwealth 1547–1640: Essays in Politics and Society Presented to Joel Hurstfield* (Leicester, 1979), 167–87; Clark, 'The Civic Leaders of Gloucester 1580–1800' in *The Transformation of English Provincial Towns, 1600–1800* (1984), 311–45. Jones's monument now looks down on the customers in the cathedral's gift shop, in the south-west corner of the nave.

[4] Suzanne Eward, ed. *Gloucester Cathedral Chapter Act Book 1616–1687* (Bristol, 2007), 23; *HMC Beaufort*, 493. Gloucestershire Record Office [hereafter GRO], GBR/B3/1 Borough

colleagues within civic government, magistrates and businessmen, and both were presumably capable of explaining why the law stood on their side. And yet the records do not show the court's justices considering whether coal really was necessary and common in Gloucester, or whether the houses on Longsmith Street really were arranged in a way that might allow Powell's smoke to damage Jones's episcopal records. Instead, they discussed those London places in which dirty trades seemed clearly illegal: Cheapside, St Paul's, and Whitehall.

The discussion in Jones's Case points to the highly political nature of nuisance law during the reign of Charles I, in particular to the law regarding smoky air around royal palaces. During these years the vague principal of fitness – the illegality of some trades in 'unfit' locations – was usefully clarified through the English crown's less ambiguous ideas about which spaces were politically significant, and about what this significance should mean on the ground. In other words the court allowed the law to follow politics. And smoky air was never more politicized in England than during Charles I's Personal Rule, the years almost immediately following Jones's Case. It was then, as this chapter describes, that earlier initiatives against smoky industry in politically important space were intensified in a way that was new and was never to return after 1640. The following chapter will examine how nuisance law was applied to and adapted by other kinds of neighbours, from the seventeenth and into the eighteenth centuries. While citizens and magistrates negotiated what was legal and desirable, the parameters of nuisance in London were most clearly established through explicitly political definitions of royal space. English monarchs decided that smoky air had no place around their person or their homes. While they were never able to achieve a real reformation of urban air, royal denunciations of coal smoke in early modern London show how pollution could be seen as incompatible with good government and political order.

II. 'THE TASTE AND SMELL OF SMOKE': ELIZABETH AND THE BREWERS

In January of 1579 Queen Elizabeth decided that sea coal smoke had become a problem, and her solution was to jail over a dozen of London and Westminster's leading polluters. This marked the beginning of royal attention to excessive coal smoke in early modern London and Westminster, attention that persisted, though with varying degrees of antagonism, into

Minutes 1565–632, 506-v. 16 feet is claimed in *Palmer*, 536. For the location of the registrar's office see the will of Elizabeth Jones, (1637), TNA PROB/11/182/119. The novelty of Powell's brewery in 1625 is corroborated in TNA E 134/10Chas1/Mich55.

and through the seventeenth century. Elizabeth's concern was not entirely novel, since almost three centuries previously Edward I had issued a series of prohibitions on coal use. These were known during the early modern period, reprinted several times by historians and antiquarians, but there never seems to have been any suggestion that they remained in force.[5] Elizabeth's pressure on industrial coal users was therefore a new departure and a response to new circumstances. In particular, it was a response to recent developments in the brewing trade that made brewhouses the largest industrial manufacturers in the early modern city, as well as its largest consumers of sea coal.

Beer brewing took off in sixteenth-century London, spurred on by a growing urban market and an influx of skilled immigrant labour. 'Dutch' migrants, originating in fact from modern Germany and Belgium as well as the Netherlands, brought expertise and ambition to an industry ripe for expansion. Before 1550 the English traditionally drank ales, which lacked natural preservatives and therefore were best produced in small batches. Beer containing hops, however, lasted longer and hence allowed producers to expand their productive capacities. Such an expansion occurred in London, especially among the Dutch brewers, and by the 1560s and 1570s rising fuel prices constituted a significant cost for large-scale brewhouses. In response they turned to the cheaper sea coal. This combination of sea coal fuel and large productive capacities made London's brewhouses its leading emitters of smoke from the 1570s.[6] By the seventeenth century brewers were always listed highly among the leading consumers of coal in the capital, and their chimneys were notorious for their thick clouds. The underworld, as described by Thomas Dekker 1613, was full of fumes 'blacker than sea coal smoke out of a brewhouse chimney', and others agreed that brewhouses were especially, even uniquely, smoky.[7]

London's leading polluters were therefore easy to identify, and their organization within the Company of Brewers made them easy to reach. On 17 January 1579 'all the ale brewers and beer brewers were summoned to the [Company's] Hall by the commandment of the Lord Mayor and the Lord Chamberlain in the Queen's Majesty's name'. Lord Chamberlain Sussex instructed the assembled brewers 'that they should not burn any more sea

---

[5] E.g., Edmond Howes, *The Annales, or a Generall Chronicle of England, Begun First by Maister Iohn Stow, and after him Continued and Augmented* (1615), 209–10; Richard Baker, *A Chronicle of the Kings of England, From the Time of the Romans Goverment* [sic] *unto the Raigne of Our Soveraigne Lord, King Charles* (1643), 137.

[6] Lien Bich Luu, *Immigrants and the Industries of London, 1500–1700* (Aldershot, 2005); Judith M. Bennett, *Ale, Beer, and Brewsters in England: Women's Work in a Changing World, 1300–1600* (New York, 1996); Cavert, 'Industrial Coal'.

[7] Thomas Dekker, *A Strange Horse-Race at the End of Which, Comes in the Catch-poles Masque* (1613), sig. D3; Cavert, 'Industrial Coal'.

coal during the Queen's majesty's abiding at Westminster'. A few days later the governors of the Brewers Company were required to identify those violating that order. They petitioned the Lord Chamberlain to reconsider the entire matter on the grounds that the brewers had no wood fuel and therefore no real alternative to coal, but Sussex refused to allow any more sea coal to be burned while the Queen was in Westminster. Fifteen brewers and one dyer were therefore arrested and imprisoned in the Sheriff of London's jail.[8] In early February the Company presented a petition to the Privy Council claiming that they burned coal 'for the commonwealth's sake' in order to preserve wood and keep down prices. Switching back would be impossible, and in any case in the depths of winter no wood could be had. They asked for the 'commandment' against brewing with coal to be rescinded, or at least restricted to 'some special days, whereof your orators may have warning'. Later that same day the council told the brewers they could 'burn out such coal as they had', and within a week they were informed that their petition had been granted. Several of them had been imprisoned, but after an ordeal lasting almost a month London's brewers were not in fact restricted any further from exercising their trade or using their best fuel.[9]

A few years later, however, the brewers once again had to make their case to an antagonistic council. In February 1586 'all the beer brewers and ale brewers which dwelleth upon the water side which burneth sea coal were summoned to be at the hall about burning of sea coal'.[10] They submitted a supplication to the Queen's council, delivered 'to the hands of Sir Francis Walsingham', again defending their coal burning on the grounds of necessity. Whereas in 1579 they had simply claimed that there was no wood available, in 1586 the brewers argued that most of London's manufacturers used sea coal and that supplying all of London's brewers with wood would deplete the city's fuel stocks. They acknowledged that Elizabeth had 'absented herself' from London because she was 'greatly grieved and annoyed with the taste and smoke' of the sea coal, which absence of course filled them, as loving subjects, with grief and sorrow.[11]

---

[8] GL MS 5445/6 under 28 January 1578/9 and 19 January 1578/9; Rep. 19, f. 412v; Letter Book Y, f. 288v.

[9] GL MS 5445/6, 5 February, 12 February, 1578/9

[10] GL MS 5445/7 (22 February 1586). There is also a reference in GL MS 5445/6 (24 July 1582) to a 'suit' to the Comptroller of the Household regarding brewing with coals, but there is no further information regarding this and the nature of the suit is not clear.

[11] GL MS 5445/7 (22 February 1586) and TNA SP 12/127/68. This letter, which includes the reference to Elizabeth's own personal disapproval of coal smoke, was calendared under 1578 in the *Calendar of State Papers Domestic* [hereafter CSPD] for 1547–80, p. 612, the editor apparently confusing it with the quite distinct petition by the brewers in January of 1578/9. This mistaken chronology has been repeated by subsequent scholarly studies and has passed into numerous popular histories of pollution. Cf. Nef, *Rise* I, 157; Hatcher, *History, Before 1700*, 438–9; Thomas, *Man and the Natural World*, 244; Te Brake, 'Air Pollution and Fuel Crises' 341.

The brewers offered the Queen a compromise by which 'two or three' brew-houses nearest to the palace of Westminster would revert to wood. It is unclear whether this concession was accepted or enacted, but no further action is known to have been taken by Elizabeth or her council for the remaining seventeen years of her reign.

These fleeting Elizabethan attempts to restrict smoky brewing reveal both the extent and the limits of early opposition to smoky air. Lord Chamberlain Sussex's decision to jail sixteen brewers, including most of London's largest beer producers, signalled the regime's seriousness, as did the brewers' offer to make concessions in 1586.[12] It is also clear, however, that this was not a general attempt to suppress smoky industry in London. Rather, the 1579 affair reveals that the regime's concern was temporary, while in 1586 the crucial limit had become spatial. In 1579 the arrested brewers were not clustered in any particular part of the capital, except insofar as brewhouses were near the river. Moreover, the petition's offer to cease brewing on 'some special days' suggests that the regime's concern was contingent rather than permanent.[13] In 1586, by contrast, the brewers offered a concession that was spatially rather than temporally specific. 'Two or three' brewers 'near unto her majesty's Palace of Westminster' would revert to wood, leaving those elsewhere unaffected. In these two episodes, then, Elizabeth's government established a principle that would remain at the heart of attempts to restrict smoky industry during the seventeenth century: that coal smoke was a problem insofar as it affected the monarch's home and those spaces that were deemed essential to monarchical display.

III. 'THAT RAGGED, POOR, AND SMOKY CASE': COAL SMOKE AND THE CATHEDRAL UNDER JAMES I

James I was less concerned with limiting coal smoke around royal palaces than Elizabeth, but he was far from apathetic to problems of urban government. He was more deeply alarmed than his predecessor about London's

---

[12] A survey of London beer brewers in 1574 lists four men brewing more than sixty quarters of malt per week, three of whom were arrested in 1579. BL Cotton Faustina C. II., f. 175–88.

[13] There are at least two – by no means mutually exclusive – possible reasons why such smoke may have been unusually problematic in January of 1579. First, the whole episode coincides very closely with the visit of Prince Casimir, Count Palatine and a leader of continental Protestantism. During his visit the regime tried to demonstrate support for German Protestants without committing much real aid. Making London sparkle may have contributed to such diplomatic theatre. In this context it may be relevant that the Earl of Sussex was not only Lord Chamberlain, and so in charge of a broad range of issues related to the queen's household, but was also one of the Privy Council's leading experts on German affairs. Second, there were fears of plague in 1578 and 1579, which often led to closer attention to cleanliness. *Calendar of State Papers, Foreign Series, of the Reign of Elizabeth, 1578–9*, 408. Thanks to David Scott Gehring for discussing Casimir's visit.

unstoppable growth and its implications for government and order, which he called 'a general nuisance to the whole kingdom'. An overgrown city, James feared, would both be ungovernable in itself and would draw the natural governors of the provinces out of the country. 'With time England will only be in London', he worried, 'and the whole country left waste.'[14] While it is clear now that this same urban growth also led to increasing coal burning, James did not understand smoke as a general urban problem. Rather, it was a problem only in very particular situations, as it had been for Elizabeth. But whereas she sought to remove smoke from her own immediate environment, by the end of his reign James became concerned to reduce coal smoke from London's most important church.

Smoke became an aspect of royal ecclesiastical policies through James's embrace of a rebuilding project for St Paul's Cathedral. Abortive plans to rebuild followed a major fire in 1561 and were again considered in 1608, but only in the final years of James's reign was a renovation programme pushed by the regime. The key spur in this seems to have been a pious layman, Henry Farley, who published poetry and prose works exhorting London's citizens and governors to redress the shameful and even irreligious neglect of Britain's 'princely church'.[15] Farley amplified the difference between the cathedral's dignity and importance and its sad material condition. It had degraded into a general public space, surrounded by stalls and defiled by dirt, dung, chatter, labour, and commerce. 'Things consecrated unto pious uses', Farley demanded in one poem, '[i]s't fit that they should suffer foul abuses? Is there no civil difference or odds, Twixt clean and unclean things, man's house and God's?'[16]

By 1620 this argument had found royal approval and a fundraising campaign was initiated with a sermon by the Bishop of London at Paul's Cross, the public pulpit in the cathedral's north churchyard. The text was set by King James himself, a clear signal that the bishop was to establish connections between appropriate regard for holy buildings and divine providence: 'Thou shalt arise, and have mercy upon Zion: for the time to favour her, yea, the set time, is come. For thy servants take pleasure in her stones, and favour the dust thereof.' Bishop King rose to the occasion, buttering up Londoners by enlarging upon their civic dignity:

If England be the ring of Europe, your city is the gem. If England the body, your city the eye; if England the eye, your city the apple of it. Here is the synopsis, and sum of the whole kingdom. Here the distillation, and spirits of all the goodness it hath. Here

---

[14] From James's Star Chamber speech of 1616. Charles Howard McIlwain, ed. *Political works of James I* (Cambridge, MA, 1918), 343.

[15] Henry Farley, *The Complaint of Paules to All* (1616), 34.

[16] Henry Farley, *Portland-stone in Paules-Church-yard Their Birth, Their Mirth, Their Thankefulnesse, Their Aduertisement* (1622), 14.

the chamber of our British Empire. Here the emporium, principal mart of all foreign commodities, and staple of homebred. Here the garrison, and strength of the land, the magazine and storehouse of the best of God's blessings.[17]

St Paul's was the temple of this unparalleled seat of empire and trade, a uniquely special building symbolizing Britain's relationship with God as well as London's standing among foreign cities and nations. A particular care for its beauty should therefore prevail. And yet King, like Farley, found dirt and decay everywhere, so he deployed his full rhetorical powers to call directly on his audience of 'so many thousands of souls' to donate generously.[18]

  Bishop King spoke in general terms of St Paul's ruin and decay, only occasionally hinting that smoke stains contributed to the facade's dirtiness.[19] Others, however, made this connection quite explicit. The antiquarian William Dugdale wrote in his interregnum history of St Paul's that Farley's 'sundry petitions' moved James's 'princely heart ... with such compassion to this decayed fabric, that for prevention of its near approaching ruin (by the corroding quality of the coal smoke, especially in moist weather, whereunto it had been so long subject)'.[20] Peter Heylyn's apology for Archbishop Laud followed Dugdale's account closely, including its stress on the agency of Farley and the central importance of the 'corroding quality of the sea coal smoke, which on every side annoyed [the stone work]'.[21] Both accounts come very close to claiming that coal smoke was the primary culprit responsible for the church's decay and even its structural weakness, but this is a claim neither made in Farley's texts nor in King's sermon. It might be tempting to conclude, therefore, that both men's focus on smoke reflected their close associations with leading figures of Charles I's government in the 1630s, during which coal smoke was more explicitly politicized. Both Dugdale (b. 1605) and Heylyn (b. 1599) came of age during the reign of James's son, and both knew and worked with many of the central protagonists of the Caroline campaign against smoke.[22]

---

[17] John King, *A Sermon at Paules Crosse, on Behalfe of Paules Church* (1620).
[18] *Ibid.*, 56.
[19] King stressed the outward appearance of the stones as powerful evidence of internal bodily decay and death, comparing them to the bloody clothes of Joseph and Julius Caesar. He compared time to a fire which consumes 'all temporal things', a fire which had left 'so many brands of disgrace upon the whole face of [Paul's]'. If this referred to stones blackened by smoke, King would then have read urban pollution as a metaphor of universal decay and impermanence. *Ibid.*, 39–40.
[20] William Dugdale, *The History of St Pauls Cathedral in London, from its Foundation* (1658), 134.
[21] Peter Heylyn, *Cypria nus anglicus, or, The History of the Life and Death of the Most Reverend and Renowned Prelate William, by Divine Providence Lord Archbishop of Canterbury* (1668), 218–9.
[22] Dugdale enjoyed the patronage of the Earl of Arundel, who sought to reduce smoky air both around Whitehall and around Arundel House on the Strand, while Heylyn's patron Laud, as will be discussed below, was closely associated with Caroline anti-smoke policy. Another

It is therefore possible that Farley, Bishop King, and James himself were all much more concerned with structural weakness and profane usage of holy space than with external smoke damage, and that Dugdale and Heylyn read the later intensification of concern over smoke back into their accounts of the royal visit of 1620.

This is possible, but the contemporary centrality of smoke to the Jacobean renovation campaign is emphasized by a remarkable set of paintings intended to illustrate the process and significance of cathedral restoration. Three paintings by John Gipkyn, now held by the Society of Antiquaries of London, are intended to represent, so Farley claimed, a 'dream or vision' which he experienced in or after 1616.[23] A first image representing a royal procession towards London is opened to reveal a diptych, with the next painting showing, according to Farley, his vision of the King and court hearing a sermon at Paul's cross. The church itself is bordered by numerous low buildings whose chimneys emit black smoke onto the church's darkened stones. This represents what Farley describes as being 'in that ragged poor and smoky case, as now you are'. The centrality of smoke to this decay is emphasized by one of the 'grievances that then were opened to his Majesty's sight and hearing … one thing written in capital letters, which was well observed on all parts, viz. VIEW, O KING, HOW MY WALL-CREEPERS, HAVE MADE ME WORK FOR CHIMNEY-SWEEPERS'.[24] The final panel shows a new structure of clean white stone and glistening ornament, what Farley described as the church 'suddenly renewed, beautifully repaired, and cured of all your evils and infirmities'.[25] Farley therefore claims responsibility for the composition, iconography, and textual adornments of Gipkyn's paintings.

It may be doubted, however, that Gipkyn's role was quite this passive, as he had worked on a mayoral pageant in 1613 which was centrally concerned with the metaphorical possibilities of smoke and darkness, particularly at sites of civic importance like St Paul's. Thomas Middleton's *The Triumphs of Truth* featured a 'mount triumphant … over-spread with a thick sulphurous darkness, it being a fog or mist raised from error'.[26] Since such a mount triumphant was also central to Farley's vision, it is likely that he was either inspired by or drawing on Middleton's text and Gipkyn's staging, or that

---

such writer was Edward Chamberlayne, who also blamed Paul's decay on 'the corroding quality of the abundance of sea coal smoke' in his 1671 *Present State of England*, but he associates its restoration entirely with Laud rather than James or Farley. Chamberlayne, *The Second Part of the Present State of England Together With Divers Reflections upon the Antient State Thereof*, (1671), 201.

[23] Henry Farley, *St. Paules-Church her Bill for the Parliament* (1621), sig. C4.
[24] *Ibid.*, sig C4v.   [25] *Ibid.*, sig. D2.
[26] Thomas Middleton, *The Triumphs of Truth* (1613), sig. C2.

Farley and Gipkyn somehow collaborated in the representation of 'Farley's dream'.

It is clear, then, that to Farley, and probably to Gipkyn as well, coal smoke contributed to St Paul's material and metaphysical decay. While Bishop King's sermon did not identify smoke as a specific problem, there is some evidence that he would have seen Gipkyn's painting regularly. As it was intended for James himself, Pamela Craig-Tudor suggests that it was instead passed along to the dean of the cathedral, since John Donne seems to refer to it in his 1631 will. If Craig-Tudor is right King might well have had time, before his death in 1621, to assess Gipkyn's portrayal of himself as the painting hung in the deanery. King's successor Bishops of London, Montaigne and Laud, would also likely have had occasion to consider the depiction of Farley's vision of the dirtied cathedral renewed.[27]

Farley was an eccentric figure whose literary advocacy of St Paul's restoration was unique, but there was nothing necessarily radical or especially divisive about James championing cathedral renovations. Mainstream Calvinists as well as the 'godly', despite their opposition to idolatry, actively built, maintained, and improved churches during James's reign.[28] Moreover the theological content of Bishop King's sermon was solidly Calvinist, as he stressed the primacy of God's power and mercy against the insufficiency of pious works. Nevertheless, the royal visit to St Paul's should be placed within a trajectory that led, by the end of James's reign, to ecclesiastical policies increasingly supporting 'the beauty of holiness'. One aspect of this shift was an association between the beauty of church fabric and decoration with the dignity of the Church as an institution. Against the perceived dangers of separatism, clerics favoured with royal patronage increasingly stressed the unifying role of the church as a national community. By the 1630s such initiatives were perceived by some as an attack on predestination and therefore on basic reformed orthodoxy, as well as being excessively ceremonial and even dangerously popish.[29] The early 1620s, then, were the turning point after which England's kings embraced a policy of ecclesiastical order

---

[27] Pamela Tudor-Craig, 'Old St Paul's' The Society of Antiquaries Diptych, 1616 (2004), 4. On Donne's death in 1631 he willed a 'large picture of ancient churchwork', supposed by Tudor-Craig to be Gipkyn's, to his successor Thomas Winniffe. Winniffe had previously been Dean of Gloucester, installed in 1624 in the presence of episcopal registrar and coal smoke victim John Jones.

[28] Julia Merritt, 'Puritans, Laudians, and the Phenomenon of Church-Building in Jacobean London', Historical Journal 41 (1998), 935–60.

[29] Kenneth Fincham and Peter Lake, 'The Ecclesiastical Policies of James I and Charles I' in Kenneth Fincham, ed. The Early Stuart Church 1603–1642 (Basingstoke, 1993), 23–50; Peter Lake, 'The Laudian Style: Order, Uniformity, and the Pursuit of the Beauty of Holiness in the 1630s', in ibid., 161–85; Nicholas Tyacke, Anti-Calvinists: The Rise of English Arminianism (Oxford, 1987).

and unity through beauty. Addressing the 'ragged, poor, and smoky case' of London's cathedral was a key moment in that development, as a series of complaints culminating in a day of political theatre revived Elizabeth's objection to coal smoke in the spaces that embodied the government's goals.

IV. 'OFFENSIVE TO THEIR MAJESTIES OR THEIR COURT': SMOKE AND MONARCHY, 1624–1640

Under James's son and successor Charles I the significance of smoky air expanded in unprecedented ways, as reforming spaces of power and display moved towards the heart of the regime's agenda. The renovation of St Paul's was revived after a period of neglect and expanded, producing Inigo Jones's Palladian west front with its huge classical portico. But concern for smoky air was centred not on the cathedral, as it had been under James, but on royal palaces, as under Elizabeth. From his first emergence as a maker of policy distinct from his father, Charles displayed a keen sensitivity to smoky air as an affront to the sorts of healthy, orderly, and beautiful spaces which he wanted Whitehall and St James to be.

Charles's push to reform the smoky air of Westminster began during the final year of his father's reign. At the parliament of 1624 Charles was a key figure, at once a champion of Protestant public opinion (through his recent and high-profile conversion to an anti-Spanish foreign policy), a territorial lord with extensive networks of personal patronage, and the heir of an aging monarch.[30] At this moment, the first when he could hope to promote legislation through his own influence, Prince Charles began his attempts to reform London's air. Both the Lords' Journal and a letter from a Member of the House of Commons record Charles as particularly interested in the passage of 'An Act concerning brewhouses in and about London and Westminster' which, if successful, would have established a non-industrial zone in all of Westminster and most of London.

The Breweries Bill, like the Elizabethan complaints of 1579 and 86, identified 'common brewhouses' as places where

the extraordinary great fires made of sea coal yield such abundance of unwholesome and unsavoury smoke and stinks as doth greatly diminish not only the pleasure and delight but the health and soundness of the said cities, a thing fit to be redressed in the chief city of this realm, where His Majesty, the prince, the nobles, and other most eminent persons of this kingdom are occasioned to be so often and so much conversant.[31]

---

[30] Thomas Cogswell, *The Blessed Revolution: English Politics and the Coming of War, 1621–1624* (Cambridge, 1989); Chris Kyle, 'Prince Charles in the Parliaments of 1621 and 1624', *Historical Journal* 41:3 (September 1998), 603–24.

[31] PA HL/PO/JO/10/4/1.

The smoke of one of London's largest industries was defined as a problem
partly because it diminished the pleasure, delight, and health of the 'said
cities' in general, but more specifically because that was where the King, his
son, the aristocracy and Britain's elite, (whom the bill describes as 'other
persons of the best rank and quality') were 'occasioned to be so often'.

The bill's provisions underlined further that it did not intend to reform
dirty air in general, nor even the air of London in general, but only of those
places that were the resorts of the 'best rank and quality'. Brewing with coal
was prohibited within one mile of the royal courts in Westminster, as well
as in that part of the city west of London Bridge and its northern extensions
of Gracechurch Street and Bishopsgate. The Tower was specifically excluded
from consideration as a royal palace, signalling that the eastern part of the
city, as well as its northern, southern, and eastern suburbs were acceptable
as places 'where their sea coal shall occasion less offence or annoyance'. The
bill would thus have created a statutory division between a clean and airy
West End and industrial, dirty areas in the East End and urban periphery,
decades before such distinctions actually came about through more gradual
processes.

The breweries bill was passed with no sign of protest in the House of
Lords, but then it stalled. 'The Commons hesitate to pass it', noted one MP
in a letter, 'as it will damage the estates of many'.[32] Some of those whose
estates were to be damaged, furthermore, were well positioned to bury
the bill. The committee to which it was referred included the burgesses for
Westminster, London, and Southwark, all of which contained substantial
brewers who would have been threatened. The London Brewers Company,
moreover, kept the brother of another MP on retainer as an advisor and,
presumably, parliamentary lobbyist. The sudden suspension of parliament
after the bill's commitment made its defeat easy for the brewers, who moni-
tored its progress carefully but did not expend large sums on the bribes that
would have been called for had they felt truly endangered.[33] The defeat of
the breweries bill was both a beginning and a turning point. It demonstrated
that the next king would make a priority of the beauty and health of his
court, 'the principal resort' of nobles and persons of quality. It was also,
however, his first and last attempt to do so through statute.

[32] Francis Nethersole to Dudley Carleton, 24 May 1624, TNA SP 14/165/34. Both author
and recipient were prominent political actors, Nethersole having served the King's daughter Elizabeth and her husband Frederick, Elector Palatine, while Carleton was ambassador
to the Netherlands. For the bill's progress through parliament, *Journal of the House of
Lords* [hereafter JHL]: *volume 3: 1620–1628* (1802), 269, 342; *Journals of the House of
Commons* [hereafter JHC] (1802), I, 790.

[33] Compare the sums of a few shillings spent on expenses related to following the 1624 bill
with the £275 distributed at court in 1637 when the company's charter was threatened. GL
MS 5442/6 (extraordinary payments 1623–4 and 1636–7).

After succeeding to the throne Charles quickly established that the decorum, order, beauty, and cleanliness of his court and capital would be central pillars of his rule. This was not merely an expression of Charles's personal interest in art and a more formal personal style than that preferred by his father; rather, it amounted to an ideological commitment. Such a vision of order required careful distinctions among persons. Household ordinances therefore defined who could attend royal meals, excluded 'beggars, idle and loose people' from entering the palace grounds at all, and demanded that those entering the chapel royal should process so as to allow rank to be easily 'distinguished and discerned'.[34] During the first few years of his reign Charles's government took many steps to reform and improve the court's organization and appearance, to prevent illegal urban sprawl in and around London, and to ensure that those areas of the metropolis regularly used by the king remained clean. Official statements connected the beauty, regularity, strength, and cleanliness of the built environment directly to the nobility of royal government. London, the crown proclaimed only a few weeks into Charles's reign, was 'our chamber and seat of our Empire [and therefore] should be kept with most state and order'.[35] It is likely, though not certain, that attention to smoky air was part of this early attempt to clean up Whitehall. The justices in *Jones v. Powell* referred to indictments made against those erecting brewhouses near to Whitehall, where the king lives. There is no other known record of cases like this during the early Stuart period, but such measures would be much more consistent with the known preferences of government under Charles than under James.[36] If these precedents dated from the first three years of Charles's reign then they suggest that his regime turned its attention to smoke almost immediately. If they did not, then the reform of urban air was entirely an initiative of the Personal Rule.

In the spring of 1629, just weeks after the dissolution of the last parliament until 1640, the king's council made its first move to suppress smoky industry in politically significant places. When the Privy Councillor William Cecil, the 2nd Earl of Exeter, sought to repress smoky brewhouses and a smith's forge near his home in suburban Clerkenwell, he pointed out that the house had been visited by the king and queen. The council arrested the

---

[34] Regulations for the Royal Household, temp. Charles I. TNA LC 5/180, f. 16, 1, *et passim*. For Charles as more concerned with order at court than James had been, see also Sir Philip Warwick, *Memoires of the reign of King Charles I* (1701), 65–6; John Chamberlain to Dudley Carleton, 9 April 1625, Thomas Birch and Robert Folkestone Williams, eds. *The Court and Times of Charles the First* (1848), Vol. I, 8.

[35] Quoted in J. Robertson, 'Stuart London and the Idea of Royal Capital City', *Renaissance Studies* 15:1 (March 2001), 39. For Caroline and earlier measures against new buildings and urban growth, see T. G. Barnes, 'The Prerogative and Environmental Control of London Building in the Early Seventeenth Century', *California Law Review*, LVIII (1970), 1332–63.

[36] *Palmer*, 538.

offenders and made it clear that such an 'annoyance' to aristocratic and sometimes royal spaces would not be tolerated.[37] Other brewhouses, according to the king's instructions to the Privy Council in 1632, were among the problems which 'annoyed and prejudiced' the houses of the royal family and so required attention.[38] From 1625 to 1632, then, smoky air was occasionally found to threaten Caroline standards of urban cleanliness and courtly decorum. But it was not until the autumn of 1633 that Charles' government initiated sustained pressure on those beer brewers who operated closest to the royal palaces in Westminster.

In September of that year the council summoned four men and charged that the king 'hath taken notice of a great nuisance and offence caused by the sea coal smoke of certain brewhouses in Tothill Street, Westminster'.[39] Their operators were informed that their businesses were displeasing to the king, against the laws of the realm and even a threat to the Stuart dynasty. The crown's use of the term 'nuisance' implied both a general and a specific meaning. Generally a nuisance was then, as now, anything displeasing, a synonym for the 'annoyance' of which Exeter complained in 1629. To displease a king was rarely good, but was not obviously illegal. Nuisance also, however, was a category of legal offence that encompassed anything preventing the use of private property or of public spaces. As the next chapter examines, these nuisances could be general and public, which magistrates were required to abate, or private, in which case victims might seek remedy through tort proceedings in royal courts. For the common law these were crucial differences, but the council's unspecific language skillfully straddled the two categories, suggesting throughout the 1630s that smoke in Westminster damaged the king in particular rather than the collective welfare, but it also acted to abate it through magisterial power. It thus used the rhetoric of a private nuisance but pursued the remedies of a common nuisance, without explicitly defending either approach. This may be seen to derive from the ambiguity of the king's person itself, which was neither wholly public nor private but always in some ways both. In court cases dealing with ordinary subjects justices recognized this, treating smoke nuisance to royal houses as limit cases, instances which were obviously nuisances and against which more ambiguous cases might be contrasted. In these senses nuisance law provided a basis for crown action, even as it itself was also shaped by the crown's claims regarding the special nature of royal space.

---

[37] TNA Privy Council Registers, [PC] 2/39, 181–2, 210, 241, 248, 276.
[38] TNA PC 2/42, 220.
[39] TNA PC 2/43, 239. For a detailed reconstruction of conflicts with Westminster and London brewers during the 1630s see Cavert, 'Environmental Policy'.

A crucial part of the council's complaint was its assertion that intensive coal smoke was a threat to health. As in the text of the breweries bill, it was again not said to threaten *public* health, but rather the royal family exclusively. The offending smoke was 'very offensive and noisome to the queen's majesty, being now with child, and may endanger the health of both their majesties and the princes their children, as their majesties have found by their own experience'.[40] Contemporary medical thought could support such a claim, and Charles's own physician William Harvey stressed the health implications of London smoke only two years later.[41] But the council cited not the textually derived expertise of physicians, but the bodily and sensory 'experience' of the king and queen themselves. The king and his heavily pregnant wife had reason to be fearful during September of 1633, both she and young Prince Charles having suffered recent illnesses.[42] These considerations are likely to have heightened concern over the healthiness of her immediate environment and to have elevated Westminster smoke from annoyance to danger.

Such short-term considerations may have helped rouse the king and his council in September, 1633, but their concern also persisted throughout the rest of the decade. From 1634 into 1638 a series of inventors and projectors received or petitioned for royal patents for a series of smokeless ovens or drying processes. The language justifying these described smoke as a serious nuisance, sometimes specifying the city of London, and sometimes the king, as its victim. A 1634 royal proclamation regarding improved furnaces, for example, justified a grant to one Thorneffe Franck through its 'lessening of the great annoyance of smoke, which is so much obnoxious to our city of London'.[43] The proclamation refers to letters patent granted to Franck the previous November, just a few weeks after the Westminster brewers were called before the council. It further established two Westminster Justices of the Peace, Lawrence Whitaker and Sir Henry Spiller, to investigate those who used Franck's invention without license. Whitaker was responsible for enforcing the council's resolutions against the Westminster brewers in 1633 and subsequently, which shows that inventions like Franck's were very much part of the regime's attempts to remove excessive smoke from Westminster. Other inventors and projectors saw this connection clearly and pitched their petitions for their own patents accordingly. One, for example, claimed in 1636 to alleviate the 'daily increase of seacoal smoke' by which London 'is

[40] TNA PC 2/43, 239.
[41] In the case of 'Old Parr', reputed to be 152 years, whom Harvey found was killed by urban pollution. See Chapter 6 below.
[42] For details see Cavert, 'Environmental Policy'.
[43] John F. Larkin, ed. *Stuart Royal Proclamations; Vol. II Royal Proclamations of King Charles I 1625–1646* (Oxford, 1983), 426.

the more and more annoyed and deformed', promising a furnace especially useful 'for brewers, especially those that annoy the Court'.[44] The language of such projectors illustrates how the regime's desire to reduce the smoke that 'annoys the court' was perceived as an ongoing policy, and one which a clever or well-connected petitioner could turn to his own advantage.

The Caroline anti-smoke campaign reached its apogee during the final years of the 1630s, when the council pursued Westminster's leading brewers with renewed determination. From 1637 brewers were called before the council again and again, bound not to use sea coal, taken to court for doing so, fined, and arrested. One described narrowly escaping out his back door and over a garden wall, only for his wife to be arrested once she assumed control of the business.[45] Such attentions were potentially ruinous whether the brewer chose to contest, ignore, or submit to them, and brewers frequently claimed to have suffered huge losses as a result of prosecutions either directly or indirectly stemming from the regime. Besides those Westminster brewers called before the Privy Council, others were pressed to cease smoky brewing by Privy Councillors who may have been eliding their public duties as a member of the king's council with their private interests and preferences as an inhabitant of a grand house in a busy city. The Earl of Arundel, for example, was a leading councillor and member of various commissions regarding urban policing and improvement, but his efforts to suppress smoky brewing near to Arundel House on the Strand seem to have stemmed from some combination of public authority and private interest. Considering brewers like these whose persecution seems in some way political, as well as those who were unambiguously charged with annoying the courts of Whitehall and St James, it is possible to identify twelve brewers in Westminster and the Strand whose operations were opposed by King Charles and his councillors. They were not all put out of business; many, in fact, were remarkably resilient and successfully resisted or evaded the council's attempts to restrict their production.[46] The Caroline campaign against coal smoke around royal palaces, therefore, was never entirely successful, but this should not surprise because neither was the politics of spatial decorum of which it was a part.

---

[44] Oxford University, Bodleian Library [hereafter Bod.] Bankes MS 11/30, warrant regarding petition of John Gaspar Wulffen. Wulffen's claim to benefit both king and brewers is again stressed in SP 16/377/61 and 16/392/19. Other such petitions include Nicholas Halse, 'Great Britain's Treasure', BL Egerton MS 1140, f. 44; Lindsay and Hobart petitions, *CSPD 1635–6*, 417 and Bankes MS 11/58.

[45] Michael and Mary Arnold, TNA PC 2/50, 343–4. Details are in Cavert, 'Environmental Policy'.

[46] Michael Arnold, for example, passed on a successful brewery to his son, also Michael, who became royal brewer and a Tory MP for Westminster under both Charles II and James II.

Attempts to prevent brewers from expelling dark clouds of smoke into the palaces and gardens of Westminster coincided with, and should be understood as a facet of, other initiatives during the later 1630s to renew the court as ideological and physical centre of government. Charles, it was reported, planned to rebuild Whitehall itself, 'in a more uniform sort', and 'with much beauty and state'.[47] This never happened, but the regime did produce both grandiose statements of intent and real, if modest, material improvements. Royal masques, a form of early Stuart drama in which the court was at once cast, setting, audience, and subject, associated royal power with beauty and order. In one masque King Britanocles emerges from a 'glorious palace of fame' to dispel 'misty clouds of error', a symbolic language quite close to that deployed by Farley and Gipkyn.[48] Besides artistic statements of royal dignity, the return of plague also pushed Charles's regime to purify public spaces, part of what the leading authority on public health measures has called 'a drive to cleanse the metropolis of public nuisances and make it a capital fit for an absolute king'.[49]

V. CONCLUSION

The campaign against Westminster brewers was also an attempt to make a capital fit for an absolute king, an aspect of a broader politics of space and royal representation. It was both a royal and a royalist policy, driven by the king himself and designed to promote the beauty, health, and symbolic power of spaces associated with his government.[50] It was not, therefore, an early outbreak of environmentalism, nor a response to pollution in general. This was quite explicit; the Privy Council suggested that Westminster's brewers

---

[47] Richard Daye to Ralph Weckherlin, secretary to the Secretary of State John Coke, 23 June 1638. *HMC Cowper*, II, 186; TNA C115/108 [letter 8606] John Burghe to John Scudamore, 24 Feburary, 1635/6. Kevin Sharpe misdated this letter to the beginning of Charles's reign in 'The Image of Virtue: The Court and Household of Charles I, 1625–1642', in *Politics and Ideas in Early Stuart England: Essays and Studies* (1989), 150; and *Image Wars: Promoting Kings and Commonwealths in England, 1603–1660* (New Haven, 2010), 223.

[48] Inigo Jones and William Davenant, *Britannia Triumphans a Masque* (1638), 20. 'Cloud' or 'mist' of error is a venerable metaphor, used by Augustine in *The City of God* in a notably anti-populist passage that contrasts the pure light of one's own conscience with 'the fond opinion of the vulgar' and the 'trivial censures of mortal men which are most commonly enclouded in a mist of ignorance and error'. Saint Augustine, *Of the Citie of God With the Learned Comments of Io. Lodouicus Viues* (1620), 31.

[49] Paul Slack, *The Impact of Plague in Tudor and Stuart England* (Oxford, 1990), 217.

[50] Peter Brimblecombe has taken testimony at Archbishop Laud's trial at face value and concluded that Laud was the originator of the anti-smoke measures of the 1630s. This is misguided, as Laud was merely one of several councillors who helped execute a policy that seems clearly to have been driven primarily by the king himself. Cf. Peter Brimblecombe, *The Big Smoke: A History of Air Pollution in London Since Medieval Times* (1987), 40–2; and Ken Hiltner, *What Else is Pastoral? Renaissance Literature and the Environment* (Ithaca, 2011), 117–8.

were welcome to brew 'in such remote places, and at such a distance, as they be no way offensive to their majesties or their court, during the time of their residence at their Palace of St James'.[51] The 1624 Breweries Act, moreover, described exactly where such places were. The health and aesthetic preferences of the people living in such 'remote' places as the East End or southern suburbs were not discussed and were probably not much considered. Charles I was therefore the most active and aggressive opponent of the entire early modern period, but this was because he found smoke to damage his vision of political and spatial order. It is significant, in this regard, that Charles's Privy Council took its first action against urban smoke just weeks after he began his Personal Rule, and that it took its last only days before it ended. For Charles, and in lesser but still important ways for his predecessors James I and Elizabeth I, coal smoke challenged conceptions of beautiful, healthy, and politically significant space. This is what made royal campaigns against smoke qualitatively different from their subjects' responses to the same problem. Both rulers and subjects described coal smoke as a nuisance, but monarchs were able to shape the nuisance law in ways that were impossible for even the richest and most powerful of their subjects.

[51] TNA PC 2/43, 239.

-5-

Nuisance and neighbours

I. BRIDGEWATER'S NEIGHBOUR

In May of 1666, the Earl of Bridgewater had a new problem. His old problems – how to pay off enormous debts, how to manage family affairs after the death of a treasured wife, how to discharge his governmental duties at a time of plague and war – did not go away, but during that summer the Earl's new problem was particularly frustrating. His London home, Bridgewater House, had as a neighbour a soap boiler whose new chimney caused smoke to cascade into the Earl's forecourt and even towards the house itself. For weeks his correspondence describes varied attempts to do something about 'the nuisance of the soap boiler, which I cannot endure'.[1]

Bridgewater's soap maker may in certain respects be compared with the Westminster brewers that so bothered Charles I in the 1620s and 1630s: both were manufacturers who used large amounts of sea coal to fire boiling vessels; both were identified as specific causes of smoke plumes that entered into gardens and homes; and both pursued their trades in ways that were legal and yet ran afoul of England's governing elite. But beyond these similarities were crucial differences. Bridgewater was a lord and grandee, but he was not the king, and Bridgewater House was a private home rather than a seat of royal government. He could not, therefore, be confident that the law stood on his side in the dispute. He was powerful, but unlike Charles I he could not order JPs to put down nuisances through their magisterial authority. His neighbour, moreover, was not obviously out of place. Bridgewater House lay just north of London's walls in the crowded parish of St Giles Cripplegate in the area known as the Barbican, over which now stand the modernist concrete towers of the Barbican Estate. The parish was poor and populous, home to tens of thousands of people including a large proportion of manufacturers.[2] Both Bridgewater himself and his father before him had

---

[1] Hertfordshire Archive and Records Service [Hereafter HALS] AH 1102, Bridgewater to John Halsey, 17 June, 1666. Also AH 1101–5, 27 May–2 July, 1666.
[2] Thomas R. Forbes, 'Weaver and Cordwainer: Occupations in the Parish of St Giles without Cripplegate, London, in 1654–1693 and 1729–1743', *Guildhall Studies in London History* 4

long experience dealing with these surroundings, preventing encroachments on their garden and forbidding tenants from practising noxious trades.[3] John Evelyn claimed that Bridgewater House gardens were among those 'planted even in the very heart of London' that only really flourished when war cut off the coal trade and thus lessened London's smoke.[4] Bridgewater House, therefore, was very far from being the sort of airy, clean, and beautiful palace imagined by Charles I during the 1630s; rather, it was a large house and garden surrounded by dense urbanization. Transforming such a neighbourhood was impossible, and Bridgewater learned that even to eliminate 'the uncomfortable company of my soap boiling neighbour' was complicated.

Despite this, the Earl was determined to escape his nuisance and in the summer of 1666 he examined what options were available to him. He began, it seems, by contacting his neighbour's landlord, the Dean and Chapter of St Paul's Cathedral. Perhaps Bridgewater was hoping that they would exert pressure on their tenant as a favour to one of England's leading Anglican noblemen. Instead they sought legal advice from Francis Phelips of the Inner Temple. According to Phelips, Bridgewater had two choices: he 'may either have an action upon the case in which he will recover damages', or he might complain to the wardmote inquest which would 'view it and present it' at their annual meeting, which would lead the ward's alderman or the Lord Mayor of London to 'order and cause a redress'.[5] Thus Bridgewater's choices, according to Phelips, were either to sue for damage in the royal courts or to seek abatement by London's civic authorities.

Bridgewater, however, chose to explore a third way: negotiation and informal pressure. He considered buying out the soap boiler's lease so long as the price was reasonable: 'that that nuisance may be removed, I should be willing to deal but not upon too hard terms'. When his legal position looked weak he lamented, 'surely the law is not so much any man's enemy but mine'. But by his next letter he was more optimistic, ready to use a strong legal case as leverage in negotiations with the cathedral. This must have worked because in early July, after at least six weeks of complaint and worry, '[I] find the Dean of Paul's speaks so well, I hope he will do as well and I shall then be well rid of my nuisance'.[6] Exactly how the Dean, the future Archbishop William Sancroft, satisfied his neighbour is not clear. Even so,

---

(1980), 119–32; petition from the brewers of St Giles without Cripplegate, 5 September 1661, Rep. 67, f. 297–8.
[3] HEHL EL 6560, 1st Earl to Richard Harrison, 17 August 1640; HALS AH 1122, answer in Chancery regarding a 1655 lease prohibiting the trades of 'Pewterer, a Brazyer, a Coper a Smyth, a Sope boyler, a Tallow Chander or Baker or any them'. Also Northamptonshire Record Office [hereafter RO] E(B) 704, lease of George March, 1666/7.
[4] Guy De la Bédoyère, ed. *The Writings of John Evelyn*, (Woodbridge, 1995), 139.
[5] St Paul's Cathedral Dean and Chapter Estates GL MS 25,240.
[6] AH 1101, 1103, 1105.

the case reveals much about how nuisances were approached in early modern London. They were, above all, aspects of relationships between neighbours, and their resolution therefore involved negotiation and contract as well as litigation and governance.[7]

Bridgewater's experience with 'the uncomfortable company of my soap boiling neighbour' is unusually well documented, but there were very many others whose thoughts were not written down and preserved, yet who were nonetheless annoyed by the smoky air in which they lived.[8] Such annoyance or concern might well have been a pervasive and widespread aspect of early modern urban life, as has recently been argued.[9] But it may be better to view nuisances as forms of deviance that were inherently limited, localized, and/or temporary. A nuisance that was allowed to continue permanently was, legally at least, no nuisance at all. Applying the term, then, implied a conflict and perhaps a debate. The common law helped, describing what nuisances were and what should be done about them. But in practice England's numerous local and central courts applied such guidance with almost infinite variability. Therefore, while Phelips was correct to advise the cathedral that excessive coal smoke was prohibited according to law, remedies available to aggrieved parties like Bridgewater were far from straightforward. Londoners probably turned to the central courts (Phelips' 'action upon the case') less often than they pursued other forms of redress, including local policing, negotiation, and attempts at prevention.[10] Residents of the metropolis wielded numerous weapons against coal smoke, an arsenal powerful enough to demonstrate that early modern people and institutions were far from apathetic to air pollution, yet also one whose weaknesses and limitations prevented systemic or long-term environmental improvement.

II. 'HAVE AN ACTION ON THE CASE'; COAL SMOKE AND THE COMMON LAW OF NUISANCE

'Anything that worketh hurt, inconvenience or damage', according to the eighteenth-century legal authority William Blackstone, was a nuisance, and nuisances were prohibited by the common law in early modern England.[11] Legal historians have generally concluded that this category did include

---

[7] The classic discussion of good and bad neighbours is Keith Wrightson, *English Society 1580–1680* (New Brunswick, NJ, 2000), 51–7.

[8] AH 1103.

[9] Cockayne, *Hubbub*.

[10] For a detailed case study of the uses of private negotiation and litigation in a later period, see Ayuka Kasuga, 'The Introduction of the Steam Press: A Court Case on Smoke and Noise Nuisances in a London Mansion, 1824', *Urban History* 42:3 (August 2015), 405–23.

[11] William Blackstone, *Commentaries on the Laws of England* (Oxford, 1765–9), 3:216.

protections against air pollution during the seventeenth century, and this protection was an established principle of English law until it was later weakened by the needs of an industrializing society.[12] The printed discussions in legal reports, cases that established precedent and therefore attract the most attention from historians of legal doctrine, did indeed establish that smoky air could constitute unreasonable and illegal damage to property and health and was therefore actionable at common law. It will become clear, however, that this story requires some important qualifications because such protections were limited in crucial ways. The broad point here is that all of the protections against excessive coal smoke offered by England's central courts rested on the assumption that such smoke was ugly, dangerous, and destructive, but that translating that belief into action was rarely easy.

Considerations of smoky air were understood to be aspects of larger questions relating to the rights of property holders. By about 1600, it was held that a property's protected easements included 'wholesome air'. This was part of general trajectory of nuisance law towards greater consideration of the potential damage to property posed by divergent land uses. Such considerations emerged from the fact that nuisance was legally an aspect of trespass, a sort of property violation.[13] Its creation in the middle ages was a response to the fact that property could be damaged as well as seized, with nuisance procedures thus offering aggrieved parties avenues for both abatement and compensation. Hales' Case of 1569 established that because houses 'cannot be without light and air', anyone deprived of such basic amenities could seek compensation.[14] In another key case of 1587 future Chief Justice Edward Coke successfully argued that 'light and sweet air were as necessary as pure and wholesome water'.[15] Aldred's Case of 1610 reinforced the importance of wholesome air despite the argument that the law cannot ensure 'dainty'

---

[12] When and how it charged during the industrial revolution, however, has been a matter of controversy. J. F. Brenner, 'Nuisance Law and the Industrial Revolution', *Journal of Legal Studies* 3 (1974), 403–33; D. M. Provine, 'Balancing Pollution and Property Rights: A Comparison of the Development of English and American Nuisance Law', *Anglo-American Law Review* 7 (1978), 31–56; J. P. S. McLaren, 'Nuisance Law and the Industrial Revolution – Some Lessons from Social History', *Oxford Journal of Legal Studies* 3:2 (1973), 155–221; Noga Morag-Levine, *Chasing the Wind: Regulating Air Pollution in the Common Law State* (Princeton, 2003), ch. 3; Christopher Hamlin, 'Public Sphere to Public Health: The Transformation of "'Nuisance"', in Steve Sturdy ed. *Medicine, Health, and the Public Sphere in Britain 1600–2000* (2002), 189–204.

[13] Janet Loengard, 'The Assize of Nuisance: Origins of an Action at Common Law', *The Cambridge Law Journal* 47 (1978), 144–66; Daniel R. Coquillette, 'Mosses from an Olde Manse: Another Look at Some Historic Property Cases about the Environment', *Cornell Law Review* 74 (1979), 765–72.

[14] Justice Monson in *A Brief Declaration for what Manner of Speciall Nusance Concerning Private Dwelling Houses* (1639), 5; also J. H. Baker and S. F. C. Milsom, *Sources of English Legal History: Private Law to 1750* (1986), 592–7.

[15] *Bland v. Moseley* in Baker and Milsom, *Sources*, 598–9. Chief Justice Wray agreed on the importance of air, citing as his authority Virgil's *Aeneid*.

pleasures such as pleasant-smelling air.[16] *Jones v. Powell*, discussed at the beginning of the previous chapter, explicitly placed sea coal smoke within the class of offences to these protections, endorsing a maxim cited in Hales' Case that each must use their own so as not to damage another's.[17] By the 1620s then, at the latest, the justices of the court of King's Bench considered excessive coal smoke to be illegal for the interconnected reasons that it constituted a serious threat to health and that it damaged household property. Coal smoke, at least on the scale produced by a brewhouse like Powell's, caused the kind of damage to the body and the estate that English nuisance law had evolved to protect since its descent from earlier protections against violent seizures of land.

Once printed in Justice Palmer's *Reports* of 1678, Jones's Case was regularly cited in discussions of nuisance law into the nineteenth century and was regularly taken to demonstrate property holders' rights against smoky industry.[18] A 1709 legal compilation directed at lay readers explained, citing Jones, that chimneys, brewhouses, or other such noxious trade were common nuisances: 'if I have any considerable offence in the smell to my health, or by the smoke to our bodies that live in the house or trees of the garden or yard, I may have an action on the case for my relief'.[19] Giles Jacob's exhaustive *Law Dictionary* included Jones's Case in a long explication of nuisance law, explaining that the defendant 'burnt coal so near the house, that by the stink and smoke he could not dwell there without danger of his health', which complaint prevailed 'though a brewhouse is necessary, and so is burning coal in it'.[20] By the middle and later eighteenth century, Chief Justice Lord Mansfield considered several cases in which the damage done by smoky air was a central aspect of claims for compensation or abatement. In 1781, a man sued his neighbour for a smith's workshop, complaining that it was a nuisance because of its loud noises and because 'great quantities of smoke and soot are blown into the rooms of the plaintiff's house'. Two witnesses testified that the smoke was 'offensive' and damaged furniture, but the Chief Justice instructed jurors to regard the smoke complaint as 'not proved' and so to separate it from the noise complaint.[21] While Mansfield weakened nuisance protections by demanding that noxious trades be clearly unfit for their neighbourhood, the key point here is that he, and indeed many of the witnesses recorded in his case notes, accepted that coal smoke

---

[16] *Ibid.*, 599–601.
[17] *Ibid.*, 595, 'sic utere tuo, ut alienum non laedas'.
[18] *Les Reports de Sir Geoffrey Palmer, Chevalier & Baronet* (1678), 536–9.
[19] *The Gentleman's Assistant, Tradesman's Lawyer, and Country-Man's Friend* (1709), 165–6.
[20] Giles Jacob, *A New Law-Dictionary* (1756), *sub.* 'Nuisance'.
[21] James Oldham, *The Mansfield Manuscripts and the Growth of the English Law in the Eighteenth Century* (Chapel Hill, 1992), I: 921–2; *Lloyd's Evening Post*, Issue 3760 (26 July 1781).

was offensive, unhealthy, and/or damaging.[22] Even the defence attorney in a 1754 case admitted that, though there were reasons to find his clients not guilty of illegal nuisance, yet 'as to the smell, the smoking chimney is bad'.[23]

The right to sue for damages caused by coal smoke, then, was clearly endorsed by printed authorities and recognized by leading justices. But to what extent this translated into actual suits brought and damages recovered is very far from clear. Printed law reports existed to record innovations in doctrine or practice, not necessarily the ways that courts were used by litigants. They are therefore not good guides to what actually went on in England's central courts. Illuminating this would require detailed examination of court records themselves, but the archives of the Westminster courts are among the most intractable and least known documents of the entire early modern period. The reason for this, paradoxically, are their importance in early modern English life: economic growth and procedural changes meant that England during the decades around 1600 produced a flood of court cases, with somewhere between 50,000 and 110,000 cases initiated in the year 1606–7 alone.[24] Because the records of these millions of early modern cases are uncatalogued, uncalendared, and are written primarily in Latin in an archaic script, they remain largely *terra incognita*. Finding records for nuisance cases therein is thus looking for needles in large and messy haystacks.

One such needle, however, has been located, which indicates that doctrines established by Jones's Case were in fact translated into real suits. In 1653 Rowland Spratt sued John Chaworth in King's Bench for converting a house outside Aldgate, near Whitechapel to the east of the Tower of London, into a glass house. Chaworth's facility, complained Spratt,

used great and excessive quantity of seacoal ... by reason whereof the sulphurous smoke and other putrefied matter ... descend[ed] into and upon the said ancient messuage, orchard, and garden of the said Rowland whereby ... [they have been] filled with noisome and filthy vapours, and the goods and utensils ... overspread with filth and thereby defiled, putrefied, and corrupted ... one hundred of the fruit

---

[22] Mansfield's approach was ambivalent, accepting a strong interpretation of the right to clean air, finding that 'it is not necessary that the smell should be unwholesome; it is enough, if it render the enjoyment of life and property uncomfortable'. This, however, was strongly circumscribed by his insistence that trades may be carried on in fit places, whatever their effects on neighbours, and that such fit places in practice constituted much of metropolitan London. Oldham, *Mansfield*, 887 and ch. 15 *passim*.

[23] Oldham, *Mansfield*, 888.

[24] Christopher Brooks estimates 54,000 for 1606, but Robert Palmer suggests 112,000 as a preferable figure. C. W. Brooks, *Pettyfoggers and Vipers of the Commonwealth: The 'Lower Branch' of the Legal Profession in Early Modern England* (Cambridge, 1986), 78; Robert Palmer, 'The level of litigation in 1607: Exchequer, King's Bench, Common Pleas' at http://aalt.law.uh.edu/Litigiousness/Litigation.html. The current author has examined c.10,000 pleas from the court of King's Bench.

trees formerly standing, growing, bearing, and increasing within the orchard and garden aforesaid are thereby withered and dead … herbs and flowers … likewise being and flourishing are also thereby dead withered and destroyed and become of no value.

A jury found for Spratt, but awarded only £40 plus costs instead of the £500 damages requested.[25] Spratt's case suggests that not only could nuisance cases be filed against those emitting excessive smoke in the London area, but that they could do so along the lines and within the limits established by earlier cases.[26] Spratt's complaint stressed both the 'noisome and filthy vapours' whereby the air of his house was corrupted, but also particular property damage that established his case as a specific harm rather than a general annoyance. These were the forms through which householders might successfully seek compensation for a neighbour's smoky chimney or furnace.

Compensation was all that the common law offered, and during most of the early modern period England's central courts had no way of requiring abatement. This changed during the eighteenth century when the court of Chancery became available for those seeking to end nuisance, or even to prevent nuisances, rather than to be reimbursed for their damage.[27] In 1736, aristocratic plaintiffs in the West End requested an injunction from Chancery against brick burners whose smoke would be a 'great mischief and annoyance' to their houses and would spoil their furniture.[28] While not seeking abatement, Lord Cheyne also hoped Chancery would aid his efforts to evict tenants who built smoky industrial facilities on his Chelsea estate in the 1670s and 1680s.[29] While the extent to which householders actually used equity courts in these ways is even less clear than it is for common law courts, Chancery did offer another option for Londoners who wanted deliverance from smoky neighbours. Since the equity court procedure to abate only developed at the end of the early modern period, however, it is likely that Londoners wanting to prevent or decrease smoke would not have turned to Chancery, but to the much more familiar officers who governed the capital.

---

[25] TNA KB27/1754, r. 282-v.
[26] In another such case in King's Bench, reported in a 1729 newspaper, the plaintiff claimed to be unable to let a house near to the 'excessive smoke' of a brewhouse in suburban Highgate. *Weekly Journal or British Gazetteer*, Saturday, 6 December 1729; Issue 236.
[27] Oldham, *Mansfield*, 885–6; P. H. Winfield, 'Nuisance as a Tort', *Cambridge Law Journal* 4 (July, 1931), 189–206; J. R. Spencer, 'Public Nuisance – A Critical Examination', *Cambridge Law Journal* 48 (1989), 55–84.
[28] *Grafton v. Hilliard*, TNA C 11/2289/87.
[29] TNA C 8/249/81; C 8/410/12.

III. 'VIEW IT AND PRESENT IT'; ENVIRONMENTAL POLICING

The central courts played a crucially important role by establishing and
defining the meaning of nuisance and explaining how smoky air might qualify as such, especially given an absence of relevant statute law. But it was
local officers and local courts that were better equipped to examine and
punish nuisances, and it was therefore these smaller and potentially more
responsive institutions that offered the best options for those upset by excessive smoke. There were, to be sure, very serious limits and weaknesses that
undermined the effectiveness of each of these courts, but before considering
these it needs first to be established how coal smoke was addressed and
redressed by officers and magistrates.

The Dean of St Paul's legal counsel advised that the Earl of Bridgewater,
if he chose not to seek compensation at law, had another option. 'Upon
complaint to your wardmote's inquest at Christmas they will view it and
present it as a private nuisance by one neighbour to another, and hereupon
the Alderman of your ward or the court of Lord Mayor and Aldermen to
order and cause a redress of the nuisance according to custom in London.'[30]
Phelips describes the wardmote court as the proper venue for 'a private nuisance by one neighbour to another', despite the insistence of common law
justices that such private complaints were properly brought before royal
courts, with common or general nuisances reserved for magisterial attention. This distinction between special or specific nuisances to one party and
general nuisances to the entire community was crucial for common law justices in determining who did and did not have standing but, according to
Phelips, was irrelevant for local courts and magistrates who seem to have
had no interest in whether nuisances damaged one or many.[31] Local courts,
therefore, could hear both the complaints regarding private nuisances, cases
that might equally have been pursued in King's Bench, and also the public
nuisances that common law courts refused to consider. The difference was
that central courts could award damages, while local courts offered the possibility of abatement or even prevention.

Local courts, then, had jurisdiction over nuisances, but whether or not
Bridgewater would have been well served by presenting his smoky neighbour at his 'wardmote's inquest at Christmas' is not all clear. The wardmote
was the annual meeting of all freemen within each of the city's twenty-six
wards, and was the occasion for a general feast as well as the election of
local officers and the presentation of nuisances for examination.[32] James

---

[30] LMA CLC/313/L/I/002/MS25240
[31] The distinction between special and common nuisances is summarized in Blackstone,
Commentaries III, ch. 13.
[32] For wardmotes in general see Archer, *Pursuit of Stability*, 83.

Howell's 1657 encomium to London's 'strict and punctual government' claimed that 'there is not the least misdemeanour or inconvenience that can be but there be officers in every corner of the City to pry into them, and find them out; but especially, the wardmote inquest'. His list of the wardmote's duties included a wide variety of offences and nuisances, but nothing related to excessive smoke or bad air.[33] Nevertheless, London wardmotes did, at least in some few cases, punish smoke. Inhabitants of St Dunstan's in the West were presented during the 1580s and 1590s for burning sea coal without any chimney, a practice both dangerous and likely to create smoke that lingered in closely built alleys and courts. Defective or smoky hearths were also examined, as in 1607 when an inhabitant of the same ward was found to have built a 'noisome' chimney.[34] In 1679, a widow's shop just north of the city's wall had a fireplace 'very offensive to the neighbourhood by reason of the smoke and perilous for fire'.[35]

Such coupling of smoke's annoyance with fire's danger was common; in some cases this may have been merely an attempt to amplify the stakes of a complaint, while in others it is likely to reflect mutually reinforcing and inextricable fears. Thus, references to the 'disquietness, terror, and annoyance' inspired by 'fires without any chimney' may also, at least to some extent, have recorded a conflict over smoke emissions.[36] Similarly, many boiling trades involved both heavy coal use and created noxious fumes, so complaints about tradesmen who melted tallow or distilled chemicals may have been reacting to both gas and smoke, rather than one or the other.[37] Some, indeed, offended in all three ways, threatening fire as well as producing fumes and smoke.[38] Some Londoners, then, did use wardmotes to address unwelcome coal smoke. But, while defective records before about 1680 prevent any clear general picture, wardmote presentments seem not to have been the preferred form of redress in most such cases.[39] Brewers Lewis Young and Thomas Mathews, for example, were both taken to court by powerful neighbours for their smoke, but were ignored by their wardmotes.[40] Perhaps

[33] James Howell, *Londinopolis* (1657), 392.

[34] GL MS 3018, 'reredoses' discussed ff. 33–69 *passim*, esp. 45v, 62v, 77.

[35] Cripplegate Ward presentments, 1680, LMA COL/AD/4/2.

[36] St Dunstan's in the West presentments, 1618, GL MS 3018, f. 100.

[37] Bishopsgate presentments 1702, LMA COL/AD/4/11; Farringdon Within presentments, 1705, COL/AD/4/13;

[38] E.g., 'Thomas Feild a Waxchandler for melting his goods in his shop having no Chimney for that purpose, which very much annoys his Neighbours and is very dangerous of Fire'. Cornhill presentments, 1708 LMA COL/AD/4/16.

[39] Only two series of wardmote presentments exist for the period before 1650, and by the final decades of the century when more survive the effectiveness of the institution in general seems to have suffered serious decline.

[40] For details see Cavert, 'Environmental Policy'.

because the court only met annually, more responsive authorities were preferable for many of those requiring abatement or prevention.

London's Court of Aldermen constituted its senate and executive council, a body of old and wealthy citizens charged with overseeing both the city's internal governance and its external relationships with crown and parliament. Despite this august status it was not above turning its attention to the minutiae of daily life including the regulation of nuisances. It investigated newly built smiths' forges, a new soap-boiling house, and prohibited two new brewhouses, all in response to complaints from unspecified inhabitants.[41] While aldermen sometimes investigated complaints themselves, there were also officers specifically entrusted with examining complaints regarding buildings, boundaries, and associated nuisances. The remit of these 'viewers' thus included disputes regarding excessive or unwelcome smoke, and they seem to have made their decisions according to a rough standard of equity and common sense rather than the learned precedents of the common law. In some cases they were implacable against nuisance. Upon a complaint from widow Anne Tanner regarding the 'smoke and steam' arising from a goldsmith's 'furnace and forge' in her cellar, for example, city surveyor Robert Hooke testified before the alderman against the unfortunate goldsmith who was then not only found guilty of nuisance and ordered to abate but was also brought before the London sessions of the peace.[42] In other cases, however, the viewers sought compromise. The chimney of a tobacco pipe maker, for example, if 'raised thirty foot in height, we are of opinion will not annoy the neighbours more than other chimneys', and raised chimneys were endorsed in other cases as well.[43]

The gulf between such commonsensical compromises and the utter illegality of heavy smoke emissions according to a certain reading of the common law is brought home by the dispute between a lawyer, Thomas Jenner, and a neighbouring soap boiler, John Peirce. In response to Jenner's complaint in 1677 the viewers found that Peirce's boilers were permissible if he raised his chimney to a height equal to Jenner's. Apparently unsatisfied, Jenner had more success a few years later after becoming Recorder, London's senior legal officer. A case, recorded as coming 'ex prosecutione' Jenner, was brought against Peirce in the King's Bench by the Attorney General in the name of the king. Chief Justice Jeffreys (like Jenner, a recent political appointee) found that Peirce was an 'annoyance to the neighbourhood', concluding that such trades 'ought not to be in the principal parts of

[41] 2 October 1621, Rep. 35, f. 275v; 29 June 1632, Rep. 46, f. 279-279v; 1 September 1629, Rep. 43, f. 275v; May–June 1674, Rep. 79, f. 215v-216, 239-239v, 349v-350; 27–29 January 1679/80, Rep. 85, f. 62-62v, 66.

[42] 14 October 1673, Rep. 78, f. 306v; January 1674, LMA CLA/47/LJ/04/44; Rep. 79, f. 111.

[43] Viewers' Reports 1674–84, LMA COL/SJ/27/467, 58.

the City, but in the outskirts'.[44] The viewers, however, were capable of much more nuance, willing both to grant and deny smoky trades a place in the heart of the capital.

Jenner was unusual, of course, as most Londoners were not leading government lawyers and could not hope for the Attorney General to take up their case. But criminal proceedings against nuisances were not uncommon, and some others did in fact appear in the central courts, as a 1678 crown case against a Southwark glass house alleged to burn 'great, terrible and large fires' and to emit 'putrified smoke ... so that the air there is rendered very insalubrious'.[45] Similar phrases were used in the case of *King v. Cole*, a soap boiler near to Haymarket and St James's Park.[46] The report of *King v. Peirce* cites two recent precedents, including the *King v. Jordan* case regarding a brewhouse in Ludgate Hill, near St Paul's Cathedral.[47] Probably more common, however, were indictments brought in the sessions of the peace by magistrates. Middlesex sessions of the peace heard cases brought by the 'commissioners' of annoyances, including a 1611 complaint against the 'filthy smoke' and hammering of a 'smithy and forge' in Grub Street, just north of the city walls.[48] Middlesex Justices of the Peace also indicted a brick kiln for 'baking them with ill-smelling seacoal'.[49] The London court could do the same, as in the case of goldsmith whose 'steam and smoke' was sent before the sessions after being condemned by the city's viewers.[50] One justice stated in Jones's Case that brewhouses constructed near to town were indicted 'in this place', but whether this referred to London or Westminster is unclear.[51] A Lambeth glasshouse was indicted during the 1680s, its owner being 'convicted and fined'.[52] At the end of the early modern period, the papers of Chief Justice Mansfield

[44] Viewers' Reports 1674–84, LMA COL/SJ/27/467, 82; Bartholomew Shower, *The Second Part of the Reports of Cases and Special Arguments* (1720), 327. Jenner went on to become an MP and a Baron of the Exchequer. In 1688, he was arrested attempting to flee with James II.

[45] *King v. Brooks*, 30 Charles II (1678), Sir John Tremaine, John Rice, and Thomas Vickers, *Pleas of the Crown in Matters Criminal and Civil* (Dublin, 1793), n.p., sub. 'Indictments and Informations for Nuisances'.

[46] *Ibid.* [47] Shower, *Second Part*, 327.

[48] Indictment of Abraham Shakemaple of Finsbury, LMA MJ/SBR 1, f. 420. Other forges were punished in Drury Lane (1614) and near St Martin in the Fields (1642). William Le Hardy, ed. *County of Middlesex. Calendar to the Session Records* (1936) II, 124; LMA WJ/SP/1642/7/2.

[49] John Cordy Jeaffreson, ed. *Middlesex County Records Volume II. 3 Edward VI to 22 James I* (1887), 304.

[50] Rep 79, f. 111; LMA CLA/47/LJ/04/44; Jenner, *Early Modern Conceptions*, 187.

[51] *Palmer*, 538.

[52] *Rex & Regina v. Wilcox*, William Salkeld, ed. *Reports of Cases Adjudg'd in the Court of King's Bench; with Some Special Cases in the Court of Chancery, Common Pleas and Exchequer* (1717), II, 458.

contain several King's Bench cases originating as indictments in and around London. In a 1776 case against a Whitechapel manufactory of spirit of hartshorn (ammonia), the long indictment included reference to 'very great and fierce fires made of sea coal ... [which contributed to] quantities of stinking smoke, fumes, and noisome stenches'.[53] On the opposite end of the metropolis in Twickenham 'fires made of sea coal' similarly contributed to 'noisome, offensive, and stinking smoke'.[54] These opposing sides of the metropolis were held to demand opposing applications of the law, as Mansfield instructed the jurors to consider that Whitechapel was already so thoroughly dirty that it must accommodate smells and effluvia that would have been illegal in Twickenham. 'Where', he asked, 'is a beneficial trade to be carried on if not in such company?'[55] This evidence of prosecutions is necessarily anecdotal, as surprisingly little is known about the policing of London and its environment before the end of the seventeenth century. These cases therefore may or may not be tips of icebergs, but they do at least indicate that London's local officers could consider excessive coal smoke worthy of redress. Mansfield's comments, however, which drew upon earlier precedent, show that the legal principles involved were ambivalent, neither straightforwardly prohibiting nor tolerating heavy smoke emissions.

The final layer of local governance in England's capital which shared this concern was the Privy Council. The council in many ways moved in an entirely separate political sphere from the tradesmen or merchants who constituted the wardmote or sessions juries. Yet this group of England's leading aristocrats and officers, despite authority to consider the weightiest questions of statecraft, were also concerned by the state of London. Such local policing was not beneath their notice, particularly considering that London served as home for both the monarch and themselves, that it hosted foreign diplomats and merchants, and that it was the financial and commercial engine of the state. It therefore was responsive to petitions from inhabitants or local governing bodies who complained about the 'nauseous and great smokes' of a new brick kiln in Spitalfields or the 'great annoyance' of a brewhouse near Piccadilly.[56] This did not mean that the council was in any sense neutral; a complaint against a Lambeth glass manufactory was quietly dropped when it emerged that the Duke of Buckingham was its proprietor, and the council paid close attention to large-scale polluters near to St James's and Whitehall Palaces.[57] But in this sense, the council shared basic similarities with London's more humble office

[53] TNA KB28/298, Trin 1776, rot. 24, cited in Oldham, *Manuscripts*, 913, where he credits Janet Loengard for the reference and a transcription.
[54] *Rex v. White* (1757) in James Burrow, *Reports of Cases Argued and Adjudged in the Court of King's Bench*, (4th ed. 1790), I, 333.
[55] Oldham, *Manuscripts*, 889–90.
[56] TNA PC 2/61, 225, 239–40; PC 2/63, 166.
[57] TNA PC 2/58, 44–5, 59, 70; see the discussion of Whitehall in Chapter 4.

holders. For all magistrates attention to excessive coal smoke derived from both common sense and common law, from calculations of private as well as public interest, from their duties as governors and their preferences as neighbours.

The point of examining these layers of government, from the annual ward-mote meetings of London's householders up through the magistrates of the aldermanic bench and sessions, culminating in the Privy Council itself, has been to demonstrate how each of these facets of the English state offered options to Londoners annoyed by a neighbour's coal smoke. The previous chapter described a series of initiatives emanating from the very centre that viewed smoke clouds as a political problem. It was a top-down policy, though one that achieved collaboration and cooperation from officers and projectors alike. Here, however, policing smoke nuisance is less top-down, though it is not bottom-up either. Rather, it has been argued that there was social depth to environmental concern, that the desire to remove or reduce heavy smoke was shared by King Charles, widow Tanner, and the unnamed petitioners of Lambeth. This desire, furthermore, was consistent with the varying conceptions of nuisance law informing the decisions of both common law justices and London's local magistrates. Smoke could be seen to damage health, devalue property, annoy the general populace, or do all of these together.

Londoners could, then, wield a variety of governmental resources against smoky neighbours. But it does not follow that there was therefore an effective regime of environmental policing. To the contrary, the fact that both citizens and officers were capable of viewing excessive smoke as a problem must be considered alongside the evident limits and weaknesses of the environmental protections in practice. The common law, we have seen, promised compensation for those suffering property damage from 'sulphurous smoke'. But it offered neither abatement nor compensation for damaged health, nor was there any procedure for class actions.[58] It may well have been an attractive tool for some property holders, but this excluded many. Moreover, we know very little about how central courts actually applied this law or how Londoners used it. More research is also needed regarding London's local courts, but there is strong evidence that smoke nuisance cases were rare. The sample of hundreds of cases brought before the London sessions of the peace between October 1738 and September 1740 do not include any related to smoke nuisance, and only one among the many hundreds of Middlesex cases examined by Jeaffreson for the reign of James I dealt with 'ill-smelling seacoal'.[59] Manorial courts leet held authority over

---

[58] Common law precedent was ambivalent regarding whether creating unwholesome air was illegal, and medical opinion was hopelessly uncertain regarding standards of wholesomeness. Oldham's useful discussion shows that Mansfield's own views were inconsistent. Oldham, 887–93.

[59] London Sessions files, LMA CLA/47/LJ/1/758–73; Jeaffreson, *Middlesex Records*, 287–314, quote on 304.

some suburban districts, but in neither Southwark nor Finsbury is there a record of nuisance presentments including excessive smoke.[60] By the years around 1700 wardmote records are repetitive and formulaic, suggesting a moribund institution less likely to search for smoke nuisances than to submit, like one ward in 1707, 'we present *omnia bene*'.[61] All of this suggests, then, that while various governmental institutions offered Londoners possible means of redress for smoke nuisances, the use of these was occasional rather than regular.

IV. 'I SHOULD BE WILLING TO DEAL'; PRIVATE NEGOTIATION

The final approach considered by Bridgewater against his annoying neighbour was private negotiation, writing 'that that nuisance may be removed, I should be willing to deal but not upon too hard terms'. Most Londoners, of course, were not earls and could therefore not simply buy out any tradesmen that bothered them. Even so, a wide variety of private agreements and arrangements regarding smoke emissions and smoke management were used by inhabitants and corporate bodies throughout greater London. Private arrangements were in most ways preferable to the court cases described above, a set of options to be tried before conflicts led to formal complaints and law suits. Legal and administrative records often acknowledge this, framing their interventions as conflict resolution coming at the end of a failed process of informal arbitration. Viewers' reports, for example, prefaced their findings with a note that the parties were not 'able to reconcile the differences'.[62] Mansfield's records indicate personal attempts to redress complaints before proceeding to law, as in the case of a man who informed a smith 'that the neighbours apprehended a nuisance. He was very civil. [He] said he would take care not to act contrary to law, nor offend his neighbours more than necessary. Since the action [was] brought I spoke to defendant, but he said the court must determine'. In this case, and doubtless in many others, court actions came after personal discussions and were only needed in the absence of agreement.[63]

One form that such neighbourly agreement could take was direct negotiation. Buying out an annoying neighbour was not possible in most cases, but there are indications that it was surprisingly common nonetheless. Bridgewater contemplated a 'deal' with his soap boiler, wishing only that the

---

[60] Southwark, LMA CLA 43/1/9; CLA/43/1/13–16; P92/SAV/1323. Finsbury, 1746–70, LMA CLA/43/4.

[61] St Sepulcre's, LMA COL/AD/4/15.

[62] E.g., Viewers' Reports 1674–84, LMA COL/SJ/27/467, p. 10.

[63] *John Hooper v. John Lambe*, London, 25 July 1781, in Oldham, *Manuscripts*, I: 922; see also *The King v. John Oliver and James Allen*, Middlesex, 13 February 1779, *ibid.*, 915–6.

terms be 'not too hard'. One of the difficulties was knowing exactly what 'hard' meant, since Bridgewater admitted that his friend and representative Halsey was a better judge than he in these sorts of 'matters of value'.[64] An aggrieved former brewer complained in 1621 that his house had annoyed the Earl of Arundel, who had 'endeavored to put down and likewise buy the inheritance thereof at some easy rate'.[65] For Bridgewater and Arundel purchase was evidently considered an alternative to legal proceedings, but it could also follow them. In 1674 parishioners of Pye Corner, just west of the current St Bartholomew's Hospital, objected to a new brewhouse. The Court of Aldermen determined that the construction was illegal, standing too close to the city wall, but nevertheless instructed the petitioners to treat with the builder and 'purchase his interest in the said tenement for the better accommodation of this matter between them, and satisfaction of all parties'.[66] Even the King sometimes did the same. In 1664, brewer Robert Breedon was summoned to the Privy Council for annoying Whitehall. This was the very brewhouse cited by John Evelyn at the beginning of *Fumifugium*, whose smoke Evelyn claimed 'filled and infested' the court, 'to such a degree as men could hardly discern one another for the cloud'.[67] Nevertheless the king found Breedon suitably deferential and so determined that he should be compensated rather than prosecuted. He therefore ordered that Breedon cease brewing and the Lord Treasurer and Chancellor of the Exchequer 'repair and indemnify the prejudice and loss he may sustain'.[68] Buying out was a useful way to reconcile all parties and it therefore sometimes appeared preferable to the cost and uncertainty of law even to those with every reason to expect victory at court.

Most private negotiations with neighbours that did not proceed to court or come before magistrates are almost impossible to access for historians relying primarily on governmental archives, but we can get occasional glimpses of the range of possible strategies for redressing nuisances through the behaviour of institutions acting in a private or non-magisterial capacity. One common tactic was evidently to complain to a landlord who might then apply pressure to their tenant to abate nuisances. Records of such complaints survive in archives of institutional landlords, but the methods must have been much the same when landlords were private individuals. Thus, a Mr Barnes complained that a house which he let from the Fishmonger's Company was annoyed by the 'steam and smoke' of a brewhouse owned by

[64] HALS AH 1101. Bridgewater to Halsey, 27 May 1666.
[65] Petition of John Taylor, 1621, PA HL/PO/JO/10/13/7.
[66] Rep, 79, f. 239-239v, and earlier petition at Rep. 79, f. 215v-216.
[67] *Writings of Evelyn*, 129.
[68] TNA PC 2/57, 188, 196, 214. This order was not pursued, and Breedon's brewhouse remained for decades.

another of their tenants. The brewer, for his part, protested that he had made a private arrangement with Barnes and so the company suffered no injury.[69] In 1611 the Court of Aldermen determined, either after a complaint or through their own observation, that their workmen used fuels that 'defaced' Christ Church, Greyfriars, and therefore ordered that cleaner-burning charcoal would be used in the future.[70] A good chimney should carry smoke away rather than belching it back into the house, and builders and householders held bricklayers and builders responsible for defects.[71] Tenants, however, complained to their landlords, as in the case of the widows of the Merchant Taylors almshouse who complained of 'the great annoyance with smoke' in their dwellings.[72]

In such cases corporate bodies, even including the governors of the corporation of London itself, acted in a primarily private capacity as landowners responsible for the good repair of their assets and so remained vigilant against nuisances arising either from neighbours or tenants. But the distinction between the private and public roles of such bodies can become quite blurry in the cases of self-governing institutions such as the Inns of Court. The benchers of Lincoln's Inn acted comparably to other private landlords in refusing to grant leases to artisans deemed prone to committing nuisances, and they considered their options with respect to 'offence given by the smoke' of a neighbouring glazier.[73] But as a self-governing corporation the Inns also policed nuisances arising from within their membership. They addressed offending chimneys and both the Inner and the Middle Temple prohibited members from burning sea coal in their rooms.[74] These orders hide a process of (perhaps very unequal) negotiation within the company. As a corporate body the internal decisions of the Inns were as private as those made within a family, but heads of private families did not claim authority by grant of royal charters. In such cases the regulation of smoke nuisances

[69] Fishmongers Company Court Book, GL MS 5570/2, f. 29–30, 45–6.

[70] 1 October 1611, Rep. 30, f. 186v.

[71] E.g., a 1669 builder's contract stipulating that chimneys should 'carry away the smoke without any annoyance', GL MS17,182; and a court case against builders failing to fulfill a similar contract, KB27/1757 (Trinity term 1653), r. 1853-v. The Bricklayers and Tilers Company fined its own members for using defective workmanship when chimneys spewed out smoke 'extremely', 22 April 1613 GL CLC/LTG3047/1.

[72] Merchant Taylors' Company Court of Assistants, 5 November 1595, GL MS 34,010/3; Brewers Court Minute book, 23 May 1581, Guildhall MSS 5445/6.

[73] W. Paley Baildon, ed., *The Records of the Honorable Society of Lincoln's Inn. The Black Books Vol. III From A.D. 1660 to A.D. 1775* (1899), 70, 99, 101, 150, 264, 268.

[74] F. A. Inderwick, ed. *A Calendar of the Inner Temple Records Vol. II, James I. (1603) – Restoration (1660)* (1898), 126; Charles Henry Hopwood, ed. *Middle Temple Records: Minutes of Parliament Vol. III 1650–1703* (1905), 1059, also 1355, 1361, 1362, 1364, 1393, 1395, 1406.

was another instance of self-government by royal command, as smoky air was redressed by a typical mixture of public authority and private dealing.

Perhaps the most important manifestation of the ways that private agreement, public authority, and social power framed approaches to urban smoke can be seen in the private contracts which excluded smoky trades from much of metropolitan London. The development of the West End – the quarter of newly built streets and spacious squares where gentry and aristocratic families increasingly possessed townhouses from the mid-seventeenth century onwards – has long been associated with its clean air. In the 1660s, Wiliam Petty claimed that prevailing westerly winds explained why 'the dwellings of the west end are so much the more free from the fumes, steams, and stinks of the whole easterly pile, which where seacoal is burnt is a great matter'.[75] Since then it has remained common sense to accept Petty's climatic explanation for the West End's the relative airiness.[76] Winds certainly helped, but contemporaries were not confident that wind alone could disperse smoky air.

Landlords, therefore, frequently prohibited or punished nuisances or noxious trades in leases and building covenants. Such restrictions certainly did not apply only to smoky manufacturing; stench, fumes, noise, and even excessive traffic were among the urban annoyances that such developers sought to avoid. But manufacturers like brewers, smiths, distillers, and makers of soap, glass, candles, and tobacco pipes, all of which were widely associated with heavy coal smoke, usually had prominent places in lists of prohibited trades. The Earl of Bridgewater himself included such restrictions in contracts with lessees in the Barbican, even evicting one under-tenant who had set up a smith's forge, a specifically prohibited trade.[77] The Earls of Holland prohibited no trades except tobacco pipe manufacture from their houses near Great St Bartholomew.[78] Very little is known about such prohibitions within the ancient walled city and its immediate suburbs, but it is clear that they were considered crucial in establishing and maintaining the desired tone in the new residential developments of the West End. Thus, the Russells' development in the area around Covent Garden after the Restoration prohibited lessees from any trade or building whereby other tenants would be 'annoyed by smoke or any noisome or stinking savour'. This clause was then usually followed with a list of prohibited shops, with

---

[75] Hull, ed. *Economic Writings*, 41.

[76] E.g., Lawrence Stone, 'The Residential Development of the West End of London in the Seventeenth Century', in Barbara Malament, ed. *After the Reformation: Essays in Honor of J. H. Hexter* (Manchester, 1980), 190; Clive Ponting, *A New Green History of the World: The Environment and the Collapse of Great Civilisations* (2011), 352.

[77] HALS AH 1122; Northamptonshire RO E(B) 704.

[78] Holland leases, TNA E 214/40; E 214/57.

smith, tallow chandler, soap boiler, and brewer all mentioned among other
offensive or undesired trades.[79] The Earl of Burlington stipulated similar
exclusions for tenants near what is now the Royal Academy off Piccadilly,
as did the Sidney Earls of Leicester for Leicester Square, the Pulteneys for
their lands just north of the current Piccadilly Circus, and Lord Cheyney in
Chelsea.[80]

The Grosvenor family used a slightly different tactic to develop its
Mayfair estates; instead of prohibiting noxious or annoying trades outright
they punished offenders with increased rent. Their fine was £30 annually
for tenants who allowed any of the denominated trades. The Grosvenors'
leases, like the Russells', varied somewhat, but a typical list included: 'a
butcher's house or shop, slaughter house, tallow chandler, melter of tallow, soap maker, tobacco pipe maker, brewhouse, victualler, coffee house,
distiller, farrier, pewterer, working brasier, or blacksmith'.[81] A penalty of
£30 was substantial compared to rents ranging from under £10 to around
£30. So while it may have been possible for an unusually successful shop to
afford this, the Grosvenors' intention seems to have been the same as that of
the Russells, Sidneys, Pulteneys, and Cheyney: to exclude smoky, dirty, loud,
and busy trades from their estates.

Through such restrictions – negotiated privately and recorded in legally
binding contracts – rich families sought to create quarters fit for rich families, urban spaces for urbane and polite inhabitants. The creation of an
entirely rich and non-industrial zone was not quite practical, as some inhabitants of Piccadilly learned in the 1670s when they appealed, unsuccessfully,
to the Privy Council against one of the two brewhouses that would eventually occupy Brewer Street.[82] But the West End was a much more uniformly
residential, elite, and non-industrial zone than anything else that London,
or any other British city for that matter, had ever seen. Its creation as such,
moreover, was the deliberate policy of aristocratic capitalists who used private contracts to exclude smoky trades in ways that the more famous protections of the common law never accomplished.

V. CONCLUSION: LIMITS AND FAILURES

Private negotiation, by contract or conversation, joined recourse to the law
and to the magistrate as the avenues available to Londoners wanting to

---

[79] LMA E/BER/CG/L22/1–27.
[80] *Survey of London: volumes 31 and 32: St James Westminster, Part 2* (1963), 9, 458; *Survey of London: volumes 33 and 34: St Anne Soho* (1966), 428; TNA C 8/410/12.
[81] Grosvenor Papers, Mayfair Building Agreements. 1720s-30s, City of Westminster Archive Centre [hereafter WAC] 1049/3/3/1.
[82] *St James Westminster*, 118–9; PC 2/63, 166, 171.

redress or prevent excessive coal smoke. That such a wide variety of legal and administrative institutions recognized heavy smoke as a public nuisance or as a threat to property suggests that the early modern period was very far from being as environmentally apathetic as many historians have suggested. Indeed this discussion has illustrated the very wide range of forms and formats that concern with coal smoke took: it was written into contracts with builders and tenants, it spurred conflicts between neighbours, it animated the action of magistrates, and it informed the common law's approach to competing land use. Inhabitants troubled or bothered by smoking chimneys thus had several options, public and private, institutional and personal, and juridical and informal. Furthermore, they took advantage of all of these, demonstrating that it was not merely kings, courtiers, and intellectuals who noticed and objected to their increasingly smoky urban environment.

Nevertheless, this discussion has also stressed that while local conflicts over smoke demonstrate something important about early modern environmental attitudes, they also show some of the weaknesses and limits of early modern environmental protections. Londoners could and did take their neighbours to court, but this was time-consuming, expensive, and only offered monetary compensation for past damages; it offered little help for those suffering impaired health, for those unable to afford access to the courts, or for those living in areas deemed fit for industrial production. London's many local courts could and did prosecute those who produced the smoke, but their officers were swamped with the impossible work of punishing urban theft, vice, violence, and negligence. Smoke, in such circumstances, was usually a low priority. The capital's inhabitants could and did come to private agreements regarding smoke, but this had very uneven effects, as the rich were naturally the most able to pay premiums for lower population densities and non-industrial residential districts. Perhaps most importantly of all, however, all of the tactics discussed above were local, limited, and specific responses to local and specific nuisances. They were largely useless in isolation against the real causes of city-wide smoke pollution. Londoners, then, had many ways to approach a neighbour emitting an unusually heavy smoke cloud, but they had no ways to redress the mundane fires in many thousands of hearths that produced the city's general haze. While early modern city-dwellers did in fact demonstrate concern for their urban environment centuries before the rise of the modern industrial city, they did not achieve any methods for systemic environmental improvement.

$-6-$

Smoke in the scientific revolution

I. THE DEATH OF AN OLD OLD VERY OLD MAN

During the 1630s the Earl of Arundel was, after only the king himself, England's greatest collector. His continental travels in company with Inigo Jones had convinced both of the virtues of classicism, and though Arundel's attempt to acquire the obelisk that now adorns Rome's Piazza Navona failed, he did fill his house on the Strand with the ancient sculptures known as the 'Arundel Marbles' as well as an exceptional collection of modern art. While Arundel retains a reputation as a key figure in the formation of English taste, it is less appreciated that he also worked practically to promote beauty, order, and decorum on the ground in London: as a commissioner of buildings he policed illegal urban sprawl; as Earl Marshall he enforced standards of chivalry; as a householder he pursued court cases against neighbouring brewers and their smoke; as a member of the House of Lords he introduced the Brewery Bill into the 1624 parliament; and as a Privy Councillor he was among the most active in pursuing Westminster's smoky brewers.[1]

Given this interest in the urban environment, perhaps Arundel should have thought twice about bringing people as well as objects to dirty London. His acquisitions abroad included the brilliant engraver Wenceslas Hollar, and his travels within England brought to his attention a humble labourer, Thomas Parr, who was said to be 152 years old. This human antiquity was acquired too, and moved from his native county of Shropshire to the capital where he was presented to the king at court during the autumn of 1635. Within a few weeks, however, 'old Parr' was dead, and a team of royal physicians was assembled to learn why. Their opinion, according to their most illustrious member, William Harvey, was that Parr had been killed not by the

---

[1] PC 2/43, 239–40; 2/47, 122; 2/52, 454; Malcolm Smuts, 'Howard, Thomas', *The Oxford Dictionary of National Biography* [hereafter *ODNB*]; Linda Levy Peck, *Consuming Splendor: Society and Culture in Seventeenth Century England* (Cambridge, 2005), 126–7, 168–71, 203–12; Chapter 4.

natural decay of his prodigious age but by a change in the 'non-naturals', – specifically, by London's smoke.

Parr, in Harvey's account, was remarkably robust. He had worked his entire life, only reducing his exertions after his 130th year. The stories that Parr had maintained an active sex life well after the age of 100 were, according to Harvey's examination, entirely plausible physically. While he needed some assistance to walk, had a failing memory, and was blind, Parr was also able to answer questions clearly, retained a muscular physique, and in sum 'might perhaps have escaped paying the debt due to nature for some little time longer' except that his journey to London had 'interfere[d] with the old man's habits of life'. Such 'habits', collectively known as the 'non-naturals' played a key role in early modern medicine theory. Parr's body had been disordered through rich food, strong drink, and, above all, a change of air. His native country enjoyed 'open, sunny, and healthy' air, but 'in this grand cherisher of life this city is especially destitute'. London, Harvey continued, was 'a city whose grand characteristic is an immense concourse of men and animals, and where ditches abound, and filth and offal lie scattered about, to say nothing of the smoke engendered by the general use of sulphurous coal as fuel, whereby the air is at all times rendered heavy, but much more in the autumn that at any other season'.[2] Harvey therefore 'judged, indeed, that he had died suffocated, through inability to breathe, and this view was confirmed by all the physicians there present, and reported to the King'.[2] Parr was buried, perhaps owing to a guilty sense that he never should have been brought to London at all, surrounded by poets and scholars in the south transept of Westminster Abbey.

Parr's life and fatal progress to London were described in a popular pamphlet, 'The Old Old Very Old Man' by the famous waterman-poet John Taylor, and memory of 'old Parr' has persisted from the seventeenth into the twenty-first centuries.[3] The oddness of Parr's story, however, seems to

---

[2] Robert Willis, ed. *The Works of William Harvey, M.D. Physician to the King, Professor of Anatomy and Surgery to the College of Physicians* (1847), 590–1; a translation of 'Anatomia Thomæ Parri annum Centesimum quinquagesimum secundum & novem menses agentis. Cum Cl. Viri Guiliullmi Harvæi aliorumque adstantium medicorum regiorum observationibus' in John Betts, *De Ortu et Natura Sanguinis* (1669), 317–25, esp. 323–4.

[3] John Taylor, *The Old, Old, Very Old Man* (1635). Laud's diary notes his appearance at court but not his death. *The Works of the Most Reverend Father in God, William Laud, D.D.* (Oxford, 1847–60), III, 225. A 1722 translation of a French work claimed that 'every one knows the story of Thomas Parr'. The original French says that James II himself told Parr's story while in exile. Harcouet de Longeville. *Long Livers: A Curious History of Such Persons of Both Sexes Who have liv'd Several Ages, and Grown Young Again*, trans. Robert Samber, (1722), 89. An eighteenth-century antiquary reported that Parr's son, grandson, and great grandson lived to be 113, 109, and 124 years old, respectively. *Additional Collections Towards the History and Antiquities of the Town and County of Leicester* (1790), 978. In 2015, Parr has a Wikipedia page, which lists numerous references to him in nineteenth to twenty-first century media and culture.

have prevented much serious attention to it from historians of medicine or environmental thought. This is unfortunate because Parr died in the middle of – and thereby quite possibly contributed to – Charles's campaign against excessive smoke in London and Westminster. His death was attributed to the same noxious smoke that Charles perceived to threaten the beauty of his court and the health of his family, and this view was endorsed by a team of experts headed by the most illustrious physician England had yet produced. William Harvey concluded that London smoke was too heavy for the lungs to ventilate the heart, without which Parr's body soon faltered. What may have been most alarming about Harvey's evaluation was that it was grounded in Parr's status as a foreigner to urban air rather than the weakness of his great age. If London smoke could so disrupt Parr's body, other newcomers to the capital might have been in similar danger. For a king with a French-born queen this would have been an unwelcome possibility, and Harvey's diagnosis could therefore have contributed to the royal campaign to reduce smoke around Westminster during the 1630s.

Parr's death and autopsy suggest that during the 1630s those most learned in medicine could consider urban smoke a serious danger. Few historians, however, have accepted this, arguing instead that smoke was among the urban nuisances which early modern medicine found *least* troubling. Environmental historians have found early modern complaints regarding smoke, above all John Evelyn's in *Fumifugium*, as primarily rhetorical and therefore not grounded in serious science. One scholar of modern environmentalism dismisses Evelyn and argues that in general his contemporaries thought smoke 'caused no harm to health', while another agrees that Evelyn had no impact because he could not back up his claims with 'medical science'.[4] Specialists in early modern culture, medicine, and natural philosophy have stressed how environmental factors were considered crucial to health, but have argued that the tradition of 'airs, waters, and places', deriving from the ancient Hippocratic corpus, focused on climate and local geography rather than the particularities of urban environments. Furthermore, they have stressed that it was rotting organic matter and noxious chemical fumes, rather than smoke, which most powerfully aroused the disgust of early modern city-dwellers. In these ways, it has been found, early modern students of medicine and natural philosophy did not consider coal smoke to be important.[5] The opposite position, however,

[4] Thorsheim, *Inventing Pollution*, 17; Joachim Radkau, *Nature and Power: a Global History of the Environment* (Cambridge, 2008), 143.
[5] Cockayne, *Hubbub*, 229, 241; Jan Golinski, *British Weather and the Climate of Enlightenment* (Chicago, 2007), 60–1 *et passim*, but see especially chapter 5, 'Sensibility and Climatic Pathology'; Glacken, *Traces on the Rhodian Shore*; Alain Corbin, *The Foul and the Fragrant: Odor and the French Social Imaginary* (Cambridge, MA, 1986), 66; Jo Wheeler, 'Stench in Sixteenth-century Venice', in Alexander Cowan and Jill Steward, eds. *The City and*

has been taken by literary critics who have found not only that concern for urban smoke existed but also that it was informed by developments in contemporary natural philosophy.[6] Ken Hiltner has stated this most categorically in his study of Renaissance pastoral literature. Early modern people, he argued, clearly 'knew' smoke to be dangerous and unhealthy, and therefore their failure to act on that knowledge is intriguingly similar to our own comparable failure.[7]

Here, then, are two completely contradictory narratives of air pollution's importance. One argues that its damaging effects were a matter of widespread knowledge by the 1660s, another that contemporary science had little place for worries about smoke. This chapter, however, will tell a story different from both of these, one in which early modern science and medicine could support claims that smoky air was unhealthy, but did not conclusively demonstrate either the scale or the nature of its dangers. The Hippocratic focus on airs, waters, and places was, in fact, quite capable of application to new urban problems, including smoky air. There was much diversity within broadly Hippocratic approaches to environmental medicine, as bad airs arose from multiple causes, had various natures, and implied divergent responses. Most importantly, the nature and function of air itself became a crucial question – in some ways even *the* crucial question – during the scientific revolution, so it is quite misleading to suggest that the enduring importance of Hippocratic thought entailed stagnation. Coal smoke, then, was indeed often seen as bad air during the seventeenth and eighteenth centuries, but in ways that changed along with broader understandings of the body and its environment. It was widely found to be medically dangerous, but in ways that did not seem certainly demonstrable nor applicable to all.

II. 'SULFUROUS COAL': SMOKE AS BAD AIR

When Harvey found smoke to have sped Parr's demise he was drawing on two millennia of medical knowledge that had long established the importance of good air as a commonplace of both learned and lay thinking. Harvey is most famous for his radical reformulation of the role of the blood, but other aspects of his thinking were quite conventional and there is little reason to imagine that either the other royal physicians who 'confirmed' his diagnosis of Parr or King Charles to whom their findings were 'reported'

---

the Senses: *Urban Culture since 1500* (Aldershot, 2007), 28; Cavallo and Storey, *Healthy Living*.

[6] Todd Borlik, *Ecocriticism and Early Modern English Literature: Green Pastures* (New York, 2011), 158–64; Diane Kelsey McColley, *Poetry and Ecology in the Age of Milton and Marvell* (Aldershot, 2007), 80–5.

[7] Hiltner, *What Else is Pastoral?* ch. 5.

would have been surprised by the idea that bad air could kill. Good air's central contribution to health was endorsed by the most popular medical reference works, which defined it through its lack of dangerous smells, corruption, and contamination: 'fair and clear … lightsome and open' or 'pure and void of infection.'[8] The key text came from the medieval regimen of Salerno, translated by Sir John Harington as 'let air be clear and light, and free from faults, That come of secret passages and vaults.' For Harington, as for many other English students of the Hippocratic and Salernitan tradition, there was little need to elaborate on the nature, constitution, and operation of good air.[9] By the late sixteenth century the importance of good air to public health was long-established, and if its precise working attracted limited attention, its sources were easily identified. The poet and pamphleteer Thomas Dekker, and most likely very many of as his readers as well, feared dangerous airs emanating 'from standing pools, or from the wombs, Of vaults, of muckhills, graves, and tombs, From bogs; from rank and dampish fens, From moorish breaths, and nasty dens'.[10]

Placing smoky air within this very common but also rather unspecific set of assumptions was not difficult and took no particular medical expertise. The text of the 1624 Breweries Bill referred to the 'health and soundness' of London and Westminster and called coal smoke 'unwholesome and unsavoury'.[11] The register of the Privy Council similarly claimed during the 1630s that brewery smoke 'may endanger the health' of the royal family, and again in the 1660s local smoke emissions were 'prejudicial to the health' of the king and queen.[12] Legal records from the second half of the seventeenth century called coal smoke 'sulfurous', 'putrefied', insalubrious', and 'unwholesome'.[13] Damage to bodily health through bad air, moreover, was expressly stated as one criterion for a complaint to be a nuisance at common law.[14] In eighteenth-century

---

[8] William Bullein, *The Government of Health* (1595), 30; Thomas Cogan, *The Haven of Health* (1634 ed.); Thomas Moffett, *Health's Improvement* (1655 ed.), 13-,29; Tobias Venner, *Via Recta ad Vitam Longam* (1650), 1, 5. The treatment of air in Italian medical texts is well surveyed in Cavallo and Storey, *Healthy Living*, ch. 3.

[9] *The Englishmans docter. Or, The schoole of Salerne* (1607), sig. A8; the original text is 'Aer sit mundus, habitabilis ac luminosus,/Nec sit infectus, nec olens foetore cloacae'. John Ordonaux, ed. and trans., *Code of Health of the School of Salernum* (Philadelphia, 1871), 56. Manuscript notes with similar content include BL Sloane MS 738, f. 115, Sloane 104, f. 23-4; Sloane 364, f. 14-5. Italian medical authors did, however, write much more voluminously on air in the sixteenth and early seventeenth centuries. Cavallo and Storey, *Healthy Living*, 70.

[10] Thomas Dekker, *Nevves from Graues-end Sent to Nobody* (1604), sig. C3v.

[11] 'An Act concerning brewhouses in and about London and Westminster' PA HL/PO/ JO/10/4/1.

[12] TNA PC 2/43, 239; 2/57, 188.

[13] TNA KB27/1754, r. 282; *King v. Brooks*, 30 Charles II (1678), Tremaine, Rice, and Vickers, *Pleas of the Crown*, sub. 'Indictments and Informations for Nuisances'.

[14] *The Gentleman's Assistant, Tradesman's Lawyer, and Country-Man's Friend* (1709), 165.

cases defendants against nuisance complaints argued that fumes and smokes were not medically harmful, a response to the close association between legal nuisance and the unhealthiness of fumes, vapours, and airs.[15] Most of the discussion of coal smoke as nuisance in the previous two chapters, therefore, was implicitly, and sometimes explicitly, also about its effects on health. It was through such medical danger, in other words, that coal smoke became a legal nuisance. But while there were specific legal implications to smoke's insalubrity, assertions that smoke could make one sick were certainly not restricted to complaints, legal reports, or any other genre. On the contrary, early modern letters often claimed that urban smoke was unhealthy. 'Pray leave the smutty air of London, and come hither to breath sweeter', wrote James Howell in 1625, and numerous seventeenth- and eighteenth-century successors agreed that London air could confidently be judged unhealthy even by those without medical expertise.[16]

One did not need medical training, therefore, to believe that coal smoke was unhealthy. But changes in early modern medical theory and natural philosophy nevertheless had profound impacts on how exactly smoke was thought to act on bodies. The nature of the air and the functions and processes of breathing were intensively studied in ways that made Harington's brief and epigrammatic definition of good air seem vague and insufficient. The new ideas, research methods, and instruments of demonstration and assent, all of which are often collectively labelled as 'the scientific revolution', changed the ways that coal smoke was said to constitute bad air. New approaches to air, in particular those grounded in alchemy and mechanistic physics, allowed physicians, philosophers, and their followers to make newly specific claims regarding how smoke threatened health and what kind of threat it posed.

Among the most important ways in which alchemical theory informed discussions of coal smoke was through the substance, and the idea, of sulfur. Harvey's use of 'sulfurous' to describe the nature of coal smoke demands examination as a term with complex early modern implications. Newcastle coal does indeed contain high levels of the element sulfur, a substance with which early modern people were quite familiar. But to read every reference to sulfurous air as a reference to this kind of sulfur and so to coal smoke is to miss out on a broader learned conversation that viewed sulfur as much more than the yellowish substance from which gunpowder was made.[17] Sulfur, according to alchemical theory, was one of the three primary materials that early modern alchemical philosophers used to explain the formation

---

[15] Oldham, *Mansfield*, 887–8.
[16] James Howell, *Epistolæ Ho-Elianæ. Familiar letters, Domestick and Foreign* (1705 ed.), 156.
[17] For early modern 'sulfur' as equivalent to the element, see especially Hiltner, *What Else is Pastoral?* ch. 5. Also cf. Brimblecombe, *The Big Smoke*; Barbara Freese, *Coal: A Human History* (Cambridge, MA, 2002), 27; Richards, *The Unending Frontier*, 235.

of matter, and was also a kind of principle of combustibility that helped
explain the transformative power of heat. Such alchemical thinking drew on
a millennium of Arabic- and Latin-language engagement with Aristotelian
philosophy, but was also newly prominent and newly dynamic during the
sixteenth and seventeenth centuries. Recent work by historians of science
has placed this alchemical tradition at the very heart of early modern science, demonstrating that figures like Boyle, Newton, and others can best
be understood as practitioners of alchemy.[18] When early modern physicians
and philosophers called coal smoke 'sulfurous', therefore, they were engaging with cutting-edge thinking regarding the nature of matter, processes of
transformation and transmutation, and the ways in which the contents of the
air acted upon the human body.

It was commonly accepted that coal itself abounded in sulfur, and its
smoke therefore partook of this constitution.[19] Evelyn complained that
London's air, which would otherwise have been pure and serene, 'is here
eclipsed with such a cloud of sulfur' so as to block out the sun. The 'clouds
of smoke and sulfur, so full of stink and darkness', were a medical as well as
an aesthetic and a political problem for Evelyn. He described the progress
of the 'gross and dense' vapours into the windpipe and so to the lungs and
thence into the heart and so on into 'the noble parts, vessels, spirits, and
humors'.[20] Despite claims that Evelyn's arguments lacked medical authority, not only had Evelyn studied medicine and chemistry, but *Fumifugium*
draws on several learned contemporaries who similarly understood smoke
through chemical categories and assumptions.[21] In addition to Hippocrates,
Avicenna, and Paracelsus, Evelyn cited Kenelm Digby, a virtuoso deeply
engaged with alchemical circles in Paris who blamed Londoners' respiratory
troubles on the coal smoke's chemical nature.[22] Evelyn also claimed to have

---

[18] Among the best recent summaries of this large literature include William R. Newman,
      Atoms and Alchemy: Chymistry and the Experimental Origins of the Scientific Revolution
      (Chicago, 2006); Bruce T. Moran, *Distilling Knowledge: Alchemy, Chemistry, and the
      Scientific Revolution* (Cambridge, MA, 2006); Lawrence Principe, *The Secrets of Alchemy*
      (Chicago, 2013).
[19] On coal as sulfurous see, e.g., Simon Sturtevant, *Metallica* (1612), 105; Gerald Malynes,
      Consuetudo, vel lex mercatoria, or the Ancient Law-Merchant (1622), 49; Robert Hooke,
      The Posthumous Works of Robert Hooke: With an Introduction by Richard S. Westfall,
      The Sources of Sciences, no. 73 (1705; reprint, New York: Johnson Reprint Corporation,
      1969), 46–56; Dud Dudley, *Metallum Martis OR, IRON Made with Pit-coale, Sea-coale*,
      &c. (1665), 10–11; George Sinclair, *Natural Philosophy Improven by New Experiments*
      (Edinburgh, 1683), 212.
[20] Bédoyère, *Writings of Evelyn*, 131, 143.
[21] Evelyn's medical, chemical, and natural philosophical studies and contacts are discussed in
      Gillian Darley, *John Evelyn: Living for Ingenuity* (New Haven, 2006).
[22] Bédoyère, *Writings of Evelyn*, 143; citing Kenelm Digby, *A Late Discourse made in a Solemne
      Assembly of Nobles and Learned Men at Montpellier in France* (1658), 39–41; Betty Jo
      Dobbs, 'Studies in the Natural Philosophy of Sir Kenelm Digby', *Ambix* 18 (1971), 1–25;

discussed smoke's role with members of the Royal College of Physicians. A marginal note, moreover, specified that his references to sulfur should be taken as its 'gross and plainly virulent vapours' rather than to the carefully prepared form endorsed by some unnamed physicians.[23] This could have been a concession to Thomas Willis, a leading physician and natural philosopher who stressed that sulfurous coal smoke was beneficial to certain patients, even while admitting its harm to others. Willis claimed that 'sulfur is justly reported by chymists to be the balsam of the lungs'. Therefore, while many consumptives 'hasten to remove out of the city-smoke into the country', Willis wanted to justify the contrary practice, explaining why 'some avoid this city as hell, [but] others fly to it as to an asylum'.[24] Other physicians, however, were closer to Evelyn's position. Writing in 1666, Gideon Harvey agreed with Evelyn that 'sulfurous coal smoke' damaged the lungs and contributed substantially to making consumption an epidemic, the 'English disease'.[25] These opposing positions seem to have less to do with sulfur's effects and more with its basic nature; as Evelyn stressed, there were multiple kinds of sulfur with multiple influences on human bodies.

Prominent as sulfur was, there were other ways in which chemical concepts seemed to explain smoke's effects.[26] After asserting London smoke's sulfurous nature, Evelyn asked whether it also contained 'an arsenical vapor, as well as sulfur', as had been argued by the Dutch physician and biblical scholar Arnold Boate in a discussion published by Samuel Hartlib. Coal, according to Boate, caused respiratory diseases 'not only by the suffocating by abundance of smoke, but also by virulency. For all subterranean fuel hath a kind of virulent or arsenical smoke, which as it speedily destroys those that dig in mines, so doth it by little and little those who use them here above'. This raised the worrying prospect that the substances causing subterranean 'damps' and leading to sudden death in miners was also slowly poisoning Londoners who were similarly exposed to coal's 'virulent or arsenical' qualities.[27] Others stressed that smoke's danger derived from

Lawrence Principe, 'Sir Kenelm Digby and His Alchemical Circle in 1650s Paris: Newly Discovered Manuscripts', *Ambix* 60:1 (February 2013), 3–24.

[23] Bédoyère, *Writings of Evelyn*, 140.

[24] Thomas Willis, 'Pharmaceutice Rationalis: Or the Operations of Medicines in Human Bodies. The Second Part' in *Dr. Willis's practice of physick* (1684), 32–3, 40.

[25] Gideon Harvey, *Morbus Anglicus: or, The Anatomy of Consumptions* (1666), 166–7. See also Harvey, *The Art of Curing Diseases by Expectation* (1689), 104.

[26] Other explorations of smoke's sulfurous qualities include Ralph Bohun, *Discourse of Winds* (1671), 202; Thomas Tryon, *Miscellania: or, A Collection of Necessary, Useful, and Profitable Tracts* (1696), 105; Tryon, *Monthly Observations for the Preserving of Health with a Long and Comfortable Life* (1688), 36, 66.

[27] Arnold Boate, 'To the Second Letter of the Animadversor', in *Samuel Hartlib, His Legacy of Husbandry* (1662), 138. For other references to coal smoke's 'arsenical' qualities see Dud Dudley, *Metallum Martis*, (1665), 61.

a salty nature. Digby considered smoke's dangerous and insinuating qualities to arise from its 'great quantity of volatile salt, very sharp' and Gideon Harvey too thought that coal consisted of 'saline corrosive steams [which] seem to partake of the nature of salt ammoniac, whereby they gnaw and in time ulcerate the tender substance and small veins of the lungs.[28] Boyle, finally, suggested that 'those numerous fires ... burning in our chimneys produce much saline smoke'.[29]

Coal smoke was also thought to offend through the mechanical effects of its weight. For some this kind of explanation was preferred to alchemical theory. Archibald Pitcairn, for example, was a Scottish physician and professor popular with English medical students at Leiden who attacked Willis's chemical reading of smoke's properties. He argued that Willis had only 'perplex[ed] the question with the fumes of sulfur, and such words', while a true explanation would have been grounded upon 'the known property of the gravity, which is greater in those particles of sea coal that pass into smoke, and are drawn in with the air, than it is in those of turf and wood'.[30] Such disputes between champions of chemical and physical approaches were probably, however, less common than eclecticism. Thus, Boyle's reference to chimney fires producing saline smoke was in the context of a discussion of air as a 'menstruum', a mixture of various constituent components acting on each other in ways consistent with the 'compressible and dilatable' qualities that his air pump experiments had famously examined.[31] Digby fused mechanical and chemical explanations, while a correspondent of John Locke, who himself espoused a chemically based theory of respiration, referred to the ill effects of London's 'heavy' smoke on Locke's asthmatic lungs.[32] In his pioneering study of consumption Richard Morton argued that coal smoke both prevented the blood's necessary fermentation and also 'stuffed' and blocked the operation of the lungs.[33]

---

[28] Digby, *A Late Discourse*, 38–9; Harvey, *Morbus Anglicus*, 166–7.

[29] Robert Boyle, 'The General History of Air,' in Michael Hunter and Edward B. Davis, eds. *The Works of Robert Boyle* (2000), XII, 31. Boyle also claimed that 'the abundant smoke of pit coal, uses to be very offensive' for use in chemical laboratories. Robert Boyle, *Some Considerations Touching the vsefulnesse of Experimental Naturall Philosophy* (1663), 176.

[30] Archibald Pitcairn, *The Whole Works of Dr. Archibald Pitcairn* (1727), 126–7; Anita Guerrini, 'Archibald Pitcairn', *ODNB*.

[31] Boyle, 'General History of Air', 12. Boyle's corpuscular theory of air was also described in 'An experimental discourse of some little observed causes of the insalubrity and salubrity of the air and its effects', Hunter and Davis, *Works*, X, 314. On Boyle's mechanism in general see Michael Hunter, *Boyle: Between God and Science* (New Haven, 2010), 116–8.

[32] E. S. De Beer, *The Correspondence of John Locke* (Oxford, 1976–89) vol. III, 592; Jonathan Walmsley, 'John Locke on Respiration', *Medical History* 51 (2007): 453–76.

[33] Richard Morton, *Phthisiologia, or, A Treatise of Consumptions Wherein the Difference, Nature, Causes, Signs, and Cure of all Sorts of Consumptions are Explained* (1694), 66, 246.

By the early eighteenth century it was increasingly common to employ both physical and chemical explanations in discussions of coal smoke as bad air. A medical treatise by Thomas Fuller, for example, discussed the implications of smoky air's pressure and weight and also argued that the lungs were damaged by the shape of sulfur atoms, with their 'acid sharp points and edges'.[34] The prominent physician George Cheyne similarly combined physical and chemical assessments of London's 'one universal nitrous and sulfurous smoke'.[35] Physician John Burton agreed that in 'large cities' (of which England had exactly one) the 'innumerable coal fires' both 'loaded' the lungs and deprived them of the 'vivifying spirit' or 'nitro-aerial particle' which chemically-informed medicine had long believed explained the role of respiration.[36] An anonymous pamphlet, long wrongly attributed to the leading Newtonian John Theophilus Desaguliers, condemned the smoke of a steam pump near the Strand both for its heaviness and sulfurous particles.[37]

The apogee of such accounts was achieved in the fullest treatment of air's medical properties produced in Britain during the entire early modern period. In his 1733 *Essay Concerning the Effects of Air on Human Bodies*, John Arbuthnot, a leading physician as well as a prominent literary satirist, synthesized the Hippocratic approach to environmental medicine with recent research into air's composition and chemistry by Boyle, Stephen Hales, Herman Boerhaave, and the secretary of the Royal Society James Jurin. Like his contemporaries Cheyne and Fuller, Arbuthnot applied chemical analysis to air's composition and mechanistic physics to its effects on the action of the lungs. He asserted that 'the air of cities is not so friendly to the lungs as that of the country, for it is replete with sulfurous steams of fuel', which both weakened air's elasticity and damaged the lungs.[38] He similarly argued that asthmatics in particular should avoid the smoke of urban fuel because it led both to 'the danger of suffocation' and to a wide variety of diseases following from 'an imperfect respiration'.[39] Arbuthnot's work was a fulfilment of Boyle's projected 'general history of the air', an attempt in the Baconian spirit to distinguish the causes and principles which led to so much variety. Medical and philosophical study of the air had certainly

---

[34] Thomas Fuller, *Exanthematologia: Or, An Attempt to Give a Rational Account of Eruptive Fevers* (1730), 85–6.

[35] George Cheyne, *The English Malady: or a Treatise of Nervous Diseases of all Kinds* (1733), 38–9; *An Essay of Health and Long Life* (1745 ed.), 11.

[36] John Burton, *A Treatise on the Non-Naturals. In which the Great Influence They have on Human Bodies is Set Forth, and Mechanically Accounted for* (York, 1738), 63–5.

[37] *The York-Buildings Dragons* (1726), 6–7. The case against his authorship is made in Audrey T. Carpenter, *John Theophilus Desaguliers: A Natural Philosopher, Engineer and Freemason in Newtonian England* (2011), 138–40.

[38] John Arbuthnot, *An Essay Concerning the Effects of Air on Human Bodies* (1733), 208.

[39] *Ibid.*, 108, 215–6.

progressed markedly from Harington's translation of the Salernitan regimen; if in 1600 one could explain air in a couplet of poetry, by the 1730s Arbuthnot required a volume of over 200 pages.

As understandings of the body, breathing, and the nature of the air changed, therefore, understandings of coal smoke's composition, operation, and significance changed with them. Evelyn was certainly not alone in his attention to urban smoke; rather, many of the leading philosophers and physicians of the seventeenth and eighteenth centuries also thought carefully about the relationship between the new urban environment and the body. They shared, in varying degrees and in diverse ways, Evelyn's concern with urban smoke. But this does not mean that they agreed that London needed to be liberated from what Evelyn called its 'clouds of smoke and sulfur'.[40] For while these authors did perceive smoke as damaging or dangerous, they did so in ways that were contained and specific. Coal smoke, they found, was not a serious problem for everyone.

This difference between Evelyn's and his contemporaries derives in part from the different conditions under which their texts were produced and consumed. *Fumifugium*'s publication history suggests that it was never meant to be sold commercially at all, but rather to circulate in a limited way as befitted a document primarily intended for the king.[41] Physicians like Gideon Harvey, Cheyne, or Arbuthnot, by contrast, wrote for a public audience that needed advice on curing or avoiding specific diseases. Such books therefore approached smoky air insofar as it contributed to therapy, stressing its importance for particular medical conditions. In such literature smoke was particularly associated with respiratory complaints, especially asthma, as well as with consumption. For example, Gideon Harvey's observation that 'at a distance London appears in a morning as if it were drowned in a black cloud, and all the day after smothered with a smoky fog' was made in a book entirely devoted to the reasons why consumptions were particular to the English.[42] The troubles of asthmatics in London's air was widely appreciated, and even the king's evil (scrofula) was said by one of Charles II's royal surgeons to be exacerbated by urban smoke.[43] In a medical

---

[40] Bédoyère, *Writings of Evelyn*, 131.

[41] Peter Denton, '"Puffs of Smoke, Puffs of Praise": Reconsidering John Evelyn's *Fumifugium* (1661)', *Canadian Journal of History* 35 (2000), 441–51, which oddly concludes that because the king was *Fumifugium*'s intended audience, historians should therefore ignore its content.

[42] Harvey, *Morbus Anglicus*, 166. See also Morton, *Phthisiologia*, 66, 246; Josiah Tucker, *Instructions for Travellers* (Dublin, 1758), 17.

[43] On asthma, e.g., John Pechey, *The Store-House of Physical Practice* (1695), 128; Richard Brookes, *The General Practice of Physic* (1754), 300–1, 319–20; Willis, Pharmaceutice Rationalis', 78–84, esp. 81; Pitcairn, *The Whole Works*, 126–7; Tucker, *Instructions*, 17; John Hill, *The Old Man's Guide to Health and Longer Life: With Rules for Diet, Exercise,*

tradition that understood diseases as discrete problems with distinct names and causes, coal smoke was repeatedly represented as a contributing factor to a contained and limited set of complaints.

It was frequently argued, therefore, that London's smoke made it a problem for certain kinds of people, but this also implied that it was quite acceptable for others. The physician John Burton, for example, argued during the 1730s that everyone would feel an 'exhilarating pleasure' upon leaving city for country air, but that it was primarily those with weak lungs who really suffered by the 'coal fires and stenches' of the city.[44] The prominent physician George Cheyne similarly emphasized smoke's particular threat to 'weak and tender people, and those that are subject to nervous or pulmonic distempers' but downplayed its importance to the rest of the population.[45] Some even claimed that habit and use made most Londoners immune to damage from their smoky environment. Foreign travellers especially remarked upon this, perceiving smoke's effects to lessen over time once they became used to it.[46] A short treatise on the non-naturals for the aged similarly noted that while London's smoky air was neither pure nor good, yet 'let not him who has attained to a healthy threescore and ten, then think of leaving London, to continue his days to a longer period. They say use is a second nature. It becomes nature itself'.[47] Arbuthnot summed up these divergent conclusions nicely. London's smoky air was 'unfriendly' to infants and children, consumptives and asthmatics should leave it for the country, and in general it promoted 'nervous' conditions. And yet, as 'the tolerance of artificial air (as that of cities) is the effect of habit' there is no indication that Arbuthnot (a sometime Londoner himself) condemned the capital's atmosphere outright.[48] The power of habit, which 'becomes nature itself', meant that for such commentators, and likely for many of their patients and readers as well, the unhealthiness of London's smoke was taken to be a problem only for the young, the old, and the weak. But for otherwise strong and healthy adults there were reasons to conclude that smoky air was not a serious threat to health.

---

and *Physick* (Dublin, 1760), 52. On scrofula Richard Wiseman, *Severall Chirurgicall Treatises* (1676), 255.

[44] Burton, *A Treatise on the Non-Naturals*, 64–5.

[45] Cheyne, *Essay of Health and Long Life*, 11; Tryon, *Monthly Observations*, 36.

[46] *M.Misson's Memoirs and Observations in his Travels over England* (1719), 38; *Kalm's Account*, 138–9.

[47] John Hill, *The Old Man's Guide to Health and Longer Life: With Rules for Diet, Exercise, and Physick* (Dublin, 1760), 20–1. See also Humphrey Brooke, *YΓIEINH, or a Conservatory of Health* (1650), 66–8. Habit as second nature is pushed especially strongly by the eccentric medical author Thomas Tryon. See *Tryon's Letters upon Several Occasions* (1700), 5.

[48] Arbuthnot, *Essay*, 208–9.

Any precise knowledge regarding this question one way or the other, however, was elusive during the seventeenth and eighteenth centuries. Medical research, according to several of its leading theorists and practitioners, had barely begun to understand the real relationships between airs and diseases. This was a broad agenda, more interested in the impact of subterranean effluvia or the steams emanated from plant matter than it was in the anthropogenic atmospheres of cities. But a key feature of this project was its emphasis on a current state of ignorance. Thomas Sydenham, the foremost champion of empirical medicine, wrote

> I have observed, with as much diligence as possibly I could, the various dispositions of diverse years, as to the manifest qualities of the air, that from thence I might learn the causes of this great variety of epidemical diseases, yet I have received no benefit thereby ... And thus it happens, there are many constitutions of years that arise neither from heat nor cold, nor moisture, nor drought, but proceed from a secret and inexplicable alternation in the bowels of the earth.[49]

Such 'secret and inexplicable' actions could best be understood, it was repeatedly claimed, by the laborious practice of making weather diaries. Detailed records of weather, in conjunction with public health information, were considered an indispensible project throughout the century after 1650. They were championed by members of the Royal Society like Boyle, Hooke, Locke, Wren, and Robert Plot during the seventeenth century, and by the Society's secretary James Jurin as well as Arbuthnot during the eighteenth.[50] Despite this enthusiasm they continually stressed the current state of ignorance. Arbuthnot still suggested in 1733 that such journals might produce useful predictions in the future, but this remained a project much as it had been for Boyle and Locke seventy years earlier.[51] For the first century during which they were championed, therefore, weather diaries and the medically useful knowledge of airs that they would facilitate remained desirable but not yet achievable. Medical knowledge regarding air's effects on human bodies thus became more complex during the seventeenth and eighteenth centuries, a process that also undermined claims to certainty. The search for demonstrable links between the almost infinite range of airs and the equally complex variations of illness and health made

---

[49] John Pechey ed. and trans., *The Whole Works of that Excellent Physician, Dr Thomas Sydenham* (1696), 4–5.
[50] *The Hartlib Papers* 2nd ed. (Sheffield, HROnline, 2002) [hereafter HP] 26/56/2B; included in Boyle's *General History of the Air*, Hunter and Davis, eds. *Works*, XII, 48–56; Jackson I. Cope and Harold Whitmore Jones, eds. *History of the Royal Society by Thomas Sprat* (St Louis, 1959), 175–7; *Life and Works of Sir Christopher Wren. From the Parentalia or Memoirs by His Son Christopher* (1903), 52–5; Golinski, *Weather*, 55, *et passim*; Andrea Rusnock, 'Hippocrates, Bacon, and Medical Meteorology at the Royal Society, 1700–1750', in David Cantor, ed. *Reinventing Hippocrates* (Aldershot, 2002), 136–53.
[51] Arbuthnot, *Essay*, 233; Golinski, *Weather*, 150.

it difficult to assign causation clearly to one factor like urban smoke. This increasing uncertainty can be seen clearly in the career of demography, a new discipline in which coal smoke first emerged as a central explanation only to recede thereafter.

III. THE DECLINE OF ENVIRONMENTAL DEMOGRAPHY

Demography offers, in many ways, the most convincing evidence that urban smoke did not stand outside of the developments associated with the new science of the seventeenth and eighteenth centuries. Its founding text, after all, stressed the particular contribution of coal smoke to London's destructive climate, a finding that has been taken to mean that early modern people 'knew' what pollution did to public health. The pioneering work of John Graunt, Ken Hiltner argued, 'offered statistical proof that linked sea coal smoke with respiratory illness'.[52] A careful reading of Graunt and of the demographers following him, however, must qualify such confidence. Coal smoke was indeed integral to Graunt's interpretation of demographic data, but his innovative statistical reasoning did not straightforwardly prove the link between London's coal smoke and its short life expectancy, nor were his assertions of their relationship universally accepted by later demographers.

In his 1662 work *Natural and Political Observations* Graunt introduced the new science of demography. He analyzed the Bills of Mortality, death registers intended to warn against plague epidemics. Realizing that the numbers derivable from the bills could illuminate questions beyond the existence of plague, Graunt compared rural parishes with urban and found that Londoners died and reproduced at different rates than inhabitants of country parishes. Londoners, he found, had fewer offspring per capita, so either the city must somehow be generally unhealthy, or there must be moral or social explanations for comparative urban barrenness. Drawing on common early modern neo-Hippocratic assumptions, Graunt associated localized regimes of health and disease with prevailing airs.[53] This tradition usually stressed the natural production of airs by local geography, but for Graunt smoke was the key feature distinguishing London's environment from others. He compared London's suburban 'country parishes' like Hackney and Newington 'with the most smoky and stinking parts of the city'.[54]

---

[52] Hiltner, *What Else is Pastoral?* 101.
[53] The Hippocratic tradition and its influence in England is survey by Andrew Wear, 'Place, Health, and Disease: The *Airs, Waters, Places* Tradition in Early Modern England and North America', *Journal of Medieval and Early Modern Studies* Fall 2008 38: 443–65.
[54] John Graunt, *Natural and Political Observations Mentioned in a Following Index, and Made upon the Bills of Mortality* (1662), 45.

He argued, along with others stressing the importance of habit, that while 'seasoned bodies' could tolerate London 'yet newcomers and children do not, for the smokes, stinks and close air are less healthful then that of the country'.[55]

For Graunt smoke contributed to a general depression of urban health, but he also muddied the waters by introducing additional considerations and so multiplying variables. There were also moral and social explanations for divergent urban and rural reproductive patterns. First, London contained fewer 'breeders' because its population was more seasonal and transient. Those 'that have business to the court of the king, or to the courts of justice, and all countrymen coming up bring provisions to the city, or to buy foreign commodities, manufactures, and rarities, do for the most part leave their wives in the country'. London mariners left their wives during their voyages while others came to London 'for curiosity and pleasure', 'to be cured of diseases', or as apprentices 'bound seven or nine years from marriage'.[56] London's role as capital, emporium, and port, therefore, produced a distinctive social profile in which there were fewer settled and cohabiting couples and more itinerant or celibate people.

Celibacy's opposite, moreover, was also damaging to reproduction. While structural features of the capital's population may have produced fewer 'breeders', Graunt also considered London to lead England in 'barrenness' for behavioural reasons. Native air is here set aside as a cause of barrenness in favour of 'intemperance in feeding, and especially the adulteries and fornications, supposed more frequent in London then elsewhere, [which] do certainly hinder breeding. For a woman, admitting 10 men is so far from having ten times as many children that she hath none at all.' Graunt then added that rural men focused on bodily work, while (male) Londoners' minds were 'full of business'. Such 'anxieties of the mind' further hindered breeding. For Graunt, then, Londoners died for reasons that can be attributed in significant part to smoky air, but their failure to reproduce resulted from other factors, including an unsettled population as well as excessive debauchery and worry.[57]

The negative impact of smoky air, in fact, was further qualified because Graunt also thought that it somehow shielded the capital from the worst effects of the plague. 'Fumes, steams, and stenches' raised mortality in normal years, but also served to lower the proportion of plague deaths compared to rural parishes. The air of London therefore appeared to be 'more equal, but not more healthful'. This, Graunt claimed, was new. Before the rapid growth of the seventeenth century London's demography conformed

---

[55] *Ibid.*, 46.   [56] *Ibid.*, 45.   [57] *Ibid.*, 46.

to rural patterns, leading Graunt to ask whether large cities must be inherently unhealthy.

> I inclined to believe that London now is more unhealthful than heretofore, partly for that it is more populous but chiefly because I have heard that 60 years ago few sea coals were burnt in London, which now are universally used. For I have heard that Newcastle is more unhealthful than other places and that many people cannot at all endure the smoke of London, not only for its unpleasantness but for the suffocations which it causes.[58]

From this Graunt concluded that London's growth must be driven by immigration from the country, meaning that the capital was an overall drain on England's population.[58] In sum, therefore, Graunt considered coal smoke to be a central factor governing London's, and therefore England's, demographic structure and trend. But his careful consideration of the many factors determining population trends – sexual practice and family formation in addition to public health – meant that Graunt's text is also somewhat ambiguous regarding the overall importance of London's uniquely dirty air.

These ambiguities were amplified by later students of demography for whom neo-Hippocratic explanations for London's population trends were less influential.[59] During the later 1690s the new demands of the post-revolution fiscal-military state prompted renewed attention to demography, carried out in particular by government official Gregory King. In an exchange with the politician Robert Harley, King argued, following what he took to be Graunt's position, that in London 'the unhealthfulness of the coal smoke' depressed procreation. Harley pushed against this, pointing out that 'several observers' had informed him that Newcastle smoke was even worse than London's and yet certain suburban parishes there 'abound in children wonderfully'. King responded by citing Graunt directly, repeating his claim that coal had originally been recognized to be offensive in London and had therefore been put under some unspecified legal restraint, and further arguing that Newcastle was in fact 'unhealthful'. King could cite no evidence for Newcastle's insalubrity, and he therefore suggested that one might, 'with some pains' ascertain the true state of Newcastle's demographic situation using reliable data. Comparing Newcastle with 'other great towns

---

[58] *Ibid.*, 70.

[59] One glaring omission from this conversation is William Petty, a virtuoso intellectual interested in the intersections of the medical sciences and the arts of governance. While he perceived London's elites to have moved into the West End to avoid the 'fumes, steams, and stinks' of sea coal, and his review of Graunt for the *Journal des Sçavans* refers to Graunt's interpretation of London's 'smoke' (fumée), Petty's own demographic work showed no interest in its influence. Petty, unsigned review of Graunt in *Le Journal des Sçavans*. 2 August 1666, 359–70; Hull, ed. *Economic Writings*, 41. For the importance of medicine and alchemy in Petty's thought see Ted McCormick, *William Petty and the Ambitions of Political Arithmetic* (Oxford, 2009).

where they burn little or no sea coal' would provide meaningful data, but such work had not yet been done. King therefore fell back on the common 'notorious' experience of the many Londoners who could not endure urban smoke.

Under pressure from Harley, King also admitted that smoke was only one of five reasons why urban conception rates were lower and why London children were less likely to survive. Despite these caveats King concluded, 'I am clearly of opinion it suffocates and destroys a multitude of infants, though perhaps in Newcastle it may not have the same operation, where the air is sharper, the town not above a 50th part of London, nor nothing near so close built.'[60] In sum, then, King reaffirmed Graunt's claims regarding the malign influence of London smoke, even as he acknowledged a number of problems with his evidence. Most notably, the mere possibility that demographic evidence from Newcastle, which King pointed out had not even been collected yet, might diverge from London's was enough to push King towards arguments that were difficult to reconcile with his emphasis on the role of the urban environment.

Given King's failure to affirm smoke's importance under Harley's pressure, it is not surprising that the other great political arithmetician of the 1690s, Charles Davenant, was similarly non-committal. Davenant's 1699 *Essay Upon the Probable Methods of Making a People Gainers in the Ballance of Trade* adopted King's and Graunt's arguments regarding coal smoke, but placed them alongside the other possible explanations of low birth rates without commenting on their relative importance as causes.[61] In Davenant's text, as in Graunt's, coal smoke joins a list which also contains urban adultery, luxury, and business. But while Graunt stressed smoky air's importance, Davenant's list made no effort to disentangle the roles of the multiple variables governing the urban demographic regime. Thus Davenant and King, the two most ambitious demographers of the 1690s, repeated Graunt's suggestion that smoky air was killing Londoners, but they displayed little conviction that Graunt had demonstrated this conclusively.

John Arbuthnot also advanced a more limited version of Graunt's argument in his 1733 *Essay Concerning the Effects of Air on Human Bodies*. While he was quite capable of using demographic techniques when he chose, in the *Essay* Arbuthnot analyzed bodies rather than populations, his focus that of the physician and natural philosopher more than the political arithmetician.[62] While the *Essay* did not use demographic methods, it did

---

[60] D. V. Glass, 'Two Papers on Gregory King', in D. V. Glass and D. E. C. Eversley, eds. *Population in History: Essays in Historical Demography. Volume I: General and Great Britain* (1965), 159–220: 164.

[61] Charles Davenant, *Essay Upon the Probable Methods of Making a People Gainers in the Ballance of Trade* (1699).

[62] For Arbuthnot's use of statistics in the debate surrounding inoculation, see Rusnock, *Vital Accounts*, 46–9.

echo Graunt's claim that urban air was particularly dangerous to 'newcomers', arguing that

> The air of cities is unfriendly to infants and children. Every animal is adapted to the
> use of fresh, natural, and free air; the tolerance of artificial air (as that of cities) is the
> effect of habit, which young animals have not yet acquired. The mortality of children
> under two years in London is not entirely owing to the small care of the brood of the
> necessitous and of bastards.[63]

Thus Arbuthnot defended Graunt's environmental argument against (or
rather, in addition to) a claim that London's high infant mortality was due
to poverty, bad parenting, and neglect.

For Arbuthnot the poor's failures to provide adequate care for children
was an insufficient explanation because it neglected Graunt's environmental argument, but by the mid-eighteenth century the clear superiority of moral and social explanations over environmental had found a
full-throated defender. Thomas Short wrote a series of works from the
1740s to the 1760s examining the relationship between air and health. He
was devoted to the Baconian goal of assembling natural histories of the air
in order to understand the progress and treatment of disease. Since Short
devoted such extensive labour to the relationships between air and health,
it is remarkable how hard he worked to refute Graunt's thesis regarding the damage caused by London's air, especially in his *Comparative
History of the Increase and Decrease of Mankind in England, and Several
Countries Abroad*.[64]

Short did not entirely deny that environment mattered. In the *Comparative
History* he listed three ways in which the nation could be made more populous. 'First, encourage marriage. Second, suppress vice and promote virtue. Third, mend the air.' For Jan Golinski, in his study of weather and
enlightenment, Short therefore serves as a prime example of how concerns
about weather and air could be translated into 'concrete programs of environmental improvement'.[65] But while Graunt stressed the environment and
Davenant was happy to list moral and material conditions alongside each
other, Short argued that the moral was clearly primary. He devoted several
pages to marriage and vice, including his thoughts on how best to punish
adultery, suppress excessive drink and luxury, and reward marriage. Along
the way he weighs in on foundling hospitals, enclosures, long leases, taxation, and care for orphans. After these detailed thoughts on the moral and
social conditions governing marriage and reproduction, Short finally turned
to the air:

[63] Arbuthnot, *Essay*, 208.
[64] Thomas Short, *A Comparative History of the Increase and Decrease of Mankind in England,
and Several Countries Abroad* (1767).
[65] Golinski, *Weather*, 158.

The last thing proposed was, that cities and great towns should (if possible) be provided with good air, water, and wholesome habitations, clean, open, wide, and well-ventilated streets, where such things can be had. Greater sickness and mortality are often charged, more than is just, on air and water; and on smoke, dirt, closeness of the place, bad effluvia, and excrements of multitudes and variety of animals, &c. Though these things want not their bad effects, yet the vices and licentiousness of the inhabitants are often more to be blamed.[66]

Then after a short list of some recent urban improvements, he concludes that despite all of these, 'Death sometimes rides triumphant, therefore the cause must lie deeper in the prevalence of vice and immorality which can only be pruned off by a vigilant, indefatigable, virtuous magistracy, having and impartially executing good laws on all sorts of offenders.'[67] For Short, then, coal smoke had no special power to explain London's divergent public health statistics; it was merely one among many dirty aspects of urban life that were themselves not nearly as important as other explanations, in particular marital customs and sexual behaviour. Mending the air, therefore, may have been moderately desirable, but only as a subordinate part of a larger programme for moral reformation. Short was not alone; others like Jonas Hanway and Arthur Young also considered the dangers of London's material environment, particularly to children, but they did so as a part of a broad-ranging call for reform that was primarily moral and governmental rather than medical or environmental.[68] A century after Graunt's *Observations* asserted that London's smoke helped explain high urban death rates, his arguments continued to be taken seriously but were also subjected to increasing scepticism.

IV. CONCLUSION

The fate of smoke within detailed examinations of London's demographic regime shares much with its place within other natural philosophical and medical conversations across the early modern period. Graunt's neo-Hippocratic assumptions were commonplace, and his novel statistical methods can therefore be seen as innovative ways to investigate and demonstrate ideas that was had been widely shared by experts and laymen alike. Graunt's demography, like alchemical or physical analyses of smoke's unhealthy properties, said with new precision and new claims to expertise some of the same things that other, non-learned voices also said in simpler

[66] Short, *History*, 35.   [67] *Ibid.*, 36.
[68] Jonas Hanway, *Serious Considerations on The Salutary Design of an Act of Parliament for a Regular, Uniform Register of the Parish Poor* (1762); *Letters to the Guardians of the Infant Poor* (1767); Arthur Young, *The Farmer's Letters to the People of England* (1768), 335, 345–9.

ways. As these new disciplines developed, however, their very complexity undermined the clarity with which one could assign a complex phenomenon to a single variable. Eighteenth-century moral reformers like Hanbury and Short could therefore be read not as ignoring Graunt's 'statistical proof' for the perils of smoky air, but as practicing a newly sophisticated kind of demography that took sociological, cultural, and governmental factors as seriously as airs, waters, and places. Arbuthnot, similarly, struggled to determine how important London's 'sulfurous steams of fuel' were compared with other sorts of effluvia and emanations which natural philosophy had only just begun to investigate. The complexity of air's physical and chemical properties had become clear, and leading natural philosophers assessed smoke's operation through those properties. But this led, not to clarity, but to greater complexity.

In these ways the discoveries and innovations of the 'scientific revolution' made it less, rather than more, certain whether and how coal smoke constituted dangerously bad air. A parallel might be drawn between Short's efforts to distinguish the importance of environmental from behaviour factors with Lord Mansfield's need to separate legal nuisance from the normal bustle of urban life. In both cases it was granted by nearly everyone that coal smoke, to paraphrase Short, 'wanted not its bad effects'. Both could point to absolute statements from seventeenth-century authorities that would seem to condemn smoke outright: Mansfield knew that the doctrine of *sic utere* protected property owners from unpleasant and harmful airs, and Short knew that Graunt, the founder of demography, had stated categorically that smoke sent Londoners to their graves. But in both cases smoke could not be abstracted from broader contexts. Bodies, for Short, were products of social and moral influences. The law, for Mansfield, could not ignore the needs of an increasingly commercial and industrial society. Both the common law of nuisance and new developments in natural philosophy therefore offered early modern English people influential languages with which to denounce urban smoke. But both were also restrained by contrary impulses. Neither law nor science offered straightforward support for claims that smoke was dangerous, an ambivalence that made it difficult for smoke's enemies to overcome the increasingly confident claims for the importance of coal.

Part III

Fuelling leviathan

−7−

The moral economy of fuel: coal, poverty, and necessity

'Week after week the winter strengthen'd,
And froze more sharply as it lengthen'd...
Beggars crept up and down, poor souls,
Cursing the price of bread and coals,
And in expressions too severe,
Damn'd those that kept them up so dear'.[1]

I. THE DANGERS OF THE CHATHAM DISASTER

In the middle of June of 1667 England was humiliated. After two years' fighting against the Dutch, when victory seemed many times to have been within reach, the enemy had sailed nearly unopposed into the Thames estuary and up the Medway, some forty miles east of London. There they quickly took key forts, penetrated the defences at Chatham, and burned or took some of the Royal Navy's greatest ships including its prize, the eighty-gun Royal Charles. It was a stunning blow to a monarchy that claimed sovereignty of the seas; the Duke of Albemarle, it was reported, wanted to die on the spot.[2] On top of the shame of defeat there was an immediate security threat as the Thames lay nearly undefended and rumours swirled of an imminent French invasion. Londoners responded with fear and anger. Samuel Pepys recorded that when the news reached the court it looked as if everyone had cried, while at home 'all our hearts do now ache; for the news is true'. Since he feared 'that the whole kingdom is undone', the navy clerk paused from his busy work to arrange for his wife and father to flee into the country with his moveable wealth, and he made his will.[3] It was feared that the royal exchequer would be shut and financiers would bankrupt. In London people

---

[1] Ned Ward, *British Wonders: or, a Poetical Description of the Several Prodigies and Most Remarkable Accidents* (1717), 44–5.
[2] Allen Brodrick to the Duke of Ormonde, 15 June 1667, Bod. Carte MS 35, f. 478v.
[3] Robert Latham and William Matthews, *The Diary of Samuel Pepys. Volume VIII 1667* [hereafter *Pepys*] (Berkeley, 1974), 262, 266.

were 'talking treason in the streets openly', and the general 'dismay' seemed to Pepys to parallel that during the Great Fire of the previous September.[4]

Among the most dangerous consequences of the Dutch raid on Chatham, according to some very well-informed men, was a dramatic surge in the price of coal. From a normal peacetime rate of around £1 per chaldron the London market soon reached £5 and even £6 per chaldron. Not only was this several times the usual level, it was higher even than the prices during the 'most bitter cold' days of the previous winter when Pepys recorded universal complaints regarding coal's 'very great price'. Then the City of London had warned the Privy Council that such prices 'will occasion the starving of hundreds of His Majesty's poor subjects unless some speedy course be taken therein'.[5] Immediate disaster was averted in March by new shipments, but there was little margin for further supply problems. After the destruction of so many riverside warehouses in the Great Fire and the interrupted wartime coastal trade, it is likely that London contained very limited coal stocks in June of 1667. Even worse, with an enemy attacking its coasts with impunity the normal summer coal fleets could not safely sail.[6] This was an immediate problem, and the Earl of Anglesey claimed that many sent their 'families' – that is, their households – 'out of town'.[7] Since great households often spent dozens of chaldrons annually such an economy was prudent, but for London's poorer households the prospect of facing a winter with empty fuel stocks was far more alarming. It was therefore the future, not the present, that was the most serious concern, and Pepys recorded that 'the great misery the City and kingdom is like to suffer for want of coals in a little time is very visible and, is feared, will breed a mutiny'.[8] Such fears of 'mutiny' were taken quite seriously, and when Anglesey – a Privy Councillor near the centre of power in 1667 – wrote to the Duke of Ormonde in July to discuss the new peace treaty, he concluded that 'the want of coals contributed as much to this peace as want of money and materials for shipping'.[9]

The claim that high coal prices might cause instability or rebellion – Pepys's 'mutiny' – and that such fears contributed directly to the government's decision to conclude a peace, point to two key aspects of the London coal economy that will be examined in this and the following chapters. First, fuel in general and coal in particular were widely considered to be basic necessities. This led fuel to hold an ambivalent place in contemporary

---

[4] Anglesey to Ormonde, 15 June 1667, BL Carte MS 47, f. 158; *Pepys 1667*, 268–9.
[5] *Pepys 1667*, 98–9, 102; PC 2/59, 336.
[6] For an examination of the ways that the coal market responded to the fortunes of war in the 1660s, and the impact such price surges had on the urban poor, see Cavert, 'Politics of Fuel Prices'.
[7] Carte 47, f. 164.   [8] *Pepys 1667*, 285.   [9] Carte 47, f. 166-v.

discussions of economic morality. On one hand it was an object of daily, even constant use, an ever-present and therefore usually unremarkable part of the background before which daily life was enacted. Unlike luxuries, consuming mundane necessities like coal did not necessarily make powerful statements, and hence were not the subject of heated polemic. But on the other hand coal's significance became subject to immediate inflation when its availability was threatened. Powerful and persistent rhetoric framed coal sometimes as the common fuel, sometimes as the fuel of the common man. Both arguments presented coal, for Londoners, as something very close to an entitlement.

Second, this language of necessity and entitlement led directly to action. If fuel was crucial for all, and in particular for London's poorest inhabitants, it followed that access to affordable firing, especially during winter, should be ensured by those in power. Chapters 9 and 10 will describe how the central state increasingly accepted that protecting and regulating the coal market should be one of its duties. Here we will see how charitable distribution of coal to London's poor remained a central feature of private philanthropy and governmental policy throughout the early modern period. Such private and public efforts, furthermore, blurred into and reinforced each other. London's governors and elites – people like Pepys and Anglesey – accepted and repeated the claim that coal was crucial for the urban poor in ways that governments could not afford to ignore. While the following chapters will develop the importance of coal to English power during the later-seventeenth and eighteenth centuries, this discussion stresses that during the first decades after coal's adoption it was described far less often as, in John Nef's phrase, 'a new national asset', than as the fuel of the poor.[10] Nef exaggerated the extent to which coal before 1650 was seen as a pillar of power. The state certainly did consider itself bound protect to London's coal supplies, but its reason for doing so was not power politics but social stability.

II. 'STARVING HIS MAJESTY'S POOR SUBJECTS': FUEL AS NECESSITY IN EARLY MODERN ENGLAND

In communities across England fuel was a vital and valued resource. In rural areas villagers often enjoyed and defended the right to gather fuel from common lands. Such rights were usually both restrictive and restricted. Access was often tied to copyhold tenure, parish settlement, or other formal means of defining community membership. Those excluded by such arrangements were thus also excluded from legal means of gathering fuel. Such customary

---

[10] Nef, *Rise*, II, part V, ch. 1.

access was also usually limited in kind and in scope, to 'lops and tops', to fuel taken 'by hook or by crook', or otherwise to specified quantities of woods, ferns, and gorse.[11] Customary access to resources like fuel, as Andy Wood and others have stressed, was almost infinitely variable and also frequently contested, forgotten, forged, adapted, and invented. The key point here, however, is that though they were malleable, customary rights were also highly valued. This is why the Star Chamber records studied by Manning as well as the Exchequer depositions analyzed by Wood are full of rural men and women who rioted or sued to protect their claims on local fuel.[12] While social historians have not made fuel provision a primary focus of their research, they agree that it was among the crucial benefits of customary use of common lands.[13]

Custom was not the only way for rural people to obtain fuel. For many, a defined quantity of fuel was included in wages for many labourers, as the Norfolk shepherd who received furze and breaks in addition to money and pasturing rights.[14] Others stole, either because they claimed rights to local resources that were found illegal by landowners and magistrates, or because they had no such claim but had no other choice. Those with their own land could supply their own needs through the annual pruning of hedgerows, but it could take miles of hedging to supply an adequate amount of fuel for a household.[15] Muldrew has stressed that such access rights were among the most important perquisites available to rural labourers and are among the reasons he finds narratives of rural poverty based merely on analysis of wages unpersuasive. Fuel rights, by the mid- and late-eighteenth century,

---

[11] Donald Woodward, 'Straw, Bracken and the Wicklow Whale: The Exploitation of Natural Resources in England Since 1500', *Past and Present* 159 (1998), 46–76: 54–5.

[12] Wood, *Memory of the People*; Richard W. Hoyle, ed. *Custom, Improvement and the Landscape in Early Modern Britain* (Aldershot, 2011); Roger B. Manning, *Village Revolts: Social Protest and Popular Disturbances in England, 1509–1640* (Oxford, 1988), 270–9.

[13] E.g., Wood, *Memory of the People*, 158, 161, 179, 184, 193; Bob Bushaway, *By Rite: Custom, Ceremony and Community in England 1700–1880* (1982), 207–33; K. D. M. Snell, *Annals of the Labouring Poor: Social Change and Agrarian England, 1660–1900* (Cambridge, 1985), 166–8, 179–80, 203; Janet Neeson, *Commoners: Commons Right, Enclosure and Social Change in England 1700–1820* (Cambridge, 1993), 158–65, 173, 279; Hindle, *On the Parish?*, 43–8, 267–8.

[14] John Walter, 'The Social Economy of Dearth in Early Modern England', in John Walter and Roger Schofield, eds. *Famine, Disease and the Social Order in Early Modern Society*, (Cambridge, 1989), 75–128: 98.

[15] Muldrew's data suggests that a double row of hedging produced about one faggot per seven yards per year, and further implies that labouring families would have consumed 2,280 faggots annually, requiring 9.5 miles of hedging. This seems impossibly high for many families, but even reducing this estimate by a full order of magnitude still supports the basic point that self-sufficiency in fuel required access to substantial woods or hedges. Craig Muldrew, *Food, Energy, and the Creation of Industriousness: Work and Material Culture in Agrarian England, 1550–1780* (Cambridge, 2011). 257, n. 179. Fuel doles distributed in faggots, it is worth noting, were usually far less than 2,280 per year.

were worth almost £2 annually to rural working families, and were therefore, in quantitative terms, a crucial component of household budgets. For
wage labourers fuel rights helped them not only to survive but increasingly
to participate in the growing consumer economy. For the very poor fuel
rights were comparable to gleaning and commoning animals: vital elements
in the economy of makeshift.[16]

The utter necessity of these fuel supplies, both to the rural poor and to
the commonwealth more generally, was perhaps argued most forcefully
by Arthur Standish in his *Commons Complaint*, first published in 1611.
While it was a commonplace to consider fuel to have a similitude with food,
both as a basic necessity and as a natural commodity, Standish argued for a
causal, even ecological relationship.[17] While others assumed that woodlands
were in competition with arable and pastoral land, Standish argued that in
fact they should be complementary.[18] This was because, crucially, some fuels
could also improve agriculture. Where wood was scarce, Standish claimed,
straw and cow dung were burned which might otherwise feed cattle and
manure arable fields. Woods, moreover, also served multiple uses. With no
adequate wood supplies sheep farming suffered for want of fencing. Hedges
and trees, moreover, could feed livestock. Sufficient trees and hedging, therefore, allowed animals to thrive and multiply and, through them, made soils
to be fertilized. The lack of this, however, had dire consequences, as livestock
would dwindle and the soil lose its 'strength'. Having witnessed and analyzed rural poverty following the Midlands revolt of 1607, Standish confidently claimed that adequate fuel was central to agricultural sufficiency and
prosperity: 'And so it may be conceived, no wood, no Kingdom.'[19] Crucially
for this discussion, his entire analysis was premised on the utter necessity of
fuel. Without wood, Standish argued, the rural population would be compelled to burn something else, even if that entailed their farm's eventual ruin.

Standish's argument for the negative feedback loop created by wood
scarcity did not apply to urban consumption, but his treatment of fuel's
importance for social and political stability did. Towns could be no less
protective of their sources for customary fuel supplies than villages and
villagers. These varied as much as rural claims, and indeed sometimes
made common cause with them. An example from the 1580s, Nef's period
of national fuel emergency, can suggest how seriously infringements on

---

[16] Hindle, *Parish*, ch.1.
[17] For wood and coal as fruits of the land, see e.g., Fynes Moryson, *Itinerary* (1617), Part III,
142, 144; Peter Heylyn, *Eroologia Anglorum. Or, An Help to English History* (1641), 295,
323, 333, 337.
[18] The competition model underlies, for example, William Harrison's discussion of woods.
Furnivall, ed. *Harrison's Description*, 336–46.
[19] Standish, *Commons Complaint* (1611), 2.

fuel supplies were often taken. In 1581, Lord Paget's tenants in Cannock Chase, Staffordshire, rioted over a variety of complaints, prominent among which were the abuse of woods caused by Paget's large-scale ironworks. A few years later the urban inhabitants of both Stafford and Lichfield claimed to be 'greatly distressed' at being barred from their accustomed access to coals from Beaudesert Park, within Cannock. A report found that the crown (which had confiscated Paget's assets) ought to 'relieve' the local poor by allowing continued access to fuel despite the damage this caused to the forest. In 1585, the new lessee of the ironworks was ordered not to spoil the forest specifically because this would cause more harm to the local poor than would be compensated by the profits arising to the crown. The crown thus accepted the claims of the local rural and urban poor to depend on access to plentiful fuel, even though these were contrary to the crown's own interest.[20] Elsewhere urban magistrates similarly asserted their rights over local fuel supplies, including Colchester's burgesses who claimed rights to the fuel from nearby Kingswood and Leicester's governors who claimed customary access to Leicester Forest, where the town's 'poor, which are not able to buy fuel, have always had much relief by gathering and fetching on their back from thence broken, dead and rotten wood'.[21] In towns and cities as in rural areas, the 'poor' were the acceptable face of the customary access to fuel rights, and it was through the rhetorical invocation of their need that conflicts regarding property and use rights were pursued.

Above such defences of customary fuel sources, however, the most important manifestation of the social and political importance of fuel supplies was the increasingly widespread practice of charitable fuel distribution. In parishes, towns, and cities throughout England private individuals acted alongside and also in cooperation with government officers to provide free or subsidized fuel stocks to the deserving poor in their community. There has been no systematic study of the geography or chronology of fuel provision in either rural or urban England, though Hindle has plausibly suggested that its frequency and scale increased into the eighteenth century.[22] What is certainly clear is that the practice was very widespread throughout the early modern period. Overseers

---

[20] Folger Library MS L.a. 305; L.a. 1031; M. W. Greenslade and J. G. Jenkins, eds., *History of the County of Staffordshire* (1967), II, 110–1; Manning, *Village Revolts*, 276–7.

[21] Janet Cooper, ed., *A History of the County of Essex* (1994), IX, 255, 258; Helen Stocks, ed. *Records of the Borough of Leicester* (Cambridge, 1923), 240.

[22] Hindle, *On the Parish?*, 268. Fuel doles were probably increasing, but were certainly not new in the early modern period. See Marjorie McIntosh, *Poor Relief in England, 1350–1600* (Cambridge, 2012).

and churchwardens in rural parishes across England provided wood or coal for their deserving poor.[23] So did urban governments, both through parochial and civic structures as well as through workhouses which heated the living and working spaces of their poor inmates.[24] In both rural and urban cases, such distributions were often funded not by local rates but by charitable bequests, usually land or money to purchase land that funded an annual fuel dole or subsidy.[25] Thus private benevolence allied with public authority provided fuel to the poor across England. Visitors to many churches, such as Great St Mary's in central Cambridge, may still read the plaques precisely recording gifts of wood or coal to local poor.[26]

Throughout both rural and urban England, then, fuel provision for the poor was an important aspect of the mutualities and obligations binding the rich and poor together.[27] In many communities the poor depended on their better off neighbours to provide charity, either formally through the parish poor rate or through informal gifts. For the rich and comfortable middling sorts such charity was among the ways in which they enacted benevolent governance in their locality and over their poor neighbours. Fuel was therefore a commodity bought and sold on open markets, but its trade was also highly charged morally, religiously, and politically. It is not surprising, then, for gardener and author Leonard Meager to argue that planting trees and hedges brought multiple benefits to England's landowners. Farmers would provide for their own household needs in anticipation of 'pinching winters', they would 'be profited in selling the overplus to the rich', and finally upon 'charitably bestowing some on their poor neighbours' they would 'for

---

[23] Examples include, among many others: Norfolk Archives, PD 629/50; GRO P170 VE1/1; Cambridgeshire RO P52/5/1; L. A. Botelho, ed. *Churchwardens' Accounts of Cratfield 1640–1660* (Woodbridge, 1999), 42, 49, 63, 96–7, 101, 106, 109; Joan Kent and Steve King, 'Changing Patterns of Poor Relief in some Rural English Parishes Circa 1650–1750', *Rural History* 14 (2003), 119–56; Hindle, *On the Parish*, 142, 267–8.

[24] E.g., 'Accounts of the Tooley Almshouse, Ipswich', BL Add MS 25,343, f. 93–169; D. M. Livock, ed. *City Chamberlains' Accounts in the Sixteenth and Seventeenth Centuries* (Bristol, 1966), 127; Gloucester Borough Minutes, GRO GBR/B3/1, f. 448v; GBR B3/2, 133, 200, 514; GBR B3/3, 303–4, 538; Cambridge parishes distributing coal include St Edward's, Cambs RO P28/4/1–2; St Botolph's, P26/5/2–5; St Andrew's P23/5/1; and Great St Mary's P30/25/1, P30/4/4. David Underdown, *Fire From Heaven: Life in an English Town in the Seventeenth Century* (New Haven, 1985), 121 on Dorchester's fuel charity as an 'implied bargain under which the poor had given up their old right to pilfer fuel from nearby woods and fields'.

[25] For the complementary of private charity, customary access, and poor law administration see John Broad, 'Parish Economies of Welfare, 1650–1834', *The Historical Journal* 42 (1999), 985–1006.

[26] Records of the St Mary's bequests administration is Cambs RO P30/25/1.

[27] The phrase is Keith Wrightson's, 'Mutualities and Obligations: Changing Social Relationships in Early Modern England', *Proceedings of the British Academy* 139 (2007), 157–94.

their Christian compassion toward them, be loaded with their blessing and prayers'.[28]

The blessings and prayers offered by the poor towards the rich for firing which must have always been deficient during the hard winters of the Little Ice Age were unlikely, in reality, to have been such a load as Meager suggested. Yet this language of benevolent provision in response to humble need was an entirely conventional way to discuss fuel provisions in early modern England. Even as England converted to a precociously modern energy regime, one that derived more heat from coal than wood by the seventeenth century, its language for discussing fuel as a commodity remained grounded in vocabularies of charity and benevolence in the face of poverty and necessity. Throughout England there was a moral economy of fuel that represented it as an entitlement, even as the actual methods of fuel production and trade were increasingly capitalist and industrial.

III. 'WANT OF COALS': THE LONDON FUEL MARKET AND THE POLITICS OF SCARCITY

Londoners did not acquire their fuel in the same ways as most English subjects. The sorts of customary access to forests, woodlands, fens, and other commons that were so important across England, as well as the wood fuels provided by enclosing hedges, were all irrelevant to a metropolis of hundreds of thousands of people. Only markets could supply a city the size of London. This had long been true, as the medieval city drew its fuel from a catchment area extending up and the down the valleys of the Thames, its tributaries, and its estuary.[29] But after its conversion to coal during the years around 1600 London depended on a long and complex chain of producers, shippers, wholesalers and middlemen. The commercialization of London's fuel market was thus not new, but the complexity and scale of it was. While inhabitants of small cities and towns like Lichfield or Leicester could gather their own fuel from their own urban hinterlands, Londoners purchased fuel that had passed through the hands of many labourers about whom they knew nothing. Its structural complexity, then, was unique, but the language with which Londoners' fuel was discussed was not. Like elsewhere in England, Londoners and their governors claimed that fuel was a universal necessity, but one of particular concern to the poor. Its provision, in London as elsewhere, was fraught with moral and political implications. The moral

---

[28] Leonard Meager, The Mystery of Husbandry: or Arable, Pasture, and Woodland Improved (1697) in Joan Thirsk and J. P. Cooer, eds. *Seventeenth Century Economic Documents* (Oxford, 1972), 186.

[29] Galloway *et al.*, 'Fuelling the City'.

economy of fuel was therefore not particular to London, but the capital's unique market and unique importance in English political life meant that the rhetoric of market morality was applied differently in the metropolis.

That rhetoric shared much of its language and assumptions with rural discussions of fuel. As in rural England, fuel in the capital was frequently associated with food, as in 1579 when the Queen's ministers ordered the city of London to manage the distribution of 'victuals, fuel, and other necessaries'.[30] In 1630 the Privy Council called both the butchers and the woodmongers to explain the high prices of meat and coal.[31] By around 1700, it was argued that overseers of the poor should legitimately provide the three basic categories of necessities: 'meat, clothing, [and] fuel'.[32] Of course such associations between food and fuel do not mean that early modern Londoners were likely to confuse bread with coal. But there was real conceptual overlap between these distinct types of necessities.

One of the terms with which contemporaries discussed fuel scarcity was 'dearth', a word typically used by historians to describe lack of food but not other commodities.[33] In the early modern period, however, because food and fuel were closely related necessities a lack of either could be called a dearth. A satirical dialogue written during the winter of 1643/4 (when parliamentary London embargoed all trade with Newcastle) claimed that the 'dearth' of sea coal was more apparent than real, as wholesalers kept huge stocks in storage while prices escalated.[34] One of Edward Hyde's informants claimed in 1653 that 'our dearth of coals exasperates' Londoners, and during the next Dutch war London's common council defined who had the right to transport coal around London 'for the benefit and relief of the poor in times of dearth and scarcity'.[35] In his descriptions of new recipe for coal briquettes, Hugh Platt repeatedly connected the 'dearth of victual' with the 'death and scarcity' of fuel.[36] In explaining why the state should require consumers to purchase his fuel, Platt referred to the current harvest crises,

---

[30] *APC 1578–80* (1974), 44. See also Rep. 20, f. 48v; LMA COL/AD/1/22 (Letter Book Y), f. 227-v.

[31] TNA PC 2/39, 35, 95–6, 823; TNA SP 16/173/63.

[32] Abraham Hill papers, BL Sloane MS 2902, f. 233-v.

[33] In John Walter's important article 'The Social Economy of Dearth', for example, the only uses of 'fuel' and 'wood' are in a discussion of the factors mitigating bad harvests. John Walter, 'Social Economy', 98. But cf. Ayesha Mukherjee, *Penury Into Plenty: Dearth and the Making of Knowledge in Early Modern England* (2014), which does consider Hugh Platt's interest in fuel.

[34] *Sea-Coale, Char-Coale, and Small-Coale: or a Discourse between A New-Castle Collier, a Small-Coale-Man, and a Collier of Croydon* (1643/4), 6.

[35] Bod. Clarendon MS 45, f. 292; Act of Common Council, 21 June 1665 in Christ's Hospital, Carrooms Memoranda, 1664–1759, GL MS 12830.

[36] Hugh Plat, *A new, cheape and delicate Fire of Cole-balles* (1603), sig A4, C1.

suggesting that governmental compulsion was reasonable because 'hunger breaketh through stone walls'.[37]

Platt's conflation of fuel and food scarcities in the service of underwriting state power was also employed, somewhat surprisingly, by the great champion of parliamentary liberties, Sir Edward Coke. During a moment of the parliament of 1628 when Charles I had very briefly convinced the House of Commons of his good will, Coke offered a conciliatory speech. In it he sought to sweep under the carpet the constitutional divisions that would soon lead parliament to present the king with the Declaration of Right. In a short-lived spirit of concord Coke asserted that the common law and the royal prerogative were not opposed at all but complementary. In an example offered to demonstrate this assertion the recent history of the coal trade was considered to show how the rule of law and accompanying property rights were, in fact, quite compatible with emergency measures justified by the royal prerogative. In a 'time of dearth', Coke explained, 'every man has a propriety in his own goods, yet a private man's corn may be brought out and a price set upon it: coals were so lately in London'.[38] The point here, for Coke as well as for Platt, was that dearths of fuel worked the same as dearths of food; in both cases physical need authorized emergency measures that could supersede normal property rights.

If a general scarcity of fuel and food were considered conceptually similar, so was the experience claimed by and for the poor during such periods. In particular, descriptions and representations of how the poor suffered during periods of fuel dearth (whether caused by poverty, cold, or high fuel prices) often focused not on being cold, but on being hungry. In 1665, the speaker of the House of Commons expressed the moving plight of 'the poor this hard season, especially those about this town who are ready to starve for want of fuel, the price of coals being so unreasonably enhanced'.[39] Two years later things were again said to be desperate. In March of 1667, a few days after Pepys noted in his diary that coal prices were four times the normal rate and the weather was 'most bitter cold', a petition by inhabitants of London protested to the Privy Council that high prices 'will occasion the starving of hundreds of His Majesty's poor subjects unless some speedy course be taken'.[40] Such claims are clearly highly rhetorical and in some cases were expressed in absurd terms considering that they masked clearly mercenary

[37] Hugh Plat, *A Discoverie of Certain English Wants, which Are Royally Supplyed in This Treatise by H. Platt of Lincolnes Inne Esquier* (1595), A3v.
[38] Robert C. Johnson, Mary Frear Keeler, Maija Jansson Cole, and William B. Bidwell, eds. *Commons Debates 1628* (New Haven, 1977–1983), III, 126–7, 133.
[39] Thomas Rugge, *Mercurius Politicus Redivivus* Vol. II, BL Add MS 10,117, f. 134. See also *CSPD 1664–5*, 154.
[40] *Pepys 1667*, 98–9, 102; TNA PC 2/59, 336.

motives. So in a 1628 petition projectors advocated royal ownership of several northern coal mines, justifying their endeavour through a rehearsal of the deprivations of the 'aged, poor, and impotent'. These and other subjects, they pronounced, were 'necessarily constrained to make no less frequent use of seacoals then they doe of the air; without which they cannot breath, and so without both, they cannot live'.[41] Towards the end of the century a broadside protested a tax on similar grounds, arguing that 'coals is a thing of so absolute necessity, that it is impossible to preserve the poor from perishing without having the same at a moderate price'.[42] Statements like these show that the rhetoric of coal as a basic necessity for the poor was available for exploitation by a variety of interests. But this does not diminish its power, especially as poor Londoners really did suffer from fuel poverty. When prices rose, many households really did have to choose between food and fuel, and they probably felt keenly the lack of both.[43]

It was therefore true that cheap fuel mattered particularly for the poor, and the rhetoric deployed in most discussions of fuel policy or governance therefore emphasized coal's status as a necessity for the poor in particular. The example of the 1628 projectors show that this language could be used for entirely self-interested motivations, but it was also often used by governors themselves who had no immediate personal or financial interest at stake. Instead, the rhetoric of coal as the fuel of the poor justified policies through which London's governors, both civic and central, sought to regulate a rapidly growing trade. Thus, in October of 1595 the civic leadership of London complained to Lord Burghley that Newcastle's Hostmen restrained the coal trade, lowered the coal's quality, and raised prices, all of which was to the 'detriment of such as use that kind of fuel being of the poorer or meaner sort'.[44] The Privy Council then repeated their language a few weeks later, condemning 'the great oppression of the poorer sort' and ordering the President of the Council of the North to investigate.[45] This 'oppression', moreover, was explicitly connected to the high food prices prevailing during the harvest crises of the mid-1590s. This was the high point of Ian Archer's 'Pursuit of Stability', when London's governors were especially conscious of the dangerous tensions within the city's social and political hierarchies.[46]

The late-Elizabethan moment of apparent social instability passed, but the language of coal as a politically sensitive fuel of the poor persisted. This

---

[41] BL Stowe MS 326, f. 6-v; the details of the project are discussed in Nef, *Rise*, II, 273–6.

[42] *Some Considerations Humbly offered to the Honourable House of Commons Against Passing the Bill for Laying a Further Duty of Coals* (n.d., c.1695).

[43] Cavert, 'Politics of Fuel Prices'.

[44] LMA COL/RMD/PA/1 Remembrancer's records, [hereafter Remembrancia], II, 105.

[45] *APC 1595–6*, 31–2; *APC 1596–97*, 27.

[46] Archer, *Pursuit of Stability*.

was not the *only* way in which coal was discussed by those considering or advocating particular policies. Another, closely related rhetoric framed coal as the common fuel of all. This was an especially useful approach for those claiming the mantle of the common good against private and particular interests. Profiteers could thus be denounced as selfish enemies of the public, as in the several cases when investigations concluded that London's woodmongers' desire for 'private lucre' was the primary cause of high coal prices.[47]

But the interests of the metropolitan poor loomed largest in sixteenth- and seventeenth-century complaints of problems with coal provisions and prices. They were invoked against the Hostmen of Newcastle and also against wholesalers within London. Thus, one William Nicholls, a petitioner to the Lord Treasurer in 1623, claimed that rich fuel wholesalers practiced frauds that devoured and ate up the poor. Nicholls drew liberally on Biblical passages to suggest that magistrates who ignored the poor risked their own souls: 'I beseech your honour, hear the cry of the poor; he that stopeth his ear at the crying of the poor, he shall also cry and not be heard.'[48] Few went so far as to threaten damnation to a potential patron, but Nicholls's association of the distress of the poor with the avarice of fuel wholesalers was conventional, even in the Houses of Parliament and the Privy Council.[49] Even when no imputations of greed were made, the suffering of the poor remained rhetorically useful. Thus in 1610, the House of Commons petitioned the king that the poor felt 'great grief' at what was, in fact, a tax on a very small proportion of the coastal coast trade.[50] Such complaints were heard most loudly and most predictably during wartime, as recurrent naval conflicts and real or feared invasions of Newcastle all drove up coal prices in the capital.[51] During the winter of 1643–4, in particular, as parliamentarian London embargoed coal from royalist Newcastle, the poor were widely depicted as the leading sufferers.[52] Throughout the early modern period,

[47] E.g., LMA COL/AD/1/25 (Letterbook AA, 1599), 285; COL/AD/1/28 (Letter book CC, 1608), 280; Rep. 70, f. 42v (1665). Similar uses in different contexts include TNA C66/2076/11 (1616); STAC 8/56/10 (1619). Wallace Notestein, Frances Helen Relf, and Hartley Simpson, *Commons Debates 1621* (New Haven, 1935), VII, 86.
[48] 'The most humble petition of William Nicholls a poor collier', KHLS U269/1 OE526, Citing Proverbs 21:13, Geneva version.
[49] William B. Bidwell and Maija Jansson, eds. *Proceedings in Parliament 1626* (New Haven, 1991), II, 127, 130; TNA PC 2/52, 465 (1640).
[50] Elizabeth Read Foster, *Proceedings in Parliament 1610. Volume 2 House of Commons* (1966), 168.
[51] For details of how sensitive coal prices were to the fortunes of war, see Cavert, 'Politics of Fuel Prices'.
[52] This is the consistent theme of *Sea-Coale, Char-Coale, and Small-Coale: or a Discourse betweene A New-Castle Collier, a Small-Coale-Man, and a Collier of Croydon* (1643/4).

then, problems with London's fuel supply were especially associated with the unfulfilled needs of the poor. This association led immediately to concerns for stability.

IV. 'WILL BREED A MUTINY': FUEL SUPPLIES AND THE PURSUIT OF STABILITY

Pepys' fear that high coal prices would lead to rebellion was born of a moment that was both unusual and yet also familiar for many Londoners. Metropolitan Coal prices were especially high in June of 1667 – perhaps the very highest ever in real terms and nearly the highest in absolute terms until the twentieth century – as the unprecedented disaster of the Medway provoked fears that the entire coal fleet would be destroyed or kept in port. And yet this unique experience fit within a pattern that was already well established by the 1660s and would continue throughout (and indeed, beyond) the rest of the early modern period. Wartime *always* caused coal prices to rise in London. This basic fact was widely, and rightly, taken to mean that the poor were especially vulnerable to wartime insecurity. Since the desperate poor were often seen as dangerous, it appeared persuasive to many contemporaries that insecure trade routes could lead directly to insecure governments. Thus, when the Earl of Anglesey suggested that 'want of coals' was a major spur to end the war in 1667 he was making an association between coal supplies and political stability that had become commonplace during the previous decades.[53]

London's first experience with the ways that naval war diminished coal supplies and threatened social disorder came during the opening years of Charles I's reign. War against both France and Spain had exposed English coastal shipping to raids, above all from the privateers of Hapsburg Dunkirk. These took a serious toll, reducing all Newcastle coal shipments by a third and raising prices.[54] Sir John Eliot claimed on the floor of the House of Commons that 'the poor of London cry out' at such deprivations. Eliot's particular agenda was to attack the failure of the Duke of Buckingham to protect the coasts as Lord Admiral, while Buckingham's defenders responded that it was hoarding wholesalers, rather than coastal insecurity, that raised prices 'especially on the poor'.[55] Whoever was blamed for them, the poor's cries appeared dangerous.

---

[53] Bod. Carte MS 47, f. 166-v.
[54] Tyne and Wear Archives, GU.TH/21/1. Export duties for years ending 1627 and 1628 were about 2/3 those for the year ending 1624. Thanks to Simon Healy for sharing images of these documents.
[55] Bidwell and Jansson, eds. *Proceedings in Parliament 1626* II, 129–30, quoting Eliot and Sir John Coke.

In response to such danger, authorities acted to protect the neediest. A survey of fuel stocks in October of 1627 found very few available supplies.[56] The following month London's Lord Mayor ordered that 100 chaldrons would be provided in small sales only to those deserving poor bearing a ticket from their churchwarden. If the intention was to ration supplies to the most vulnerable and worthy, the actual effect was to demonstrate widespread desperation among the urban poor. Assembling those deserving of charity at two central distribution points was found impossible, 'the officers for delivery out of the coals not being able to perform the business, nor resist the violence of the poor'.[57] To disperse the multitudes and prevent such 'violence', a committee of aldermen divided up the available supplies between the twenty-six wards for more local distribution. High prices persisted, and through to the war's end the poor were still found dangerous. In March of 1630 the admiral tasked with protecting coastal trade was ordered to convoy Newcastle's colliers. To underline the urgency of the order it was mentioned that 'the cry of the people of London for want of coals is very great. They have been with great clamour and noise to the Lords [of the Admiralty].'[58] Finally, the naval wars of the 1620s also prompted the first mention of the coal trade as an object of strategic consideration by foreign enemies. In 1628, an agent of Buckingham's circulated a forged discussion by foreign Jesuits in which they discussed how they hoped to cripple England. Among their observations was that 'London is as it were besieged for want of fuel'.[59] This was a forgery but it still indicates how already during the late 1620s coal scarcity produced genuine disruptions. The poor battled for cheap supplies and angrily voiced their demands to state officers, all of which were considered to aid the foreign enemy in its Machiavellian plans to sow domestic discord and thereby ease eventual conquest.

The forged agenda of a non-existent foreign Jesuit during the 1620s became, during the unprecedented conditions of the civil war, an important aspect of actual state policy. Whereas coastal trading was harmed during the 1620s, in 1643–4 it was almost entirely stopped.[60] In January of 1643 Parliament ordered a halt to trade with royalist Newcastle, citing its benefit

---

[56] TNA SP16/83/34.   [57] LMA Rep. 42, f. 13v, 19v.

[58] Sir Edward Nicholas to Sir Henry Mervyn, *CSPD 1629–31*, 222.

[59] Sir John Maynard, *The Copy of a Letter Addressed to the Father Rector at Brussels* (1643).

[60] Nef, *Rise*, II, 286; Hatcher, *History, Before 1700*, 489; Ben Coates, *The Impact of the English Civil War on the Economy of London, 1642–50* (Aldershot, 2004), 144. In June of 1643 the Committee of the Navy ordered all shippers to be bound not to trade to Newcastle on the grounds that some 'have presumed to lade coals' at the north-eastern ports, suggesting at least some smuggling. *An Order Concerning the Price of Coales and the Disposing Thereof, within the City of London, and the Suburbs, & c.Die Jovis 8. Junii 1643*.

to royal coffers.[61] Despite that ordinance's claim that there were 'sufficient coals' for the kingdom available elsewhere, it began measures to supervise, regulate, and expand London's fuel supply almost immediately. By May a committee officially concluded that coal from Scotland and Wales were insufficient, and that 'only Newcastle coal will be able to furnish the City of London'. The 'most likely' way to achieve this supply, it continued, was by 'reducing of Newcastle by force, either by sea or land or both'. It therefore suggested a loan, to be repaid at 8 per cent interest, to fund the £50,000 which was 'conceived needful' for the conquest of Newcastle.[62] When these proposals were enacted by parliament a few days later they were again justified through the irresistible and dangerous needs of the urban poor: 'the City of London, and all the great part of this Kingdom are like to suffer very deeply in the want of that commodity, so absolutely necessary to the maintenance and support of life, and which is like to be of very dangerous consequence in the influence which it may have upon the necessities of the meaner sort'.[63]

By October, with the conquest of Newcastle not imminent and winter approaching, parliament again invoked the urgent necessity and potential violence of the poor to justify a new approach to fuel provisioning. This ordinance again justified itself through the lack of 'sea coal, the want of which in winter may prove almost an insupportable misery, especially to the poorer and meaner sort of people'. Their 'insupportable misery', indeed, had already become dangerous: 'the common sort of people have of late destroyed, and still are destroying great store of timber trees being urged unto by necessity to procure to themselves fuel'. It therefore authorized the organized plunder of delinquents' woods within sixty miles of London, from which London's 'poorer sort of every parish to be first served'.[64]

London's civic and parliamentary leadership therefore explicitly claimed that fuel scarcity would lead to disorder, and they made it clear that preventing this was a priority. These claims were also picked up and shaped to fit the agenda of parliament's enemies. In May of 1643 the royalist press jumped on the likelihood of coming fuel scarcities in London, claiming that prices were already high and growing higher daily 'which makes the people apt to mutiny and the prime Members of the Houses to fear

[61] C. H. Firth and B. S. Rait, eds. *Acts and Ordinances of the Interregnum, 1642–1660* (1911), I, 63.
[62] LMA Jour. 40, f. 60.
[63] Firth and Rait, eds. *Acts and Ordinances* I, 171.
[64] *Ibid.*, I, 303. The collection of turf and peat the following summer was to proceed in similar ways, *ibid.*, I, 481–3; Jour. 40, f. 101–2. 'Delinquents' were enemies of the parliamentary cause.

the consequences'.[65] A Scottish traveller noted in 1643 that Londoners, despite the 'daily relief they get from Scotland' constantly complained of expensive coal and 'heavily bewail the loss of their advantageous Tyne'.[66] The parliamentarian press was concerned to stress the relief offered by the eventual conquest of Newcastle in 1644, anticipating at what 'easy rates the City will suddenly be provided with coals'.[67] The Scottish army appreciated in 1644, as it had in 1640, that one way to Londoners' hearts was through their fireplaces, publicizing their efforts to provide coal for the city's poor immediately upon their conquest of the Tyne.[68] The Venetian ambassador, finally, repeated the conventional wisdom regarding coal's political implications. By October of 1643 it was 'beginning to be wanted', causing 'a great outcry among the people'. The following summer he predicted riots in London if the Scots did not free its coal supplies, and by October 1644 he turned to cipher (indicating news of particular sensitivity) to predict that the Scots control of the Tyneside mines gave them 'such an advantage over London' that future divisions with their English allies would be likely.[69]

By the end of 1644, then, it had become clear to Royalists, Parliamentarians, and neutrals, as well as to English, Scots, and foreigners, that London's coal supply both depended on and had implications for its political stability. Scarce or expensive coal elicited grumblings, complaints, and, it seems likely, frequent outbreaks of minor and contained disorder – whether the 'violence of the poor' recorded in 1628 or the unauthorized plunder of woods and fuels noted in 1643–4. There is, however, no evidence that London ever actually experienced a fuel riot. Nevertheless, the lesson taken from the experience of the civil war by contemporaries was not the remarkable stability of urban society in the face of scarcity and danger, but rather the fragility of order in the metropolis.

From the 1640s forward it was widely assumed that coal scarcity would make the poor suffer, and that in response basic bonds of civility and obedience would dissolve. Thus, during the 1650s London's aldermen feared that a delayed spring shipment would bring 'a great clamour and disturbance amongst poor of the City for want of that necessary fuel', while a correspondent of Edward Hyde's warned 'our dearth of coals exasperates [the people], and I assure you, if the Dutch keep them from us, we shall

---

[65] *Mercurius Aulicus*, Thursday 25 May 1643 (Oxford).
[66] William Lithgow, *The Present Surveigh of London and Englands State* (1643), sig. A3v.
[67] *Exact Diurnall*, 11 July 1644. Also *Weekly Account*. 23 October 1644.
[68] LMA Rep. 57/2, f. 29-v; *JHC* III, 715. For 1640, John Rushworth, ed. *Historical Collections. The Second Volume of the Second Part* (1721), 1259–60.
[69] *CSPV* 1643–7, 30, 116, 152.

shortly cut one another's throats'.[70] The Venetian ambassador concluded in the spring of 1653 that the recent arrival of a coal fleet had prevented 'universal consternation' and an 'insurrection' in the capital.[71] The agendas of these three observers were quite different, as the aldermen hoped to avoid instability, the royalist exile Hyde wanted the opposite, and the Venetians had no such clear stake – but each offered the same analysis of coal's relationship to the governability of the poor. In 1667, as we have seen, Pepys feared rebellion, while a few years later a projector cited the recent experience of 'the City's incredible want of coal, near to the hazard of an insurrection'. Like the Earl of Anglesey, he argued that 'the consideration of this want' contributed greatly to 'contracting of that speedy peace'.[72] Those seeking instability perceived the same associations: in both 1708 and 1715 Jacobite plotters and their allies were stunningly optimistic in their beliefs that choking London of Newcastle coal would hasten regime collapse. In 1708 it was asserted that without coal London would suffer greatly within six weeks, while in 1715 Lord Mar suggested that an invasion of the north-east had a chance of success, since thereby the coal trade would be obstructed, 'which would either induce London to declare or at least distress the Government'.[73] Throughout the eighteenth century, long after the living memory of the civil war, or even the Dutch wars, had passed, London's stability still seemed to depend on coal. In the event of a hard winter and naval war, one historian wrote in the 1730s, 'poor people of London would not only be in danger of perishing, but the government would run a great hazard of being distressed by an insurrection'.[74] Even as late as the American Revolutionary war John Paul Jones argued that just a few frigates could destroy the 'coal shipping of Newcastle etc., which would occasion the utmost distress for fuel in London'.[75] All of these claims depended upon some very dubious assumptions about the poor Londoners; fuel riots in the capital seem not to have happened at all, and even if they had the urban poor were not an angry mob likely to topple a regime. Despite this, the persistent and pervasive rhetoric claiming that fuel scarcity caused disorder was not utter invention.

[70] Rep. 64, f. 131v (17 April 1656); London newsletter, 15/25 April 1653, Bod. Clarendon MS 45, f. 292.
[71] *CSPV 1653–4*, 63.
[72] Francis Mathew, *A Mediterranean Passage by Water, from London to Bristol* (1670), 8.
[73] Nathaniel Hooke, *The Secret History of Colonel Hooke's Negotiations in Scotland, in Favour of the Pretender; in 1707* (1760), 4; John Gibson, *Playing the Scottish Card* (Edinburgh, 1988), 98–9; HMC Stuart, I, 521–2; Daniel Szechi, *1715: The Great Jacobite Rebellion* (New Haven, 2006), 87.
[74] Thomas Salmon, *Modern History: or, the Present State of all Nations* (Dublin, 1739) IV, 215.
[75] John Paul Jones to the American Commissioners at the Court of France, Passy, June 5. [ver. July 4–5?] 1778. *The Adams Papers Digital Edition* (2008).

V. CONCLUSION: THE USES OF SCARCITY

Fuel as necessity, scarcity as dangerous, government as protector of the vulnerable: this chapter has examined a conventional rhetoric. Its power drew in part from deeply laid foundations of British culture, including the special status of the poor in a Christian polity and the assumption that good government produced prosperity. But the specific power of the poor's association with urban fuel supplies, an association often appropriated by governmental and commercial interests, must also be explained through the lived experiences of early modern Londoners. For precisely because they were entirely dependent on commercial markets for fuel, the London poor endured real hardship when their complex supply networks were interrupted. Historians have generally underestimated the importance of fuel in the budget of poor people, but during typical winter months poor households probably spent over 10 per cent of their total income on fuel. When prices rose steeply budgets suffered; if prices tripled (as they often did during wartime) a middling household on a fixed income would have had to cut its food bills by over 35 per cent in order to continue buying the same amount of coal, while for a very poor household the picture is much worse.[76] Complaints that the poor could 'starve' for lack of coal, therefore, were not mere rhetoric. During the rare but recurrent periods of high prices, London's poor really did face the choice of buying food or fuel.

Such extreme fuel scarcity, because it was episodic, was probably felt all the more keenly by London's poor. All pre-modern societies lacked fuel and energy compared to contemporary standards, but it was also the subjective, lived experience of *relative* scarcity that mattered. Feeling cold was doubtless a very real physical sensation, but recent research has also suggested that the perception of fuel scarcity would have produced dramatic mental, and therefore probably social, cultural, and political effects. Social scientists Sendhil Mullainathan and Eldar Shafir have argued that perceived scarcities of anything, whether food, companionship, or time, lead people to focus their minds in comparable ways. The object of desire takes up increasing amounts of their conscious and also subconscious mental capacity, as the singled-minded pursuit of the scarce good necessarily requires the sacrifice of attention elsewhere and even the sacrifice of basic cognitive competencies.[77] The very poor, they suggest, need to focus, worry, and obsess about the short term just to survive until the next paycheck or meal. Their research studies the twenty-first-century world, but it makes sense to imagine that during periods of fuel scarcity or high prices London's poor were not only

---

[76] Cavert, 'Politics of Fuel Prices'.
[77] Sendhil Mallainathan and Eldar Shafir, *Scarcity: Why Having Too Little Means So Much* (New York 2013).

very cold, but were also very preoccupied with the immediate problem of how they might become and stay warm in the immediate future. Depictions of the poor as obsessed with cold and fuel may therefore not be mere rhetorical or literary tropes. The urban poor probably did, as Ned Ward suggested, spend considerable time 'cursing the price of bread and coals'.[78] Elites and governors were well aware of this, and feared its implications.

[78] Ward, *British Wonders*, 44.

−8−

Fuelling improvement: development, navigation, and revenue

I. HOUGHTON'S ENGLAND: COAL FIRES AND GHOST ACRES

The mid 1690s were hard times. Disastrous harvests, expensive foreign wars, and currency devaluations strained the financial capacities of families, corporations, and the English crown itself. For John Houghton, such problems underscored the importance of the gospel of economic improvement that he had been preaching for years through his *Collection for Improvement of Agriculture and Trade*.[1] Every week Houghton provided his readers with commodity prices, advertisements, and suggestions for how England could be made more productive, more integrated, and more competitive. In his vision of a transformed national economy coal held a central place. Houghton recorded over 5,000 separate coal prices from across England, Wales, and Scotland, he distinguished the various types and grades of coals found throughout Britain, and he described the ever-increasing number of industries that these fuels supported. Most provocatively, Houghton argued that the nature and extent of the coal supply meant that many of England's landowners could and should cut down their trees.

For most of the seventeenth century, deforestation was an approaching danger, but for Houghton it was an elusive goal. In 1611 Arthur Standish had argued 'no wood, no kingdom', and innumerable statements, including royal patents and statute law, agreed that woods provided both necessary fuel and the navy's wooden walls.[2] But now Houghton awaited impatiently the day when the English would 'come to be of the humour of destroying our wood'.[3] He defended this heretical notion for many years, announcing in 1701 that after eighteen years of hearing all that had been said against him he was as sure as ever that woods 'threaten the public and private wealth'.[4]

---

[1] These were published from 1692 to 1703, but were often very similar to his earlier *A Collection of Letters for the Improvement of Husbandry and Trade*, published from 1681 to 1685.
[2] Standish, *Commons Complaint* (1611), 2.
[3] John Houghton, *A Collection for Improvement of Agriculture and Trade*, n. 174 (29 November 1695). Reprinted as *Husbandry and Trade Improv'd: Being a Collection of Many Valuable Materials Relating to Corn, Cattle, Coals, Hops, Wool, &c.* (1727), I, 449.
[4] N. 483 (24 October 1701), Houghton, *Husbandry and Trade Improv'd*, III, 176–7.

He offered twin justifications. First, forestry was simply not as valuable as other forms of agriculture, and so anyone who sold off trees would receive an immediate windfall of cash which could then be invested at interest, as well as enjoying continually productive acres for arable farming.

Second, and at least as importantly, this change was not simply more profitable; it was more commercial. Without local wood supplies all would be driven to work harder and to engage more thoroughly in the market. Labour would be disciplined and transport costs would fall as estates carried grain to market and returned with fuel. This entailed greater coastal traffic, which in turn implied more trained mariners and more ships, growth which Houghton eagerly quantified. Cutting down trees therefore produced a more efficient and more productive economy, one that could support more people, employ more labour, produce more food and manufactures, and pose a greater naval threat to belligerent neighbours. All of this, Houghton stressed, was because coal and trees were fundamentally different. From the perspective of the political economist ('by considering the nation as one man, in one joint-stock') coal was free because it required no imports and used no land. Wood ground, therefore, was doubly undesirable. First, because it rendered land unavailable to commercial agriculture, and second because it did not promote commercial integration – 'such ground is in effect lost because it hinders the burning of coal'.[5]

Houghton's argument was intended to shock his contemporaries, but it surprises the modern historian for a quite different reason. By arguing that coal use freed up additional land for productive agriculture, and that this process would produce positive feedback loops of fuller employment, more efficient labour, rising population, falling transport costs, greater domestic demand and, therefore, a growing manufacturing sector, Houghton anticipated one of the most influential recent interpretations of English economic history. Scholars interested in why England became the first industrial economy (whether their object of comparison is Tudor England or eighteenth-century China) have argued that coal provided significant 'ghost acres'.[6] Since all production in what E. A. Wrigley calls an 'organic economy' derives from the land and from the energy provided by photosynthesis, England's embrace of a subterranean fuel source meant that it both enjoyed more energy than other early modern economies *and* it required much less land for its production.[7] With coal,

---

[5] Houghton, *Collection of Letters*, II, 56, and 50–73 *passim*.
[6] Wrigley, *Energy*, 38–9; Pomeranz, *Great Divergence*, 275–7.
[7] Wrigley importantly stresses, however, that this distinction is only meaningful in the short term. In the longer term fossil fuels like coal are not free energy but rather capital deposits, accumulated over geological time, and can only be drawn down rather than replaced. Chapter 2 considers these issues in more detail.

according to this view, English agriculture was freed to produce more food, which – exactly as Houghton predicted – allowed for demographic growth, urbanization, expanding consumption, and a uniquely expansive manufacturing sector.

The other crucial contributions of coal to English industrialization – to prompt and to power the great new technologies of the nineteenth century – were not part of Houghton's vision. But historians stressing the unprecedented industrial potential of steam power, as well as those scholars who focus on showing the minor role of steam in early industrial factories, both risk missing the ways that coal, already for Houghton and his contemporaries writing during the pre-industrial 1690s, seemed to provide English manufacture with a crucial competitive advantage. Coal, Houghton noted, was used 'in most mechanic professions (except ironworks) that require the greatest expense of fuel; witness the glass-houses, salt-works, brick-making and malting'.[8] He did not belabour this point, however, perhaps because others were explaining coal's centrality to English, and in particular to London's, manufacturing sector well enough already. As parliament considered heavy taxes on coal in 1695–6, a series of petitions and pamphlets complained that London's artisans 'cannot carry on their trade without great quantities of coals', and that unless fuel taxes were lowered 'manufacture will be ruined and lost to his nation'.[9] Already in the 1690s, then, the developmental potential of 'ghost acres' as well as the industrial pay-off for cheap energy had already found their champions.

Houghton was not quite content, however, to stress such commercial and industrial benefits. Improvement was a long-term project but England also faced more immediately pressing concerns, so Houghton rehearsed the long-familiar claim that a buoyant coastal coal trade contributed crucially to national security. More shipping required more ships and more seamen, both of which were easily converted into naval assets during wartime. This association between maritime trade and power was, Houghton had suggested in 1683, clear to everyone, as a flourishing coal trade 'might frighten our neighbours from making war with us; and ... bring a further advantage of an honourable lasting peace'.[10] Because it contributed so directly to military power as well as to general improvement, Houghton stated in 1695 that 'if I would wish a tax upon anything it should be upon the wood of our

---

[8] N. 241, 12 March 1696/7, Houghton, *Husbandry and Trade Improv'd*, II, 155.
[9] 'A Petition of several Dyers ... in the City of London, and several Places adjacent', *JCH XI*, 394; see also 'A Petition of the Masters, Workmen, and Servants, of the Glass-houses in Southwark, *ibid.*, 391; 'A Petition of the Feltmakers, Dyers, Smiths ... of Southwark', *ibid.*, 390; and 'A Petition of the Lord Mayor, Aldermen, and Commons, of the City of London', *ibid.*, 412.
[10] Houghton, *Collection of Letters*. II, 56.

own growth and not upon coals, which, I am sure, hinders our navigation'.[11] The petitions presented to parliament that same winter of 1695/6 agreed, but they failed to convince the legislature to ignore such a lucrative source of revenue. Whether coal taxes should be raised (to the benefit of the state's finances) or lowered (to the benefit of industrial consumers and coastal shippers) was not resolved during the 1690s nor indeed thereafter. But for all sides Houghton's arguments for the transformative power of coal's energy and of the commercial coal trade were increasingly persuasive. Whereas discussion of London's coal supply before the Restoration of 1660 tended to focus on the desperate needs of the urban poor, thereafter it became increasingly devoted to the possibility that expanding coal consumption would facilitate improvement and power.

II. 'MECHANIC PROFESSIONS THAT REQUIRE THE GREATEST EXPENSE OF FUEL': COAL AND THE MANUFACTURING ECONOMY

During and after Houghton's lifetime coal remained the fuel of the 'poor', but the nature and purposes of much of this description changed profoundly. The needs of the deserving poor, the householder or head of family, were still invoked in discussions of fuel policy, and civic and parochial authorities still distributed free or subsidized fuel. Yet during the late-seventeenth century this discourse changed in important ways; the 'poor' invoked in discussions of fuel policy were increasingly understood to be labouring tradesmen and manufacturers, important primarily as contributors to the national economy rather than as heads of households or potential threats to social and political stability. The importance of such people to the state was increasingly represented as their productive potential rather than, or in addition to, their capacity for disorder or their claims on paternalist care and good governance. Coal remained in many senses the fuel of the poor, but gradually the 'poor' were seen less as vulnerable and threatening subjects than as factors of production. Through this shift the interests of London's poor artisans and its rich producers could be put in alignment, as if their similar dependence on cheap fuel overcame their social differences and rendered them equally members of the manufacturing sector.

The extent of this rhetorical shift can be illustrated by beginning with earlier claims that poor and rich customers engaged with the fuel market entirely differently. A 1653 pamphlet, lampooning the frauds and manipulations of fuel wholesalers, contrasted modest artisans with great manufacturers. The former were a 'beggarly rabble of poor silk weavers, button makers,

[11] N. 174, 29 November 1695, Houghton, *Husbandry and Trade Improv'd*, I, 454.

taylors, ragmen, bonelace makers, tobacco pipe makers, wash women, [and] scolding oyster-wives', who bought their fuel in small quantities, 'seldom ... above half a bushel at a time'. Such characters, claimed the pamphlet's woodmonger, were served by the chandler, while he reserved for himself the larger fry: 'we deal with brewers, and dyers, and merchants, and rich citizens, some of whom fetches in forty, fifty, or a hundred chaldron at a time'.[12] For all its polemics this pamphlet represents fairly accurately what, as we saw in chapter 2, were major differences between how the rich and the poor engaged with the market and acquired their fuel. The pamphlet's point is that the wholesalers are capable of fleecing each class equally, but in making this allegation there is no suggestion that poor pipe makers have anything in common with rich brewers. Wealth defined how one engaged with the market, not the uses to which customers put their fuel.

From the 1690s this distinction became decreasingly significant as pamphleteers, lobbyists, and political economists saw cheap fuel as an essential contributor to London manufacturing. This association of coal with industry was not new; as we have seen coal always was and was seen to be an industrial fuel, and it was the brewers, glass-makers, and brick-burners that were most often said to consume large amounts of fuel and emit large amounts of smoke.[13] The novelty arose, rather, from the debates surrounding new taxes, instituted first after the Great Fire of London in 1666 and growing in particular during the 1690s. These high taxes were repeatedly debated in parliament in ways that invited public lobbying efforts. The pamphlets and petitions created for such debates argued for lower coal taxes, and in so doing they addressed the relationship between coal and industrial production that earlier authors and government officers had often commented upon but had not developed with any sophistication. From the 1690s the state's new tax demands meant that many trades and industries were obliged to defend their contribution to the greater good.[14] Within this context claims for the London coal market were both newly broad-ranging and newly specific.

Coal, it was now increasingly argued, was essential for urban artisans and manufacturers. In some cases these trades were specifically enumerated or their interests singled out by defenders. Thus, advocates of the London

---

[12] Well-willer to the prosperity of this famous Common-wealth, *The Two Grand Ingrossers of Coles: viz. The Wood-Monger, and the Chandler* (1653), 10. A half-bushel was 1/72 of a chaldron.

[13] See Chapter 2 and Cavert, 'Industrial Fuel'.

[14] Jason Peacey has argued that the English Revolution prompted new levels of engagement, including sophisticated lobbying, with parliament. The methods used during the 1690s were not new, therefore, but the state's new revenue demands did allow them to be applied to a new class of industries and trades. Peacey, *Print and Politics*.

glass-makers, dyers, and sugar-boilers during the years around 1700 complained that high coal prices hurt their interests.[15] A better tactic, however, was to associate the specific interests of such particular trades with the collective (and therefore, it was claimed, more public and common) interest of urban manufacturers in general. According to one such pamphlet, the coal trade benefitted 'the brewer, the glass-maker, the salter, the smith, and all manufacturers and workers with fire'.[16] Others found it more powerful simply to gesture towards the scale and importance of 'our manufactures depending upon sea-coal'.[17]

The primary reason for coal's essential role in London manufacturing, it was increasingly argued, was the highly competitive nature of international trade. London's coal was taxed more heavily than British coal exports, a policy repeatedly denounced as a ludicrous subsidy of foreign manufacturers. Because so much production aimed for export markets, any source of comparative advantage was critical and any policy that counteracted such an advantage misguided. Many agreed with a pamphlet of the 1690s that claimed 'foreigners will be able to out-work us and afford several commodities, as dying of all sorts of woollen manufactures, iron-work for shipping, all sorts of wrought iron, glass, and refined sugar, &c. (which are manufactured by our own coals) much cheaper in all foreign markets than we'.[18] 'Foreign artificers', claimed another pamphlet, 'have been highly encouraged to set about their endeavors to beat us out of our manufactures depending upon sea coal'.[19] While the tax structure remained unreformed these complaints continued. By the 1740s fuel prices in and around London had caused, one argued, 'the languishing condition of many manufacturers', while another rued that French manufacturing had to become 'equal to, if not superior to Great Britain; as sugar-baking, and making glass bottles, and many other valuable branches of commerce too plainly witness'.[20]

Another argument was less prominent at the time but appears quite prescient in light of England's subsequent pattern of industrialization. The most elaborate defence of coal's benefits from the pamphlet literature of the 1690s is an anonymous tract called *The Mischief of the Five Shillings Tax*

[15] *The Case of the Glass-Makers in and About the City of London* [n.d., *c.*1711]; also *Reasons Humbly Offered Against the Bill for Laying Certain Duties on Glass Wares* (n.d. *c.*1700); *Reasons, Humbly Offered to the Honourable House of Commons, by the Dyers, Against Laying a Further Duty upon Coals* (n.d. 1696?); *CSPD 1700–2*, 559.

[16] *An Account of a Dangerous Combination and Monopoly upon the Collier-Trade* (1698), 8.

[17] *The Mischief of the Five Shillings Tax upon COAL* (1698), 23 and *passim*.

[18] *Reasons Humbly Offer'd to the Honourable the House of Common, Against the Bill for Laying a Duty of 5s per Chaldron upon Coals* (n.d. *c.*1690s)

[19] *Mischief*, 23.

[20] William Bowman, *An Impartial View of the Coal-trade* (1743), 22–3; Thomas Lowndes, *A State of the Coal-trade to Foreign Parts* (1745), 7.

upon COAL. In addition to rehearsing the dangers of foreign competition as well as coal's benefits to navigation, it makes an unusual set of predictions regarding the future of English industrial geography. High prices in and around London will 'draw most sorts of manufacture made with sea coal into the bishopric [of Durham] and Northumberland, or into the other coal countries'. Since coal was naturally so much cheaper near the mines, any policy exacerbating that differential only further threatened the viability of London industry and hence of the coastal coal trade. 'I do appeal to common sense', it argued, 'to pronounce the fate of all our manufacture made with sea coal under these certain advantages the north have above the south countries. They will most certainly by these means be run out of the south into the north; nay, they have already got footing there.'[21] For this pamphleteer, economic historian Robert C. Allen's argument regarding the industrial benefits of cheap energy would be unsurprising.[22] *The Mischief* did not, of course, predict the invention of the steam engine, but it did quite clearly associate cheap fuel with industrial competitiveness, and it concluded that Londoners could be put out of work by northerners as easily as by the French or Dutch.[23]

During the opening decades of the eighteenth century, arguments regarding coal's importance to urban manufacturing were regularly elaborated in print and in parliament. Printed tracts continued to defend the various interests of mine-owners, shippers, middlemen, and consumers, but some of these grew into minor treatises defending to role of coal in Britain's political economy.[24] One 1739 tract, for example, gave a detailed account of a trading cartel which was said to be 'to the no small discouragement of a great number of manufacturers, such as dyers, sugar-bakers, brewers, distillers, glass and salt makers, smiths, founders, &c and to the great oppression of the poor'.[25] Another extended discussion described the 'great quantities' of coal used by 'sundry of our manufactures' while yet another explained precisely which industries suffered from competition with foreigners who imported cheap British coal.[26] These arguments

---

[21] *Mischief*, 19–21.

[22] Allen, *British Industrial Revolution*.

[23] The importance of cheap fuel for manufacturing competitiveness elsewhere in England was also recognized in 'Essay on Ways and Means to Raise New Funds' (1757), BL Add MS 1215, f. 232v; *An Account of the Constitution and Present State of Great Britain* (1759), 4.

[24] The pamphlet literature includes *The Case of the Owners and Masters of Ships Imployed in the Coal-trade* (1729); *Reasons Humbly Offer'd for Continuing the Clause Against Mixing at the Staiths the Coals of Different Collieries* (1711); *Reasons Humbly Offered; To Shew, That a Duty Upon In-land Coals, will be no Advantage to His Majesty* (1725?)

[25] George Nixon, *An Enquiry into the Reasons of the Advance of the Price of Coals, Within Seven Years past* (1739), 10.

[26] *A Letter to Sir William Strickland, Bart. Relating to the COAL TRADE.* (1730), 17; Lowndes, *A State of the Coal-Trade.* See also Bowman, *An Impartial View.*

were increasingly sophisticated examinations of energy's role in political economy, but their language was not the preserve of an enlightened
philosophical elite. Nearly identical arguments were offered to parliament
by manufacturing lobbying groups. In 1738 London's court of common
council identified both 'the manufacturers and poorer sort of people' as
the chief sufferers from high coal prices, but within a few months the
manufacturers moved clearly to the fore.[27] A petition was submitted to the
Commons on behalf of the collective interests of the 'consumers of large
quantities of coals'.[28] A few weeks later London's civic government submitted a supporting petition, noting specifically the expenses incurred by
'those who use a large amount'.[29] The whole debate, according to Horace
Walpole, seemed driven by 'the manufacturers'.[30]

By the middle of the eighteenth century, then, London coal consumption
was frequently and increasingly associated with industrial manufacturing. The needs of the urban poor were still invoked too, but often in ways
which conflated their interests with those of large-scale industry. This conflation reached its early modern apogee in the work of two of the eighteenth
century's leading political economists. Josiah Tucker found the tax on the
domestic coal trade 'absurd'. 'The taxes upon the necessaries of life', Tucker
wrote in 1750, 'are in fact so many taxes upon trade and industry; and such
must be accounted the duties upon soap, coal, candles, salt and leather.'[31]
To tax coal was to tax life, which raised the cost of labour and therefore the
cost of making everything and of doing all manner of business. Elsewhere,
however, Tucker assessed coal quite differently, not as a necessary of life
but as a 'raw material'. When exported, the French and Dutch use this raw
material 'in the manufacture of such articles as are intended to prevent the
sale of English merchandise'. Tucker therefore recommended doubling the
export duty in the interest of 'the manufacturers of the City of London, the
metropolis of the whole kingdom'.[32] Tucker thus considered coal important
for manufacturing because it was a raw material and also because it was a
necessity of life and therefore an ingredient in the cost of labour – two quite

---

[27] Jour. 58, f. 82v.
[28] 28 February 1738/9, *JHC XXIII*, 263. The petition was on behalf of 'the glass-makers,
brewers, distillers, sugar-bakers, soap-boilers, smiths, dyers, brick-makers, lime-burners,
founders, and calico-printers, and other traders and inhabitants in and about the Cities of
London and Westminster who are consumers of large quantities of coals, imported from
Newcastle and the Northern parts of England into the ports of London'. Robert Walpole's
copy is CUL MSS Ch(H) Political Papers, 80, 64.
[29] *JHC XXIII*, 305.
[30] *HMC Buckinghamshire*, 28.
[31] *HMC Townshend*, 375; Josiah Tucker, *A Brief Essay on the Advantages and Disadvantages,
Which Respectively Attend France and Great Britain, With Regard to Trade* (1749), 23.
[32] Josiah Tucker, *The Case of the Importation of Bar-Iron From Our Own Colonies of North
America* (1756), 28–9.

different claims made in different contexts, but both assimilating coal burning into industrial production.

While Tucker offered these two approaches to coal in different works written years apart, Adam Smith drew on Tucker's thinking to elide these into one discussion. Coal, wrote Smith in *The Wealth of Nations*, was 'a necessary of life' just like – following Tucker – soap, candles, salt, and leather. It therefore was such an important component of the cost of labour that 'all over Great Britain manufactures have confined themselves principally to the coal countries; other parts of the country, on account of the high price of this necessary article, not being able to work so cheap'. Having defined coal as a necessity, Smith then shifts, in the following sentence, to consider it as a production factor: 'In some manufactures, besides, coal is a necessary instrument of trade; as in those of glass, iron, and all other metals.' He then notes the oddity of taxing exports more lightly than the domestic trade.[33] For Smith and for Tucker, then, coal's role in heating the hearths of labouring men and their families were analytically equivalent to its role in heating furnaces. Both mattered as, and only as, costs of industrial production in a world of vigorously competitive trade.

III. 'FRIGHTEN OUR NEIGHBOURS': THE NURSERY OF MARINERS

Houghton's chief arguments in favour of the coal trade were its commercial benefits, but he also identified an associated military consideration. Greater coastal trade required more shipping and more mariners, a growth that he suggested would double the number of ships then supplying London with both coal and timber. This, as anyone who had ever seen the early modern Thames would have known, was a prodigious fleet, and the prospect of its doubling was therefore enough to 'frighten our neighbours from making war with us; and with that, bring a further advantage of an honourable lasting peace'.[34] While Houghton's elaboration of the commercial advantages of subterranean fuel was original and insightful, the association with naval power was entirely commonplace. Its familiarity made it all the more useful for his purposes as it buttressed his unusual claims with quite conventional arguments. Houghton's predecessors had argued for many decades that the coal trade was vital for English shipping, and his immediate successors would develop this further to explain how exactly the coastal coal trade could improve English national security during the newly large-scale warfare after 1689.

From the 1650s England fought three naval wars against the Dutch and then a longer and more costly series of conflicts with France, even as the extent

---

[33] Adam Smith, *An Inquiry into the Nature and Causes of the Wealth of Nations*, R. H. Campbell and A. S. Skinner, eds. (Oxford, 1976), 874–5.
[34] Houghton, *Collection of Letters*. II, 56.

of its overseas trading interests also expanded. Each of these wars required the sudden enlistment of many thousands of men, and while any able-bodied man might make a passable soldier, naval service required special training. The duties of an 'able seaman' or a gunner certainly required experience and skill, but simply to endure the motion of a ship without sickness often required experience at sea. This could be a serious matter in a crowded ship; one author claimed in 1693 that seasick 'landmen' on board naval vessels 'empty themselves every way, either between decks or in the hold, to the great annoyance of all the rest of the crew, who are hereby exposed to so many diseases as have appeared in the late years' expeditions of the royal fleets'.[35] Sailors thus needed both 'sea-legs and sea-stomachs' but becoming acclimated to life on board took time. The Royal Navy, however, required new men immediately upon the outbreak of war – 20,000 were already enlisted by the winter of 1689/90 and 33,000 a few months later.[36] Without manpower the millions spent on ship construction were useless and the decisions of diplomats, administrators, and officers impossible to execute. For many early modern commentators England's coastal shipping therefore served an essential service to national security, and the coal trade was regularly cited as the foremost apprenticeship for future mariners.

Expanding English shipping and the supply of trained mariners were goals of royal advisors from the late sixteenth century, and under the early Stuarts discussions of the coal trade often stressed its service to the general advancement of shipping.[37] Such early statements, however, were usually restricted to claims that coal shipping contributed to maritime capacity, with few extending this to suggest that 'navigation' translated directly into naval power.[38] From

---

[35] Anon, *Proposals to Increase Seamen* (1693), in J. S. Bromley, ed. *The Manning of the Royal Navy: Selected Public Pamphlets 1693–1873* (1974).

[36] The maximum enlistment was over 45,000 men, reached both during the mid 1690s and 1706–11. Daniel A. Baugh, 'The Eighteenth Century Navy as a National Institution 1690–1815' in J.R. Hill and Bryan Ranft, eds. *The Oxford Illustrated History of the Royal Navy* (Oxford, 1995); N. A. M. Rodger *The Command of the Ocean: A Naval History of Britain 1649–1815* (New York, 2006), 206.

[37] Examples include 'Reasons against impositions to be layed upon Coles' (*c.*1590) BL Lansdowne MS 71/13, f. 24; John Keymer's project, KHLS U269/1 OE1515; BL Lansdowne MS 169/54; Robert Kayll, *Trade's Increase* (1615), 10–2; Commons' grievances, 1624, SP 14/165/53, IX; Statements by the Bishop Neile of Durham, Bidwell and Jansson, eds. *Proceedings in Parliament 1626*, I, 239, 241; 'Petition of masters of owners of ships trading to Newcastle for coals to the council', SP 16/180/58, (calendared 1630, dated by Nef to 1638); Maija Jansson, ed. *Proceedings in the Opening Session of the Long Parliament House of Commons* (Rochester, 2000), I, 82, 86.

[38] Exceptions include the 1618 finding by the Privy Council that regulating the coal trade 'had relation to the state of the navy and the puissance of the kingdom by sea, whereof this trade to Newcastle is a principal nursery', APC *1618–9*, 276; 'The French and Dutch growing daily more powerful at sea it is high time for our King to increase his Royal Navy ... it is very probable the coal trade from Newcastle may answer all ends aimed at by the fishing, especially that of breeding up seamen the bulwark of this Kingdom.' Bod. Rawlinson MS A 185, f. 309-v (*c.*1628?).

the mid-seventeenth century this changed as the coastal coal trade was increasingly as seen as an essential basis for England's unprecedented number of sailors in the royal service. The regular and frequent runs from the Tyne and Wear valleys along England's eastern coast to London (and, to a lesser extent, to other south-eastern ports) actually did dominate England's coastal trade, and the labour involved was increasingly seen by administrators, reformers, and pamphleteers as a way to produce the naval expertise that would be needed to rule the waves.

A thriving coastal coal trade, it was argued, yielded the trained and experienced men necessary to provision the capital, defend the realm at sea, and expand England's maritime trade throughout the world. *The Mischief* pamphlet developed these claims in particular detail, asserting the unique contribution of the coastal coal trade to national security.[39] First, the collier trade constituted the largest sector of England's shipping trade by capital value and employed the most men, and was therefore 'the true parent and support of our navigation. It is the greatest, the most constant, and almost universal nursery of our ships and sailors.' Foreign trade, this author (inaccurately) claimed, was relatively insignificant because it employed less tonnage.[40] Oversees shipping was a drain on manpower, with the profitable trades to the East and West Indies and Africa among those where 'English bodies' withered under 'the salt meat, the stench water and the hot climates'. Shipwreck near such hostile shores meant certain death, but in the coastal trade a storm simply washed English sailors onto their native beaches, 'into their mother's lap, where their lives are saved'. To neglect the collier trade, this pamphlet argued, was to neglect navigation itself, the danger of which was amply demonstrated by England's early history of successful foreign invasion by Saxons, Danes, and Normans. 'A flowering collier trade', in sum, was essential for national security and a secure foundation for empire; it would make England 'safe and easy at home, and formidable abroad'.

The author of this pamphlet certainly stretched his case as far as it would go, but other sober and relatively disinterested observers agreed that with proper management the collier trade really would be the best method to turn English boys into trained mariners. A puzzling dilemma arose, however: the coastal collier trade could not be suspended during wartime without compromising the very interests of poor householders and industrial manufacturers described above. It was therefore at once a long-term producer of potential sailors, and yet also the navy's direct competitor in the wartime

---

[39] *Mischief.*
[40] Hatcher indicates that during much of the seventeenth century the coastal coal trade constituted a majority of English domestic shipping tonnage. Hatcher, *History, Before 1700*, 470–1. Its value, however, was dwarfed by foreign trade, and by 1700 it was no longer true that foreign trade used less tonnage than the coal fleets.

labour market. Projectors, reformers, and statesmen considered this a problem that could be solved. Some advanced proposals founded on the belief that the collier fleet could supply sailors to the navy and also keep down its own labour costs and thus the price of fuel.[41] In 1703 desperate MPs, including future Tory leader Sir Robert Harley, argued that encouraging the coal trade was among the least bad of the many problematic proposals for manning the navy in the aftermath of a devastating hurricane.[42] At their most ambitious such projects framed the coal trade as a kind of machine to produce sailors, 'ever [to] be preserved, and perpetually increased'.[43] From the 1690s onwards the claim that the coal trade supplied the nation's commercial and naval ships was common in petitions to and speeches before parliament.[44]

Despite such optimism, resolving the wartime tension between preserving the coal trade and supplying the navy was nearly impossible. Naval administrators and officers, even when they accepted the political and economic importance of London's coal supply, nevertheless viewed colliers with a predatory eye. During the Second Dutch War of 1665–7, for example, some of the navy's leadership could barely restrain themselves. Admiral Sandwich recorded seeing eighty coal ships in May 1665 off the Suffolk coast. Despite the 'proclamation of the King's to secure their men, it was deliberated … whether we should instantly seize [them]'.[45] The officers decided against impressing in regards to 'the expectation of the City of London from this collier fleet' but a few weeks later Sandwich recorded no such scruples. 'A very great fleet of colliers passed by bound for London, out of which our ships took men (notwithstanding the proclamation prohibiting it) and indeed there was necessity for it, we expecting a battle daily and wanting seamen much.'[46] That same month the Duke of Albemarle wrote

---

[41] Proposals include speeches by Col. John Birch and George Downing, 1671 and 1675, Anchitell Grey, *Debates of the House of Commons, from the year 1667 to the year 1694* (1763), I, 441; III, 333; *An Answer to the Coal-Traders and Consumptioners Case* (1689); Robert Crosfield, *England's Glory Revived* (1693) in Bromley, ed. *Manning of the Royal Navy*, 6; *Proposals to Increase Seamen* (1693), in *ibid.*, 13–4; Sir William Hodges, *Great Britain's Groans: or, An Account of the Oppression, Ruin, and Destruction of the Loyal Seamen of England* (1695).

[42] Minutes of a debate by a parliamentary committee on manning the fleet, 27 November 1703, and committee of the whole house, 4 December 1703. R. D. Merriman, ed. *Queen Anne's Navy: Documents Concerning the Administration of the Navy of Queen Anne, 1702–1714* (1961), 184–7.

[43] *A Plan for the Better Regulating the Coal Trade to London, by Preventing the Fluctuation of the Price* (n.d., c.1750), 4.

[44] E.g., *JHC XI*, 382; *JHC XII*, 587, 609; *JHC XIV*, 310; *Vol. JHC XVI*, 531; *JHC XXI*, 474, 516; *JHC XXIII*, 309.

[45] R. C. Anderson, ed. *The Journal of Edward Mountagu First Earl of Sandwich Admiral and General at Sea 1659–1665* (1929), 212.

[46] *Ibid.*, 223

to Sandwich from Whitehall that none should be impressed from colliers because a fuel shortage would 'occasion great disorders' among the city's poor.[47] The next year, however, as Albemarle joined Prince Rupert in command of the navy during a busy campaign, they eagerly preyed upon the collier fleets as a means to replace casualties.[48] The colliers, for their part, knew how the navy viewed them and one admiral described to Samuel Pepys the 'constant fraud' through which they avoided the press.[49]

While there was real tension and competition in practice, then, it was nevertheless agreed by many that the coastal coal trade did – or could – provide the navy with the men it needed to pursue England's increasingly ambitious foreign policy. Moreover, they may in fact have been right. Some concerned with manning England's navy envied the French system, by which a national registry of mariners allowed bureaucrats to identify thousands of trained sailors for quick enlistment. But continuous coastal trade during wartime may have provided Britain with more sustained and renewable sources of necessary manpower than the French method.[50] Little wonder, then, that many believed throughout the early modern period that the coastal coal trade was of such strategic benefit that the use of coal mines close to London was deliberately suppressed. There were some, according to Edward Chamberlayne in 1683, who believed that Blackheath, near Greenwich just outside of London, contained abundant coal, but that it would be to 'the great prejudice to navigation ... in regard this colliers trade between Newcastle and London is the greatest nursery of seamen we have'.[51] Blackheath did not contain coal, but the persistence of this belief, which still needing refuting in the nineteenth century, indicates how strong was the view, during the century after 1650, that the coastal coal trade could

---

[47] Albemarle to Sandwich. 2 November 1665, Bod. Carte MS 75, f. 389.

[48] J. R. Powell and E. K. Timings, eds. *The Rupert and Monck Letter Book 1666* (1969), 66, 71, 131, 140, 285.

[49] Robert Latham, ed. *Samuel Pepys and the Second Dutch War: Pepys's Navy White Book and Brooke House Papers* transcribed by William Matthew and by Charles Knighton, (Aldershot, 1995), 134.

[50] For registration schemes see Captain George St Lo, *England's Interest; Or, a Discipline for Seamen* (1694) in Bromley, ed. *Manning of the Royal Navy*. For comparative effectiveness Rodger, *Command of the Ocean*, 210–2.

[51] Edward Chamberlayne, *The Present State of England. Part III and Part IV*, (1683), 28. For this belief see also *Mischief*, 6; *The Present State of Great Britain, and Ireland ... Revised and Completed to the Present Time, By Mr. Bolton*. (10th ed. 1745), 102; *The Foreigner's Guide: Or, a Necessary and Instructive Companion Both for the Foreigner and Native* (1729), 12; Madame van Muyden, ed. *A Foreign View of England in 1725–1729; The Letters of Consieur Cesar de Saussure to his Family* (1995), 195. During the summer of 1643 the City of London in fact funded searches for coal supplies in Blackheath, and finding coal in Kent was still a goal of geologists in the nineteenth century. Rep. 56, f. 198v, 207v, 237; W. Stanley Jevons, *The Coal Question: An Inquiry Concerning the Progress of the Nation, and the Probably Exhaustion of our Coal-mines* (1906), 52–4.

supply the capital's enormous market, and still supply men to a maritime nation and a naval state.

IV. 'UPON COALS': THE COASTAL TRADE AND THE SINEWS OF POWER

There were many throughout early modern England who agreed with Houghton that taxes on coals were imprudent because they hindered navigation. And yet such taxes persisted. Even Adam Smith conceded that while coal taxes increased manufacturing costs their existence was not as 'absurd' as Tucker alleged. 'Such taxes, though they raise the price of subsistence, and consequently the wages of labour, yet they afford a considerable revenue to government, which it might not be easy to find in any other way. There may, therefore, be good reasons for continuing them.'[52] Finding new taxes was, as Smith observed, no more 'easy' in the eighteenth century than it has become since. London's coal imports were therefore taxed from 1600 onwards, and with particular severity from the 1690s. This high taxation and the debates it sparked were the essential context for the pamphlet literature that celebrated coal's benefits. Coal's vital contributions to employment, trade, and naval power were asserted loudest by those most upset with its continued taxation. But such taxes contributed to expanding national power in ways no less important than trained mariners and integrated commerce. By the early eighteenth century, London's coal consumption offered the British crown a large and dependable source of revenue, an essential precondition for its creation of the fiscal-military state.

From the first explosion of London's coal consumption under Queen Elizabeth the English state had sought ways to turn the trade to its own benefit. The incorporation of the Newcastle Hostmen in 1600 brought the crown a duty of a few thousand pounds annually, and monarchs also repeatedly used the lucrative posts in the London coal meters' office as a free source of patronage.[53] In the 1630s Charles I pursued projects to turn the coastal coal trade into a royal monopoly that would increase crown revenue, and during the interregnum parliamentary ordinances established various new duties on London's coal imports.[54] At the Stuart Restoration

---

[52] Smith, *Wealth*, 875.
[53] F. W. Dendy, ed. *Extracts from the Records of the Company of Hostmen of Newcastle-upon-Tyne* (Durham, 1901); Healy, 'Tyneside Lobby'. For recurrent attempts by the crown to secure patronage of sea coal meters' places Rep. 20 f. 458; Rep. 24, f. 162; Rep. 27, f. 45v, 180; BL Lansdowne MS 152, f. 42.
[54] Oxford Bod. Bankes MS 37/34; TNA SP 16/447/88; TNA PC 2/49, 581, 592–3; Nef, *Rise*, II, part V, ch. 4.

in 1660, therefore, significant taxes on the Newcastle-to-London fuel trade had already been levied for sixty years.

Despite such duties, coal was not a continuously important source of state revenue until the Great Fire of London of 1666, after which it was seized upon as a convenient and logical source of funds to rebuild the City. The Rebuilding Act of 1667 placed a one shilling duty on every chaldron imported into the port of London. This was not a great burden, constituting a price increase of about 4–5 per cent, but it was renewed and expanded repeatedly until, by the early eighteenth century, total coal duties stabilized at around ten shillings per London chaldron, twice the original price at the pit-head, as much again as the cost of shipping 400 miles by sea, and 30 per cent or more of the final cost to the London consumer. Coal taxes, therefore, were indeed a significant ingredient in the price of fuel during the eighteenth century, just as pamphleteers often complained.[55] But these high and continuous duties made the coal trade a vital element of Britain's fiscal-military state. An act of 1694, for example, authorized a loan of £564,700, largely on the security of new coal duties, stipulating that the money raised should go towards the navy.[56] Another tax of 1709 funded borrowing through a state lottery, which duty was then made perpetual in 1720 to contribute to the South Sea Fund.[57] These taxes raised between £190,000 and £250,000 per year from the accession of George I in 1714 to 1750, comprising around 4 per cent of Britain's total tax revenues.[58] This was a large amount of money, and one whose collection was in many respects easier than most other contemporary revenue sources.

The demand for coal was considered steady, natural, and inevitable, meaning that while fashion, innovation, or foreign competition might drastically change the landscape of commerce for commodities like textiles, coal duties were quite likely to remain stable for years into the future. In economists' terms, coal demand was relatively inelastic; in contemporary terms it was a necessity rather than a luxury. Such dependable demand meant that coal taxes were exactly the kind of revenue that the British state required

---

[55] William John Hausman, 'Public Policy and the Supply of Coal to London, 1700–1770' (PhD thesis, University of Illinois at Urbana-Champaign, 1976), 161 and *passim*; M. W. Flinn, *History 1700–1830*, 283. For the progress of coal's various taxes, see T. S. Ashton and Joseph Sykes, *The Coal Industry of the Eighteenth Century* (Manchester, 1929), 247–8.

[56] 6&7 W&M, *c*.18, sec. 17, 21. *The Statutes of the Realm* (1819), VI, 604–5.

[57] 8 Anne *c*.10. *Statutes of the Realm* (1822), IX, 205–16. For other loans raised on coal duties see P. G. M. Dickson, *The Financial Revolution in England: A Study in the Development of Public Credit, 1688–1756* (1967), 72, 206, 417. For a pamphlet war regarding the fate of a £1,750,000 loan raised on the security of coal duties, see William Pulteney, *Some Considerations on the National Debts, the Sinking Fund, and the State of Publick Credit* (1729), 7, 79; Robert Walpole, *Some Considerations Concerning the Publick Funds: The Publick Revenues, and The Annual Supplies, Granted by Parliament* (1733), 42–8.

[58] Hausman, 'Public Policy', 165.

as security for its increasing debt. As John Brewer has pointed out, Britain's 'financial revolution' was 'contingent upon the belief among its creditors that it had the capacity and determination to meet its payments'.[59] Many economic historians have focused on the constitutional aspects of this belief, arguing that parliament was seen to be (and also in fact really was) more likely than absolute monarchs to honour debts and hence to attract credit.[60] But many contemporaries also saw the nature of the tax itself, and the nature of the trade from which it derived, as crucially important. Excising basic staples like coal inspired trust precisely because they derived revenue from consumption that was widely seen to be necessary and unavoidable.[61] Londoners needed coal, and this need could be translated into a revenue stream that could then be manipulated to suit a variety of financial needs and goals.

As historians have recognized, there was an inherent tension between the state's fiscal interests and the associated, yet distinct, goal of economic growth. Smith's observation that high fuel costs could restrain competitiveness was a commonplace for those who argued against England's high internal but low export duties. They claimed that Londoners suffered by such taxes, suggesting a zero-sum world in which taxes drained the pockets of poor consumers and hard-pressed manufactures. But many commentators, petitioners, and advocates also realized that such a pessimistic position was unhelpful during a period of persistent fiscal emergency. They therefore argued that properly managed regimes of taxation could maximize benefit to both subjects and treasury. One tactic to this end was to claim that coal demand was, in fact, elastic, stressing therefore that high taxes depressed consumption and hence were counterproductive. A 1743 tract by William Bowman painted the suffering of those who scrounged out miserable fuel supplies because they were too poor to accommodate rising coal prices:

How many poor families here in London, not to speak of the inland poor furnished from London with sea coal if they can any way purchase it, now make a shift with little or no firing, burn bones, or eek out the small parcel of coal they are able to purchase, gather in the streets peas cods and bean shells in the summer, which they

---

[59] Brewer, *Sinews of Power*, 88. See also William J. Ashworth, *Customs and Excise: Trade, Production and Consumption in England 1640–1845* (Oxford, 2003).

[60] Douglass C. North and Barry R. Weingast, 'Constitutions and Commitment: The Evolution of Institutions Governing Public Choice in Seventeenth-Century England', *Journal of Economic History* 49 (1989), 803–32. Important critiques include Julian Hoppit, 'The Nation, the State, and the First Industrial Revolution', *Journal of British Studies*, 50:2 (April 2011), 307–31; D'Maris Coffman, Adrian Leonard, and Larry Neal, eds. *Questioning Credible Commitment: Perspectives on the Rise of Financial Capitalism* (Cambridge, 2013).

[61] Coal taxes were administered by the customs because they were shipped along the coast, but their place in contemporary economic thought shares much with excised goods like beer and soap.

lay by and burn in the winter, mixed with an handful or two of coals, who would ...
[were this tax taken away] ... pass the winter much more comfortably.

Their desire for 'comfort' would lead directly to increased consumption. If
current taxes were removed, Bowman argued, one-third more coals would
immediately be brought to market. Labourers would consume more fuel,
which would then expand the trade and thus employ many poor people.
These former paupers would at last become excise-paying consumers, buy-
ers of 'malt liquors, spirits, tobacco, sugar, soap, candles, &c. which their
want of employ will not now allow them, [which] will bring in consid-
erably more than the government now receives by the coal-tax'.[62] Here
an untaxed and therefore flourishing coal trade spurs productive labour
and hence rising (taxable) consumption. Coal, for such authors, was a
commodity whose trade the government could expand or contract at
will through taxation. In this vision the most prudent course would be to
remove taxes entirely so as to drive broad-based growth in other related
sectors – to use cheap coal's benefits, in effect, to create a market for other
more expensive goods.

Bowman's rhetorical tactics illustrate that, by the mid-eighteenth century,
a stand against coal taxes was only tenable if it could demonstrate how this
policy would raise, rather than lower, royal revenue. Others made similar
claims in less developed ways, arguing that their preferred policy would
ultimately help the trade and hence improve the fisc.[63] The demands of the
state could not be brushed aside lightly, so once coal taxes became a central component of royal revenue all commentators were compelled either to
accept this regime or to show what could replace it. Bowman's response was
to replace coal taxes with growing consumption of excised goods, a program based upon yet more coal consumption. Whether as direct or indirect
contributor to national revenue, therefore, the coastal coal trade to London
was universally seen as an indispensible asset to a state with enemies abroad
and debts at home. Bowman, finally, also stressed that his proposal would
aid 'our only nursery for seamen' and hence Britain's interests as a maritime
and naval power.[64] This was a quite typical example of the ways that the
three agendas discussed in this chapter, the commercial/industrial, naval,

---

[62] William Bowman, Merchant, *An Impartial View of the Coal-Trade* (1743), 16, 14, 28.
[63] E.g., 'A Scheme for Erecting a Magazine for Coales', (*c.*1710) BL Add MS 29,948, f. 174-v;
    Philalethes, *A Free and Impartial Enquiry into the Reason of the Present Extravagent
    Price of Coal; Shewing The great Inconveniences which arise from thence, especially to the
    Manufacturers of the City of London, and Adjacent Counties* (1729), 4; *The Case of the
    Glass-Makers, Sugar-Bakers, and Other Consumers of Coals* (1740); *A Plan for the Better
    Regulating the Coal Trade to London, by Preventing the Fluctuation of the Price* (1750?), 2.
[64] Bowman, *Impartial*, 17, also 27.

and fiscal, were often in practice seen as overlapping, mutually constitutive, and complementary.

V. CONCLUSION: NOURSE AND MANDEVILLE ON IMPROVEMENT AND ENVIRONMENT

This chapter has argued that according to numerous commentators writing in a variety of genres, London's coal consumption and the coal trade on which it depended were crucially important for national prosperity, naval power, and fiscal health. These arguments were not, it should be stressed, uniquely applied to the coal trade. Far from it; much of their power arose from their conventionality, from the easy consensus over the desirability of prosperity and power. Contemporaries knew that claims for coal's unique ability to multiply commercial links or to produce expert seamen were often special pleading. As Charles Davenant pointed out in 1696, 'there is hardly a society of merchants that would not have it thought the whole prosperity of the Kingdom depends upon their single traffic'.[65] Many readers could certainly, therefore, view with a sceptical eye the most extravagant claims made by the lobbyists and pamphleteers who argued for coal's importance. Strong arguments existed against most of the rhetoric discussed above: advocates of the English fishing fleet claimed that it produced many more mariners than the coal trade, while debates over the hugely important textile manufacturing sector could be carried on with no reference to fuel policies at all.

But while claims for coal's importance were not universally accepted, neither were they merely the extravagant claims of a few lobbyists. The importance of cheap fuel to manufacturing was increasingly seen as a basic aspect of London's general interest. By 1729 petitions to parliament, for example, were submitted not by 'one society of merchants' but in the name of all the industrial 'consumers' in the metropolis, and these petitioners were backed by the authority of the civic government itself.[66] Through commentators like Tucker and Smith coal was a part of some of political economy's central debates. A few years after Smith's *Wealth of Nations* parliament debated the importance of the coal trade during the war against the American colonies and their European allies. In the Lords strong claims were made for the coal trade as the nursery of mariners and as vital to manufacturing, with the Duke of Bolton going so far as to claim 'the coal trade was of the most material consequence to the metropolis. Its very existence might be said to depend upon it.'[67] This was polemic, as Bolton was fighting a political battle

---

[65] Charles Davenant, *Discourses upon the Publick Revenues, and Trade of England* (1698), 30.
[66] *JHC XXI*, 366, 368; Rep. 133, p. 181–2, 190–1, 266, 292, 347–8.
[67] William Cobbett and John Wright, *The Parliamentary History of England, from Its Earliest Period to the Year 1803*, 36 vols. (1814), XX, 998, 999–1002.

against the government and its conduct of the war. But his battle drew on very persuasive rhetoric, as William Pitt discovered a few years later when he was forced to withdraw a plan for new coal taxes in response to massive public opposition.[68] The point here is not who won which political debate, but more generally to illustrate that by the late seventeenth and throughout the eighteenth century claims regarding coal's importance were increasingly made and were increasingly hard to ignore.

Many, indeed almost all, of these claims were made without reference to coal's by-product, the smoke clouds visible over London. But those who were interested in smoke, and especially those who argued that London might have less of it, found themselves confronted by the arguments summarized above. In some such cases coal's status as an unassailable national asset is only hinted at, as in the anonymous treatise *Orvietan* in which most Londoners are said to find coal 'the only cause of polluting our air'. The pamphlet then argued that despite this awareness of coal smoke's unpleasant and unhealthy qualities, reform was impossible. The capital's citizens, it argued, believed that real improvement in public health will never occur 'except (say they) you can procure first a final prohibition for the importing hither any more coals from Newcastle, which is not to be expected'.[69]

Such a prohibition of the coal trade was not, in fact, expected by anyone at all, even the author of the early modern period's only attempt to argue that London actually could give up its coal supply. Even John Evelyn's *Fumifugium* did not really make that argument, claiming instead that domestic consumption was so insignificant that the air would clear once industrial burners were expelled from the city. But Timothy Nourse's *Essay Upon the Fuel of London*, apparently written during the 1690s, did – or very nearly did – claim that London could switch to wood fuels and hence do without the sea coal which he, like Evelyn, blamed for a series of environmental, medical, aesthetic, and moral disasters. To make his case Nourse not only needed to show that sufficient woods could be planted, but that losing the coal trade would be no great problem. Remarkably, his argument granted all of the claims discussed in this chapter, and then proceeded to show why none of them were any real impediment to his scheme. The many 'smoky occupations … furnace-gentlemen … [and] sons of vulcan' who used coal for industrial production might be relocated south of the river, where their continued coal use would not hurt the bulk of the metropolis. Two thousand four hundred poor families would be put to work managing the forests newly planted on former wastes and commons. 'As touching the breeding of seamen by the coal voyages', Nourse granted "tis certain that

---

[68] Jonsson, *Enlightenment's Frontier*, 181–2.
[69] *Orvietan*, f. 8v.

some advantage doth accrue to this nation by that means', but he stressed that cultivating coastal fishing could do the same, and further argued that continued consumption elsewhere meant that in any case the coastal trade would continue. The taxes laid on coal could continue until they had finished funding the building of St Paul's and thereafter might easily be transferred to wood. The urban poor might still receive some sea coal doles, 'without any considerable annoyance to the city'.[70] Nourse, therefore, who came closer than any other early modern author to arguing that London might actually abandon its coal fuel, felt obliged to show how this plan would damage neither the city's poor relief system nor its industrial production nor the nation's supply of trained mariners nor its revenue.[71] He even agreed with Houghton regarding improvement and increased employment, only he did so through the more conventional method of advocating the transformation of waste and commons into productive land. Altogether his tract devotes much more space to commercial, naval, and fiscal improvement than to the harms caused by smoke, demonstrating that by the end of the seventeenth century redressing London's 'one great nuisance' would demand a substitute 'new national asset'.

This leads, finally, to the author who saw most clearly into the relationship between London's economy and London's environment. If coal burning was the basis for the poor's comfort, the artisan's comparative advantage, the navy's manpower, and the treasury's solvency, but if coal's smoke was also the immediate cause of the shopkeeper's 'tarnished and sullied' wares, of 'stinking and smoky' rooms, of ladies' decayed and deformed complexions, of sunless urban skies, blackened gardens and walks, and of unhealthy babies, then there was a cruel paradox at the heart of urban life.[72] The great philosopher of such paradoxes, Bernard Mandeville, put this succinctly:

If we mind the materials of all sorts that must supply such an infinite number of trades and handicrafts as are always going forward, the vast quantity of victuals, drink and fuel that are daily consumed in it, the waste and superfluities that must be produced from them ... it is impossible London should be more cleanly before it is less flourishing.[73]

[70] Timothy Nourse, 'An Essay upon the Fuel of London', in *Campania Fœlix. Or, a Discourse of the Benefits and Improvements of Husbandry* (1700), 351 (ver. 355) for industrial production, 352 (ver. 356) and 355 (ver. 359) for poor families, 357–9 (ver. 361–3) for breeding seamen, 353 (ver. 369) for taxes, 361 (ver. 365) for charity.
[71] Evelyn was convinced that the 'ingenious' Nourse had demonstrated that this scheme was 'not impossible', but he prefaced this endorsement with an acknowledgement that this was contrary to the interests of 'our Newcastle trade and seminary of mariners'. Evelyn, *Silva, or A Discourse of Forest-trees, and the Propagation of Timber in His Majesty's Dominions*, (4th ed. 1706), 265.
[72] Nourse, *Fuel*, 363–4 (ver. 367–8).
[73] Bernard Mandeville, *The Fable of the Bees: and Other Writings*, ed. E. J. Hundert (Indianapolis, 1977), 22.

The suggestion that business, production, and population caused urban problems was far from new, but Mandeville's innovation here was to accept, and even to celebrate such problems as signs of underlying success.[74] If consumption, employment, and growth were good, as surely they were, then it was foolish to regret their inevitable by-products. The implication here, which Evelyn and Nourse had worked so hard to avoid, is of course that societies can choose to be clean (and simple and virtuous) *or* they can choose to be 'flourishing'. Because, as Mandeville stressed, we are so full of desires, everyone would naturally choose the latter. For early-eighteenth-century London society, and perhaps by implication for all people everywhere, human nature would inevitably push societies towards economic growth and accompanying environmental degradation.

[74] For Mandeville's thought and its influences, E. J. Hundert, *The Enlightenment's Fable: Bernard Mandeville and the Discovery of Society* (Cambridge, 1994); John Robertson, *The Case for Enlightenment: Scotland and Naples 1680–1760* (Cambridge, 2005), 261–80.

—9—
Regulations: policing markets and suppliers

I. THE QUEEN'S SPEECH: 'SOME REGULATION'

On 9 November 1703, Queen Anne addressed the Lords and Commons to declare her priorities for the new session of parliament. The dominant themes, predictably, were war with France and the rage of party. The former, hoped Anne, would soon be brought to 'a glorious and speedy conclusion', and she therefore asked the houses to vote the necessary supply. The latter theme was addressed obliquely but unmistakably through the Queen's wish for 'peace and union' among her subjects and the avoidance of 'heats or divisions'. During her short speech, whose printing required only two pages, Anne avoided all other issues except two. First, she hoped that the Royal Navy could achieve an easier and cheaper way of recruiting manpower. Second, she asked for action regarding the high price of coal.

Anne was hardly unaware that her capital's coal consumption had environmental and medical consequences. During the reign of her father, James II, she had let it be known that she considered London's air to threaten the health of her young children.[1] Within a few years the new king, her brother-in-law William III, agreed regarding his own health, which he found severely compromised by the capital's smoke. By 1703 Anne's husband, George of Denmark, had already begun to suffer from the severe asthma that would contribute to his death in 1708.[2] Anne was therefore typical of the elites who, as we will see in this book's final chapters, considered London's air as something to avoid. It is therefore contradictory, and yet not at all surprising, that she wanted the power of the English state deployed so as to ensure that London's coal burning would continue unabated. The coal trade, for the reasons described in the previous two

---

[1] Edward Gregg, *Queen Anne* (New Haven, 2001), 51; 4 June 1687 N. S., Bonrepaux to Seignelay TNA PRO 31/3/170.
[2] Discussed in Gregg, *Anne*, 161–2, 279–80. For her concern over the effects of London air on her husband and children, see also James Anderson Winn, *Queen Anne: Patroness of the Arts* (Oxford, 2014), 139, 191, 194.

chapters, was for the Queen an object of governmental concern, regulation, and protection.

Anne wanted action, wanted for 'you to make some regulation for preventing the excessive price of coals', and she recommended two distinct approaches. The first, those convoys that the Queen herself claimed to have 'taken particular care to appoint' will be the subject of the next chapter. These convoys, Anne claimed, had succeeded in protecting imports but not in lowering prices, which 'gives great ground of suspicion there may be a combination of some persons to enrich themselves by a general oppression of others, and particularly the poor'. This combination of unnamed persons therefore 'will deserve your consideration how to remedy'.[3] The remedies the Queen hoped for were neither easily nor quickly achieved, but her government did take a series of measures, some of which were new and others long-familiar, to restrain prices and promote the consumption of coal in London. Indeed it may be this combination of approaches, none of which were applied with great rigour but all of which played their part, that has led historians to underestimate the state's vital contribution to the continued prosperity of the London coal trade.

II. CIVIC AUTHORITY AND THE STATE

For London's civic governors the central state became an increasingly essential resource in their efforts to protect and manage the coal market, as the city's focus changed during the seventeenth century from the direct oversight of the urban fuel market towards lobbying the crown and parliament. For London's ruling elite of Common Councilmen, Aldermen, and Lords Mayor it was clear that a healthy coal trade, meaning abundant supplies and low prices, could best be promoted and protected by the state. Despite their authority over local markets, and despite their increasing unease over fixing prices or otherwise directly regulating trade, London's governors never lost the conviction that the crown and parliament played important roles in establishing and maintaining the urban coal market.

The City of London had long held authority over its fuel market. Like other local governments in medieval and sixteenth-century England, London supervised prices, quality, and the practices of fuel sellers.[4] These powers were asserted with particular vigour, and were unusually visible, precisely during the years when Londoners could no longer acquire sufficient affordable fuel supplies from their traditional sources and therefore

---

[3] *Her Majesties Most Gracious Speech to both Houses of Parliament, on Tuesday the Ninth Day of November 1703* (1703), 4.
[4] This chapter summarizes material in Cavert, 'Villains of the Fuel Trade'.

turned to Newcastle coal. During the 1570s and 1580s, London's government showed itself very concerned with fuel, supervising prices, shaming fraudulent traders in the city's most public spaces, and improving the machinery of charitable stocks for the poor. Quickly, however, it became clear that the new seaborne coal trade rendered these local initiatives insufficient. The coal trade involved a longer and more complex supply chain and several distinct interests, including miners, mine-owners, shippers, river labourers in both the Thames and Tyne, wholesalers, and finally London distributers, chandlers, or chapmen. This complicated, inter-regional trade could be subject to monopolies, inefficiencies, bottlenecks, and dangers at several points, so by the 1620s the king's Privy Council and parliaments emerged as the key loci of governmental oversight for London's intricate fuel supply system.

London's governing elites were allies, not victims, of this process, as the state was more a useful resource than an imposition to men who had ability to influence its policies. London's governance of its urban fuel market was therefore exercised in collaboration with a state whose policies it did not control but could often strongly influence. This dynamic first emerged during the late 1620s when, as was described in Chapter 7, naval war with Spain allowed the privateers of Hapsburg-held Dunkirk to attack coastal shipping and so hurt London's coal supplies. Exports declined sharply and by the summer of 1627 London's Court of Aldermen expressed its concern with consequent high prices.[5] During August of that year it ordered an investigation into the practices of fuel traders, suggesting that wartime high prices were the result of fraudulent practices. They summoned fuel traders and fixed prices at their normal peacetime rate.[6] These were familiar responses to heightened fuel prices, for which London's magistrates needed recourse to no superior authority. Within a few weeks, however, they decided that this was insufficient and that governing the capital's fuel market required help from the state. By the end of September the Aldermen concluded that coal ships were deliberately waiting off Harwich (about eighty miles up England's eastern coast) because they feared losing men to impressment in the Thames. The capital's coal supplies were being interrupted by decisions taken far from the capital itself by ship captains who themselves were responding to royal policies. The aldermen therefore turned to the state, ordering a committee to address the Privy Council and requesting an order to send the collier fleet into the Thames.[7] The council proved entirely receptive, reinforcing the Corporation's authority over fuel

[5] Tyne and Wear RO, GU.TH/21.
[6] Rep. 41, f. 314, 325v, 334; Jour. 34, f. 160.
[7] Rep. 41, f. 344v-345. Also Rep. 42, f. 1-1v.

prices, ordering a survey of existing fuel stocks, and requiring the collier vessels to come immediately into the port of London.[8]

This was mandatory trade, an infringement upon property rights that was raised and defended the following spring by Sir Edward Coke as a textbook example of how the royal prerogative might legitimately supersede the normal protections of common law.[9] Because London suffered a 'dearth' of coals, argued Coke, the principal of equity as embodied in the Royal prerogative and as executed by the Privy Council temporarily trumped the principal of ownership as protected by the common law. Perhaps it is not coincidental that the instance of extra-legal prerogative power that Coke singled out for approval was one explicitly requested by the City of London. Since Coke's purpose was to downplay the tensions at the heart of the English constitution, the council's actions in this case could be spun to stress their devotion to the common good. It had responded to a humble request from the city governors and applied its power in the interests all the poor consumers of London. Protecting the coal trade could unite the interests of urban consumers, civic magistrates, and the royal council; this, for Coke and doubtless for London's civic authorities as well, was exactly how England's collaborative and participatory government was supposed to work.

Of course, relations between the city and crown were not very often so comfortable. Royal efforts to supervise or protect the coal trade were rebuffed by a city that suspected the motives of a crown that was constantly searching for new revenues. In 1630, London's Lord Mayor James Cambell, who had led both governmental and charitable efforts to secure and distribute fuel during the later 1620s, sought assistance from the crown even as he simultaneously kept it at arm's length. He argued that a system for coastal convoys should not be undertaken by the city alone but rather was 'fit to be carried by a royal hand'. This was at once an attempt to co-opt the naval resources of the crown for the benefit of the capital and simultaneously an effort to deflect some of the cost of that protection away from the London Corporation.[10] A few months later when the council ordered Cambell to report on the high fuel prices caused by trade combinations, Cambell and his colleagues had no desire for either help or interference from the centre, and so responded calmly that all was well.[11] Throughout the following decade this tension persisted, as London's governors were wary of the

---

[8] *APC 1627–8*, 99. The survey found only 283 chaldrons of coal on London's wharves, TNA SP 16/83/34.

[9] Johnson *et al.*, eds. *Commons Debates 1628* III, 126–7, 133. See also the discussion of Coke in chapter 7.

[10] TNA SP 16/166/19, 5 May 1630.

[11] TNA PC 2/39, 811, 823; PC 2/40, 34–5, 96, 177; TNA SP 16/173/63.

consequences if the coal trade came more fully under the direct protection, supervision, and even management of a grasping crown.[12]

The closest collaboration between London and the state, but one that emphatically did not produce the constitutional unity that Coke hoped for, was reached in 1643–4 as London's coal supply became a central component of economic warfare. Even as parliament closed London's coastal coal trade it simultaneously allowed the city an unprecedented level of influence regarding its fuel policies. During the summer of 1643 Parliament set coal prices in London and then, when it became clear that new fuel sources were required, it authorized the plunder of Royalist woods throughout south-eastern England for London's benefit.[13] This was justified through the familiar picture of poor men driven to desperation by lack of fuel, which 'misery' and 'danger' the act sought to prevent. From the royalist perspective this was particularly absurd, the woods of innocent people 'to be cut down to prevent spoil'.[14] This, for the Royalist press, was typical parliamentary 'nonsense', but it becomes easily explicable in the context of the close collaboration between Parliament and London's leaders and their agreement that the state must do all it can to protect London's fuel supplies.[15]

The entanglement of London's fuel trade with civil war during the 1640s is an extreme example of a more general process: throughout the seventeenth and eighteenth centuries, as maintaining London's fuel supply became a political issue of national significance, mobilizing state resources on its behalf also meant engaging in the rivalries, factions, and parties that beset high politics. This had begun in the early seventeenth century when parliament first considered the regulation, taxation, and protection of London's coal. Thereafter any attempt to reform some aspect of London's fuel market was complicated both by the complex nature of the trade – in which mine owners, Newcastle merchants, masters of coasting ships, London wholesalers, and finally London consumers took up shifting alliances with or against other interests – as well as by the difficulty of successfully managing parliamentary inquiries.

An example of both of these dynamics can be seen in the partial success achieved by the City of London in its efforts to punish mine owners for an illegal combination in 1711. During the late spring and summer of 1710 London's Court of Aldermen investigated a combination among ships'

---

[12] Nef describes the crown's attempts during the 1630s to manage and profit from this trade, *Rise* II, part V, ch. 4.

[13] Firth and Rait, eds. *Acts and Ordinances* I, 84, 303–5.

[14] *Mercurius Aulicus*, 11 October 1643.

[15] For the speed with which policies of the Court of Common Council became embodied in parliamentary ordinances, see Jour. 40 f. 60 and Firth and Rait, *Acts and Ordinances*, I, 171–4.

masters 'to advance the price of coals', and within two weeks had the matter before the Privy Council.[16] Investigations of any one aspect of the coal trade were always liable to creep into others, and by July the Privy Council's attention also encompassed northern mine owners.[17] Some of these owners had, only a few weeks before, entered into a formally organized cartel to regulate the supply of coal from Newcastle with the goal of preventing a glutted market and a fall in price.[18] The mine owners were identified by the ship owners as well as by striking keelmen (river labourers) in the Tyne as the true origin of difficulties in the coal trade.[19] By the following winter it was no longer shipmasters but this 'Regulation' that was particularly attacked in parliament with a bill against combinations in the coal trade. London's coal-burning industrialists, calling themselves 'the principal consumers of coals in and about the cities of London and Westminster', lobbied for the bill's passage. Mine owners, however, in particular William Cotesworth, worked hard to deflect blame from their group.[20] Cotesworth admitted the Regulation's existence but denied that it was an illegal combination in restraint of trade. Partially through his efforts the bill that was eventually passed into law condemned the Thames wholesalers along with the mine owners, which Cotesworth and his allies considered a major victory.[21] The law's basic goal, the source for its support in the city and in both houses, was to provide the London market with abundant coal at a reasonable price. But its eventual form, which addressed both the actions of the Thames wholesalers and their rivals the mine owners, showed how the messy and unpredictable process of parliamentary politics could hijack a bill and steer it away from the intentions of its initial champions. London's lobby was very powerful indeed, but it always competed for influence with other interests.

The London lobby thus did not control the crown or parliament, but nor did it shy away from attempting very ambitious interventions into the national coal industry for its own benefit. The uses of the state, as perceived by some of London's large-scale manufacturers and civic magistrates, emerge particularly clearly in the 1729 investigations into combinations

[16] LMA Rep. 114, 220–1, 235, 241–2, 261–2, 304–6, 313–4; TNA PC 2/82/2, 587–8; PC 2/83, 10, 12, 24, 26–7, 35, 49.

[17] TNA PC 2/83, 26–7.

[18] Edward Hughes, *North Country Life in the Eighteenth Century: The North East, 1700–1750* (Oxford, 1952), 168–9, and 151–257 for the coal trade in general. The Regulation began in 1708 but its minuted meetings began in June 1710.

[19] J. M. Fewster, *The Keelmen of Tyneside: Labour Organization and Conflict in the North-east, 1600–1830* (Woodbridge, 2011), 61–3, which corrects some of Hughes's discussion.

[20] *Journal of the House of Lords: volume 19: 1709–1714* (1767–1830), pp. 311.

[21] Hughes, *North Country*, 170–90, esp. 186–7. See also J. M. Ellis, ed. *The Letters of Henry Liddell to William Cotesworth* (Leamington Spa, 1987), 12–40; H. T. Dickinson, ed. *The Correspondence of Sir James Clavering* (Gateshead, 1963), 115–22.

among the mine owners and shippers. Mine owner George Liddell learned at the end of January 1729 from City MP Alderman John Barnard that there were 'great complaints in the City' against methods to drive up prices and that therefore 'the City design to bring it into Parliament'. A week later newspapers complained of the 'great prejudice' these combinations caused to the poor and called for 'parliament's consideration', which they duly received within a few days. Edward Wortley Montagu, MP and major mine proprietor, declared that he would give £10,000 to avoid such a complaint in the Commons, and mine owners seemed to make some progress negotiating directly with London's aldermen, yet they were not able to avoid a parliamentary investigation.[22]

The agitation against the combining mine owners seems to have been led by the 'great consumers' of coal, industrialists who spent many hundreds of pounds annually on fuel. In late March they planned, according to Liddell, to hire ships and attempt to buy coals in the Tyne directly from the 'fitters', wholesalers blamed for constricting trade. If the fitters would refuse to supply them the Londoners could claim a monopoly in restraint of trade and seek to cancel the charter upon which the Newcastle Hostmen depended for their trading rights. London's aldermen considered mine owners to be in league with the fitters, and they therefore also planned to petition parliament to 'lay open wayleaves'. Controlling these passages over private lands was an essential way for mine owners to restrict competitors' access to river shipping and thereby to limit the total coal output. Depriving the Hostmen of their charter and transforming wayleaves into public roads would have transformed England's coal trade at a stroke.[23] As in 1711, the progress of the bill was not smooth, as owners, shippers, wholesalers, and others sought to deflect blame onto their rivals.[24] The key point here, however, is that Parliament's investigations into the coal trade, which Wortley found so threatening, were advanced in the name of the London consumer by city MPs described as 'the two tribunes of the London plebeians'.[25] The justifications for such investigations had changed little over the previous century;

---

[22] Hughes, *North Country*, 218–9; *London Evening Post* Tuesday, 28 January 1729.
[23] Hughes, *North Country*, 223–5. For similar violations of property rights in the interest of general improvement, Julian Hoppit, 'The Nation, the State, and the First Industrial Revolution', *Journal of British Studies*, 50:2 (April 2011), 307–31. London's goal was to remove restraints on trade and thereby lower prices. Mine owners, however, refused to accept that this would work, arguing instead that unlimited production would glut the market, causing prices to plummet and small mines to bankrupt, which would then leave the trade entirely in the hands of a few large proprietors. Thus, the mine owners argued that an unlimited trade led not to lower prices but ultimately to harmful monopolies.
[24] *JHC XXI*, 369–73.
[25] Eveline Cruickshanks, 'Micajah Perry', *History of Parliament: The House of Commons 1715–1754*, at www.historyofparliamentonline.org/volume/1715-1754/member/perry-micajah-1753.

coal was still the fuel of the poor, of industry, and a source of royal revenue. But to translate this rhetoric into legislation entailed engaging in increasingly elaborate political manoeuvres in order to wield or to hinder the broad powers of the state.

III. 'HOW TO REMEDY': THE STATE'S TOOLS FOR MANAGING THE COAL TRADE

Despite repeated royal and parliamentary attention to the coal trade it was never made a state monopoly, never controlled directly by the crown, and large fortunes were made throughout the period by mine owners and wholesalers alike. There were important limits to what the state did, but this does not mean that state power was irrelevant, weak, or absent from the coal trade in early modern London or England more broadly. A brief examination of its powers and tools can reveal how the state set rules and enforced provisions in ways that made London's coal market possible and, often, helped determine its extent, nature, and role in the urban economy.

The actions that came closest to direct regulation were probably the infrequent but recurrent emergency measures deployed in wartime, including price maxima and mandatory trade. During the 1630s the Privy Council set prices as part of a broader effort to assume direct control of the coastal coal trade. The council justified this by reference to existing high prices and frauds, but there was also a clear fiscal motive.[26] More often, however, prices were set in response to short term problems, particularly trade disruptions caused by war. Early in 1643 a parliamentary ordinance, framed in collaboration with the London authorities, set coal prices within and around the capital.[27] A statute was also enacted in 1665 authorizing JPs to set prices, and this remained a basis for further regulation during later wartime conditions.[28] Such conditions also called for the mandatory trade celebrated by Coke in 1628; ships were again ordered to proceed forthwith into the Thames in 1666 and prompt trade was made a condition of government protection in 1672.[29] In 1704 the council did the same, this time to break up a 'combination' of shippers delaying at Yarmouth.[30] In such cases the

---

[26] TNA PC 2/49, 72–3, 155–8 (1638); PC 2/50, 29 (1639).

[27] *An Ordinance of the Lords and Commons Assembled in Parliament, That no Wharfinger, Woodmonger, or other Seller of New-Castle Coales, within the Cities of London and Westminster, or the Suburbes thereof, shall after the making hereof sell any New-Castle Coales, above the Rate of 23s. the Chaldron, and after the first of Aprill next, above 20s at the most.* (1643),

[28] 16 & 17 Car 2, cap. II, *Statutes of the Realm* (1819), Vol. 5, 552–3; TNA PC 2/76, 107–8 (1695); BL Add MS 32,693, f. 74–5; PC 2/95, 597–9 (1740). For the civic government of London as the origin of the 1665 bill, Rep. 70, f. 97.

[29] TNA PC 2/59, 154 (1666); PC 2/63, 232–3 (1672).

[30] TNA PC 2/80, 124–5, 422–4; *CSPD 1704–5*, 21–2.

same policy that Coke described as emergency prerogative powers could also appear as the entirely regular execution of laws against combination in restraint of trade. One final version of mandated trade was the requirement that coal merchants sell directly to the poor rather than restricting their goods to small dealers who would then further increase prices. This was often coupled with price maxima, so that dealers were required to sell small quantities and at limited rates.[31] In such cases trade was both mandatory and regulated, as merchants were officially permitted neither to withdraw from the market nor to allow prices to track demand.

One of the oldest functions of the English crown was the regulation of weights and measures, and its execution in early modern London was considered to have an important role in producing trustworthy markets.[32] At the most basic level standardized measures were necessary to make markets possible and avoid confusion, but once such units are established their policing was in practice a key way to prevent fraud. Short measures that gave customers less than they had paid for was an easy way for fuel merchants to augment profits, and the punishment of such fraud was widely seen as basic duty of magistrates. In London this was carried out by officers of the city, who occasionally punished fraudulent traders by placing them in the pillory, a classic way of shaming them before neighbours and customers.[33] Standard bushels were to be available in every parish so that citizens could confirm that purchases were not deficient, and the city's coal meters were responsible for examining imports to establish their precise volume.[34] The units of trade were therefore subject to examination from a variety of civic officers and courts who enforced official standards and punished those guilty of fraud. Persistent fraud, moreover, was grounds for the state to wield further powers.

The most important of these for the urban coal trade was probably the trading privileges granted to companies, both in London and in the north-east. In London the coal trade was dominated by the Fellowship of Woodmongers from their incorporation in 1605 until they were dissolved in 1667.[35] Their dissolution was part of a broader set of initiatives meant to restrain fuel prices during the second Dutch War, as insecure shipping and

[31] LMA Rep. 28, f. 146 (1608); Rep. 56, f. 209-290v (1643).
[32] For an overview of the association between sovereignty and weights and measures, especially but not exclusively in coinage, see Diana Wood, *Medieval Economic Thought* (Cambridge, 2002), ch. 4.
[33] Cavert, 'Villains of the Fuel Trade'.
[34] Rep. 35, f. 215-215v (1621) for measures in each parish. The duties of sea coal meters were stated in a statute of 1746, 23 Geo. 2, cap. XXXV, *Statutes at Large* (1765), XVIII, 500-9.
[35] Hylton B. Dale, *The Fellowship of the Woodmongers: Six Centuries of the London Coal Trade* (n.pub. info, c.1923). The Woodmongers archives are lost except for Bod. Rawlinson MS D 725B. For their dissolution *JHC VIII*, 671, 676; *JHC IX*, 39-41.

loss of stocks in the Great Fire of 1666 drove prices far above normal levels. London's wholesalers again saw privileges come and then go early in the eighteenth century, as the lightermen (operators of the 'lighters' that brought coal from ship to shore, who therefore were in a position to manipulate prices) gained and lost corporate privileges. In 1700 they were incorporated as members of the Lightermen's and Watermen's Company, but in 1730 the methods that they had used were made illegal.[36] The importance of charters and privileges are especially visible in the case of those lacking them, such as the Thames coal heavers who repeatedly petitioned the Privy Council for incorporation and repeatedly were denied. The city consistently opposed the proposal presumably because it would have incorporated a body of labourers who could have joined the mine owners, Tyneside fitters, coastal shippers, and London lightermen, as those with an interest in heightening – as well as power to do so – the prices of the capital's vital fuel supplies.[37] The coal heavers of the Thames, like their fellow workers in the Tyne, had cause to regret their powerlessness.[38] They were exploited by gangmasters until a strike in 1768 led to disorder that was suppressed by legal and extra-legal violence. The dispute culminated in the execution of seven heavers, all but one of whom were Irish, who were killed before 'an incredible number' of viewers in the heart of the east end, as well as 'a prodigious number of peace officers'.[39] The coal heavers were kept without corporate privileges and so without protection against exploitation and abuse for the same reasons that the Woodmongers lost their charter in 1667 and that the lightermen were stripped of their role in the market in 1730: because London's governors and their allies in Westminster perceived them as threats to cheap and dependable coal supplies.

The parliamentary investigations described in the previous section could appear to wield formidable powers. Their accomplishments, however, are less clear, as trade combinations of all sorts were repeatedly identified throughout the early modern period and beyond into the nineteenth century. Neither parliament nor the royal council, it must be admitted, ever

---

[36] Hughes, *North Country*, 260–1; 3 Geo. II, *c.*26, *The Statutes at Large, from the Second to the 9th Year of George* II (Cambridge, 1765), 170–9

[37] TNA PC 2/74, 87 (1672); *CSPD* 1696, 338; Rep. 102, p. 230–1 (1698); PC 2/78, 26 (1700); *CSPD* 1703–4, 360; Jour. 110, f. 82 (1706); Rep. 114, f. 5 (1709).

[38] The 'keelmen' who worked in the Tyne attempted to establish control over a charitable hospital that could serve as a strike fund. The battle over the hospital, therefore, was a battle over labour in the Tyne. *The Case of the Poor Skippers and Keel-men of New-Castle, Truly Stated* (1711?); *A Farther Case Relating to the Poor Keel-men of Newcastle* (1711?); Fewster, *Keelmen of Tyneside*, ch. 2.

[39] White, *London in the Eighteenth Century*, 243–5, quotes on 244. For coal heavers' earlier attempts to improve pay and conditions, Ashton and Sykes, *Coal Industry of the Eighteenth Century*, 205–6.

succeeded in ending combinations and cartels, and historians have therefore reasonably concluded that state regulation was weak in this regard. However, there is also evidence that such combinations were severely constrained by the state, both because they feared governmental exposure and because their very illegality froze them out of some important legal protections. In 1710–11, for example, the Regulation of Tyneside mine owners perceived that they could not lean too heavily on their allies among the Thames fitters because the fitters could easily reveal kickback payments that parliament would take to be 'plain proof of a designed monopoly'.[40] Nor could they use royal courts against members who removed themselves from their agreement. Liddell wrote 'I don't see that the owners ever durst venture to sue their covenant, least it be thought an illegal one and then it turns upon themselves.'[41] Without the ability to enforce contracts the Regulation was an entirely voluntary union, no more able to compel membership than were the congregations of dissenting churches. Unwelcome attention ended the Regulation, albeit only temporarily, in 1711, and future cooperation was always undermined by the inherent difficulty of maintaining voluntary cooperation among individuals and families that were also commercial, social, and political rivals.[42] While the state never ended combinations, then, it did significantly limit, undermine, and threaten them.

The state's ability to assess the coal trade, to evaluate its prosperity or to know when it suffered from restraints, was unmistakably derived from the gathering of taxes. Taxing created knowledge about what would now be called the economy, and adjustments to taxation always brought winners and losers. This was clear to all, as is evidenced by the grand claims made about the damaging power of new or heightened taxes in every session of parliament. The power to tax entailed the power to destroy. But that destruction could also be creative, as tax policy fashioned the conditions under which certain trades floundered and flourished.[43] Taxing London's coal consumption, therefore, was both the foundation and a key goal of most of the above interventions in the trade. Taxes produced the data and expertise necessary to investigate a complex trade, and taxes were among the most important reasons why that trade was as indispensible to the state as it was to London's consumers. Coal duties therefore provided the state with power over London's coal consumption that was potentially complete, but was in practice constrained by the need to avoid serious disruptions. Lowering or removing coastal taxes, for example, would have removed one of the key factors that, during the eighteenth century,

[40] Ellis, ed. *Liddell Letters*, 31.   [41] *Ibid.*, 12–3.
[42] Hughes, *North Country*, 176, 180, and *passim*.
[43] A theme in Ashworth, *Customs and Excise*.

pushed much manufacturing industry out of London and into the north, and would also have benefitted London consumers and the shippers who supplied them. Pamphlets for and against coal duties debated whether consumer demand was elastic or inelastic, whether or not high taxes suppressed consumer demand.[44] In practice duties settled at about the level that appeared to maximize revenue, high enough to bring in substantial income but not so high as to depress trade.[45] The fiscal imperative thus reinforced the goals of social stability and political-economic improvement, all three of which agreed that London's coal consumption should remain as high as possible. This policy was not contested, but if it had been taxation would have offered a convenient and effective tool to alter or end London's coal trade. If, for example, parliament had preferred to keep England's coal in the north, perhaps in order to stimulate local industries, or if it had decided to promote woodlands or coal fields closer to London, simple adjustments to tax levels could have accomplished their goals. By the eighteenth century, taxation through the customs and excise meant that London's coal consumption, like so many other aspects of the national economy, were subject to state management.

Finally, in addition to the these methods of managing the practice and scale of the coal trade, the state also retained one powerful way to manage its social and political significance. Through charity London's civic and royal governors were able to mitigate the effects of price escalations on the most deserving poor and to present themselves as concerned and benevolent magistrates. Unusually hard frosts or scarce supplies regularly prompted intensified efforts to administer charitable distributions of fuel, either free or subsidized, and to be seen doing so. During the late 1620s, for example, Alderman James Cambell not only administered the city's charitable fuel fund, he also seems to have pushed London finally to accept a gift to that fund of £300 from the estate of his own father.[46] King James I had granted the city the right to import 4,000 chaldrons without paying the 1s. customs duty, which in practice seems to have meant a grant or rebate of £200 annually for charitable fuel provision. During wartime London took particular care to ensure that this sum was in fact deployed for the intended purpose and that the city got the best possible deal for its money.[47] During the maritime war of the 1660s the city not only tried to manage its own fund of

[44] See Chapter 8.
[45] The tensions between fiscal and social policies are explored, from an economic perspective, by Hausman, who counts eighty separate statutes dealing with the coal trade between 1695 and 1770. Hausman, 'Public Policy', 45.
[46] Rep. 42, f. 17v-18; Rep. 43, f. 186.
[47] Rep. 62, f. 301v; Rep 63, f. 182, 445; Rep. 64, f. 18, 71v-72 (1653–5); Rep. 70, f. 97 (1667); Rep. 78, f. 97; Rep. 70, f. 97 (1673–4).

fuel for the poor, but also ordered the various livery companies to supply themselves with fuel totalling 7,500 chaldrons, presumably to insulate the market from price fluctuations and to provide for the poor.[48] During harsh winters additional coal was distributed, with over 1,300 people receiving coal doles in Westminster alone during the winter of 1739/40.[49] Through such distributions London's governors sought to, and it seems to a significant degree actually managed to present themselves as good governors. This was especially the case because charity was organized and distributed in conjunction with other measures.

IV. COMPLEMENTARY POLICIES

Ostentatious charity, investigations into trading privileges and practices, charters invalidated, weights and measures policed, and price maxima set: the crown and its officers could do all of these things in order to protect London's coal supply, and when it used one of them it was more likely to use many. The powers described here were not exclusive but complementary, and there was no ideological or practical reason not to resort promiscuously to many of them at once. The second Anglo-Dutch maritime war during the 1660s provides a good example of such a practice. The company of Woodmongers lost their charter for unreasonable trading practices, and profiteering fuel merchants were brought before the Privy Council for examination and punishment.[50] Parliament passed a bill detailing proper fuel measures and authorizing London magistrates to set coal prices, which they duly did within a few weeks.[51] City of London magistrates sought to reinvigorate existing traditions of providing cheap coals to the poor, and King Charles directed naval vessels to travel to Newcastle specifically to supply London's charitable fuel stocks.[52] Finally, the king and Privy Council, anticipating a major coal dearth, prompted a search for inflammable turf and peat in Windsor Forest and another search, conducted by John Evelyn himself, along 'both the sides of the River of Thames, and up into the Meadows and pasture grounds'.[53] This

---

[48] John Noorthouck, *A New History of London, including Westminster and Southwark* (1773), 222.

[49] WAC E3107.

[50] *Some Memorials of the Controversie with the Wood-Mongers, or, Traders in Fuel;* (1680); *JHC* IX, 40; TNA PC 2/58, 91, 103, 122; PC 2/59, 478, 581, 591; Rep. 70, f. 41-44v; Rep. 72, f. 128v. Among the fuel merchants so admonished was Sir Edmund Berry Godfrey, whose death in 1678 became a central part of the Exclusion Crisis.

[51] 16&17 Chas. II, cap. 2; *JHC* VIII, 582, 611–2; Rep. 70, f. 97; *CSPD 1664-5*, 262.

[52] Rep. 70, f. 108v, 143v, 156v; Rep. 72, f. 2, 14v, 29v-30; Guy De la Bédoyère, ed. *Particular Friends: The Correspondence of Samuel Pepys and John Evelyn* (Woodbridge, 1997), 50.

[53] Rep. 72, f. 130-130v; PC 2/59, 468; BL Add MS 78,393, f. 18.

was a diverse set of policies, meant to stabilize London's fuel supply in several ways. None of them would have had any hope of success, however, were they not joined by the single most important policy at the English crown's disposal: direct military protection by the Royal Navy.

$-$IO$-$

Protections: the wartime coal trade

I. THE QUEEN'S SPEECH: 'CONVOYS FOR THAT SERVICE'

When Queen Anne spoke to parliament in November of 1703, she wished for peace and unity, victory and glory, but offered clear positions on government policy regarding only two issues: parliament should man the fleet and it should reduce the high price of coal. Lowering fuel prices entailed, as the previous chapter has described, surveillance and intervention in the urban market, which was justified because the crown had already, as Anne put it, 'taken particular care to appoint convoys for that service'. This claim for particular care was especially credible because the official at the head of the Royal Navy was Anne's own husband, Prince George of Denmark, the Lord High Admiral of England. The Queen's care, and that of her husband and of her government, in directing the navy to protect the ships of the coastal coal trade, justified further investigations into trade combinations; since lack of convoys were not causing high prices, something else was to blame. Her language suggested that convoys and coal prices were related, but convoys and naval manning were not. Governmental discussions of what to do about coal during wartime did tie them together, however, repeatedly and unavoidably, throughout the War of the Spanish Succession and other wars both earlier and later. The basic question was whether government should protect the coal trade and so keep supplies abundant and prices manageable, or benefit from it by harvesting the mariners trained up by its ships for service in the Royal Navy. Manning the fleet was not a separate issue from convoying colliers, but another aspect of one problem: that the trade of coal to London both depended upon and propped up the power of the navy.

Deploying its naval vessels to escort the collier ships was the most important way that the state facilitated and protected London's coal consumption during the early modern period. It demonstrates how significant this trade was perceived to be by London's civic authorities, by the crown, and also by England's enemies who saw coal vessels as both attractive prey and as convenient ways to foment political chaos in the capital. This belief is noteworthy given the attention devoted by scholars to coal's place in eighteenth- and nineteenth-century industrialization. Some economic historians have questioned the importance of cheap coal to English growth, suggesting along the way that Baltic wood might have supplied England with fuel at only insignificantly higher costs.[1] There are serious problems with this proposition, including the unsupported assumption that all existing Baltic forests were economically accessible and that forestry in those regions could have easily been scaled up so as to meet Britain's enormous demand. Londoners, as these scholars suggest, could probably have made do with less fuel, through some combination of efficiency and suffering. But if Londoners c.1700 were to receive the same amount of heat from Baltic wood as they did from English coal, then some 5,000 shipping journeys would have been required to transport the one million tons of wood the 1,200–1,800 miles from Russia, Finland, or Sweden into the Thames.[2] It would also have meant travelling through the home waters of the Swedish, Danish, and Dutch navies, a consideration that would have demanded constant diplomatic balance to avoid endangering London's fuel supplies. Even had such a substitution been economically feasible, therefore, contemporaries had good reason to think of London's coal supplies as much in strategic terms as in economic. The difficulty of securing such a fleet would have made the wood substitution hypothesis laughable to any early modern Londoner, as experience showed that protecting the real collier fleet sailing along England's home coasts was quite challenging enough.

II. CONVOYING COLLIERS

The application of English naval power to protect coastal coal shippers began with its first prolonged naval war of the early modern period, the Anglo-Spanish war of 1625–30.[3] Early in that conflict it was recognized that the primary danger to English shipping came from the privateers of Dunkirk, a port still under Hapsburg allegiance and well positioned to harass shipping along England's southern and eastern coasts. The Bishop of Durham worried that the Tyne was vulnerable, commissioners of the navy sought opinions on how to protect the coastal trade, members of both houses of Parliament complained of the Dunkirkers' damages, and the Privy Council agreed that the coal trade along England's eastern coast

---

[1] Gregory Clark and David Jacks, 'Coal and the Industrial Revolution, 1700–1869', *European Review of Economic History* 11 (April 2007), 39–72; Mokyr, *Enlightened Economy*, 102; McCloskey, *Bourgeois Dignity*, 187.

[2] By c.1700 the average coal ship departing from north-eastern England for London carried c.250 tons. Hatcher, *History, Before 1700*, 473–4.

[3] For modest precedents beginning in the 1590s, see Nef, *Rise*, II, 263.

had to be defended.[4] The state papers of these years are full of evidence of the need for convoys and full of projects describing how they might be funded, but they say little about how convoys actually worked. They were certainly deployed, however, apparently under the leadership of Captain James Duppa at least during 1626–7. Duppa was a brother to the future bishop of Winchester, an investor in the Virginia company, an owner or partner in a large London brewery, had some sort of interest in the coastal coal trade, and had served at sea in the early 1620s.[5] He was ideally placed to undertake the protection of coastal shipping on behalf of the crown, but did so for his own benefit and under his own initiative.[6] Duppa was thus a projector, that is an entrepreneur acting with the authority and assistance of the crown, but, according to many observers, one more concerned with private than public gain.[7]

J. U. Nef saw the convoys very differently, as another attempt by the crown to assert its direct control over the coal trade, royal 'interference' through protection.[8] He thought that the convoys of 1627 and 1628 had been discontinued because of political opposition by coastal traders, but though there does seem to have been a disruption in 1629 convoys were again deployed in 1630.[9] This opposition he interpreted as part of a broader struggle between the crown and the 'rich merchant class', which he thought included coastal shippers, who feared for their lucrative government loans should the crown use the coal trade to become financially solvent. None of this is credible, but Nef was right to notice the political stakes of the coal convoy during these years. The royal officer in charge of protecting England's coasts during wartime was the Lord High Admiral, the Duke of Buckingham, who was impeached during the parliament of 1626 and denounced as 'the grievance of grievances' in 1628. Buckingham claimed that he required more revenue in order to pay for convoys, and the crown did indeed attempt to fund them with non-parliamentary duties.[10] MPs saw

---

[4] Richard Neile, Bishop of Durham to the Council, 27 October 1625, TNA SP 16/8/48; Commissioners of the Navy to the Council, 14 January 1626, SP 16/18/59; Bidwell and Jansson, eds. *Proceedings in Parliament 1626*, I, 239, 241–2; II, 126–31; *APC 1625–1626*, 295.

[5] For his family and background see Sir Gyles Isham, ed. *The Correspondence of Bishop Brian Duppa and Sir Justinian Isham, 1650–1660* (Northampton, 1955), 192–3.

[6] Duppa's letters to Edward Nicholas, Buckingham's secretary for naval affairs, show him quite proactive in promoting his services and quite free with tactical advice. E.g., TNA SP 16/23/87, 16/95/74.

[7] For Duppa's project to police brewing, see Peter Clark, *The English Alehouse: A Social History 1200–1830* (1983), 175.

[8] Nef, *Rise*, II, 265–6.

[9] For convoy activity and administration, *CSPD 1627–8*, 216, 219, 222; TNA SP 16/95/36, 16/95/74, 16/101/30; PC 2/38, 447, 563; *CSPD 1629–31*, 254, 261, 343.

[10] Bidwell and Jansson, eds. *Parliament 1626*, I, 241–2, 537, 548; TNA SP 16/62/56, 16/68/34I, 16/88/71, 16/94/56.

such initiatives as precedents that might have allowed the crown to fund itself without a parliament. Indeed they may well have been right, as it has been argued that these attempts to fund convoys by non-parliamentary taxes did in fact lead directly to the hugely controversial Ship Money project of the 1630s.[11] During the later 1620s, then, both the funding and the administration of the coal convoy were highly politicized, as parliament refused to grant extraordinary funds to Buckingham or to the king.

During subsequent naval wars convoys for the coastal coal trade seem to have worked much better because they gradually became integrated into other operations of the Royal Navy. This is not at all to say that all went smoothly or that convoys invariably succeeded. The new Commonwealth government moved quickly to protect the coastal trade in 1649 when it arranged for vessels to scour the waters off the north-eastern coast for pirates, but coal shipping was nevertheless vulnerable until the Navy was able to establish supremacy over the remaining Royalists.[12] During all three of the naval wars against the Dutch of the 1650s, 1660s, and 1670s, very many English ships were lost despite the convoys deployed to protect them.[13] Some historians have been highly unimpressed with the state's record during this period, and indeed a few fourth- or fifth-rate convoys were useless when, as in 1666, the very formidable Dutch battle fleet prowled off the English coast.[14]

Under such difficult conditions, however, and despite the many other trades with claims on naval protection, convoys did help secure a substantial supply of fuel. On 9 October 1666, for example, Secretary of State Williamson was informed that the coast off Hull was infested with three Dutch 'pickeroons' [i.e., pirates] who had already taken four colliers, the sort of loss stressed by Willan's study of the deficiencies of the coastal convoys.[15] On the same date, however, Williamson also learned that 'the collier fleet', accompanied by its convoys, was in the Tyne waiting for good winds and a clear coast, and on the following date, the tenth, he was told by an informant in Suffolk that forty empty coal ships were returning

[11] Andrew Thrush, 'Naval Finance and the Origins and Development of Ship Money', in Mark Charles Fissel, ed. *War and Government in Britain, 1598–1650* (Manchester, 1991), 133–62.
[12] *CSPD 1649–50*, 160, 165; Jour. 41, f. 5v-6; Bernard Capp, *Cromwell's Navy: the Fleet and the English Revolution, 1648–1660* (Oxford, 1989), 61–6.
[13] J. R. Bruijn, 'Dutch Privateering during the Second and Third Anglo-Dutch Wars', *Acta Historiae Neerlandicae: Studies on the History of the Netherlands* 11 (1978), 79–93, which counts at least twenty-eight colliers taken during the Second Dutch War, and at least fifty-nine in the third.
[14] T. S. Willan, *The English Coasting Trade 1600–1750* (Manchester, 1967 orig. 1938), 23–7 finds the Restoration regime guilty of abandoning a successful convoy system. Movements and operations of the Dutch and English fleets in 1666 are described in Rodger, *Command of the Ocean*, 65–79.
[15] *CSPD 1666–7*, 189.

north under the protection of two men of war. That same informant had written on the fourth that 100 ships were proceeding southwards towards the Thames, and the following day a further correspondent stated that 'a great fleet of London colliers' was to the north at Lynn, protected by four armed ships.[16] Thus, there were probably four distinct large groups of coastal coal ships moving simultaneously along the eastern coast, despite the Dutch threat, and at least three of them were said to be under armed guard. This commitment to protecting the coastal trade even continued into the Third Dutch War of the 1670s, when the civic government of London agreed with the Privy Council that it would raise its own convoys. Despite this, the Royal Navy also escorted colliers during this conflict, so London's municipal convoy was rather a complement to than a substitution for the Royal Navy's protection.[17]

By the French wars of the reigns of William and Mary, and Anne coal convoys seem to have become more regular and more effective. While Newcastle's coastal coal shipments for the three years of the Second Dutch War were only 55 per cent of what they had been during the previous three years, during the War of the League of Augsburg (1689–97) they totalled 81 per cent of what they had been during the peaceful reign of James II. London's imports during the latter period were even steadier, constituting 87 per cent of their tonnage during peacetime.[18] This was a remarkable accomplishment, for while the English Navy fared better against the French after 1689 than it had against the Dutch in the 1660s and 1670s, French privateers still took thousands of English vessels during the wars with Louis XIV.[19] During the 1690s there was a nearly constant supply of convoys, usually three or four though sometimes as many as five ships dedicated to that service during the summer months.[20] As in the 1670s, these ships reported to the admiralty, whose papers show not only the movement of the ships specifically assigned to the coal trade but

---

[16] *Ibid.*, 181, 184, 189, 191.
[17] For arrangements and payments PC 2/63, 232–3, 240; Jour. 47, f. 265, 291v; LMA COL/ CHD/PR/3/11. For the admiralty's oversight of collier convoy activity, TNA ADM 106/283, f. 91, 324. For naval vessels protecting coal fleets, and for Secretary of State Williamson's informants regarding the movements of colliers and their convoys, including evidence of effective protection, e.g., *CSPD* 1673, 1, 44, 75, 90, 200, 203, 205, 328, 365, 382, 421, 484.
[18] Derived from Hatcher, *History, Before 1700*, 489–91, 502. Newcastle totals indicate periods from Christmas 1661–67 and 1685–97, while London figures run from Michaelmas 1685 to Michaelmas 1697. The London data for James's reign includes the year 1687–8 during which imports were substantially higher than any other year during the seventeenth century. Without this outlier there is no substantial decrease in London coal consumption in the 1690s compared to the 1680s. London's growing population, however, would mean that there was a small decline of coal consumption per capita during these years.
[19] Rodger, *Command of the Ocean*, 158, 177.
[20] 'An Account of the Number and Rates of her Majesty's ships as were appointed Convoys to the Colliers in the last war', HL/PO/JO/10/6/44, f. 96.

also the occasional assistance of other naval vessels that escorted coal fleets on their way to or from other stations.[21]

By the time Queen Anne told her parliament of her special care for convoys, therefore, such protection was already a long-established and expected duty of the state during wartime. The high priority placed on protection for the coal trade was made evident when other traders tried to increase protection for their own shipping at its expense. 'The Newcastle convoy could not be touched', Russian merchants were told in 1703, 'the City of London having applied that their coal trade may be taken care of, and complained of the dearness of coals', and in 1706 it was noted that no ships may be 'diverted from that service without the Queen's order'.[22] As in the 1690s, under Anne the navy deployed 2–5 ships specifically for the coal trade, besides others whose stations might also provide some additional protection for the eastern coasts.[23] Some of these ships were relatively light cruisers of the fifth and sixth rate, but fourth-rate ships of the line like the Portland, Dartmouth, and Warwick, each of which convoyed colliers during the summer of 1703, carried over fifty guns. Though the entire naval fleet included over 120 ships of the line and over sixty cruisers, five ships was nevertheless a significant proportion. It was the largest single naval commitment to protecting a domestic trade and an allocation of resources not far below those deployed in defence of more profitable branches of overseas commerce. English strategy during the War of Spanish Succession was to occupy French armies on as many fronts as possible, requiring the navy to blockade both France's Atlantic and its Mediterranean squadrons, to support allied operations in the Mediterranean, the Iberian Peninsula, and in the Netherlands, to protect English commercial interests in the North Atlantic, Caribbean, Mediterranean, Indian Ocean, and Russia, and finally to protect the southern and western coasts of England and the coasts of Ireland from privateers.[24] Given such diverse and dispersed responsibilities, the allocation of even a few fifty gun ships was a significant extension of military power. The English state devoted significant resources to protecting

---

[21] E.g., TNA ADM106/489, f. 45, 54, 85, 88, 114, 118, 138, 231, 360; TNA ADM106/493, f. 84, 95–125 *passim*.

[22] John Hely Owen, *War at Sea under Queen Anne 1702–1708* (Cambridge, orig. 1938, 2010), 57–8.

[23] 'An Account of what Convoys have been appointed in the Year 1703, for securing the Importations of Coales to the City of London', HL/PO/JO/10/6/44, f. 84–6; 'A list of H.M. ships and vessels in sea pay at home as are employed as convoys or cruisers', TNA SP 42/7/49.

[24] Rodger, *Command of the Ocean*, 164–80, 608, 612; John B. Hattendorf, *England in the War of the Spanish Succession: A Study in the English View and Conduct of Grand Strategy, 1702–1712* (New York, 1987), 168–72 *et passim*.

London's fuel supplies during its many wars, and became increasingly good at doing so, in large part because it seemed clear that the alternatives to such protection were dire.

III. THE PRICE OF INSECURITY

When wartime convoys failed, or seemed as though they might fail, the London coal market responded quickly. The inverse was also true, that secure coasts led prices to moderate. This was seen clearly in the Dutch wars, beginning in the 1650s. In March of 1653, a newsletter writer described a straightforward relationship between the military situation on the coast and the price of coal in London. In April he wrote that a convoy of twenty ships had been sent to fetch 120 colliers, but 'if the Dutch have taken our colliers as it's reported we must pray for fair weather, for coals are already at £5 the chaldron'.[25] In following weeks the bad news continued, as reports circulated of coal fleets sheltering in ports to avoid enemy attack. The effects of the morale of the capital, and hence the stability of the regime itself, were claimed to be severe:

Our dearth of coals exasperates [the people], and I assure you if the Dutch keep them from us, we shall shortly cut one another's throats; they are above £6 the chaldron, and scarce any to be had for money. A merry fellow the last week went through the City crying coals at 3*d.* a bushel, and as the people gathered about him to know where, he told them at Rotterdam stairs, and so went on not being taken notice of by any of our officers.[26]

These letters, intended for the royal court in exile, must be read with some suspicion due to the evident pleasure with which they describe London's unstable condition. Yet sources from their political enemies made similar assessments, as London's court of Aldermen agreed that without adequate convoys London's fuel prices would rise dangerously.[27] The Council of State collected almost daily intelligence regarding the collier fleets in the spring of 1653.[28] The Venetian ambassador also described, as his predecessor had in 1643–4, a straightforward relationship between a successful convoy by 'a squadron of war ships' under Vice-Admiral Sir William Penn and lowered coal prices. With a coal fleet hemmed in by the Dutch in April prices tripled, but after the fleet's arrival in May they quickly declined back to only half their recent height.[29]

[25] London newsletter, 1 April 1653, Bod. Clarendon MS 45, f. 222.
[26] London newsletter, 15 April 1653, Bod. Clarendon MS 45, f. 292. See also f. 221v-3, 284-v. 3*d.* per bushel would equal 8 shillings per chaldron, far cheaper than coal had been within any Londoner's memory.
[27] Rep. 62, f. 232, 268v.   [28] *CSPD 1652–3*, April and May, *passim.*
[29] Steve Pincus, *Protestantism and Patriotism: Ideologies and the Making of English Foreign Policy, 1650–1668* (Cambridge, 1996), 175–6; *CSPV 1653–4*, 60, 63.

This process occurred again with the Second Dutch War just a decade later, as the outbreak of war once more saw prices rise immediately and military reversals prompted further spikes. The most dramatic such fluctuation, not only of that war but probably of the entire history of the London coal market, occurred in 1667. In March of 1667 Samuel Pepys recorded cold temperatures and high prices, but during the third week of April a 'great fleet', reported to contain from 300 to 500 sail of colliers, proceeded under convoy of six ships to London, where prices quickly plummeted to 23s.[30] Within two months, however, they had quintupled, as the sensational Dutch descent on Chatham in June raised the possibility that England's eastern coasts would be undefended for the rest of the summer. Peace soon moderated prices again, showing that it was not immediate necessity but rather the anticipation of future scarcity that pushed prices up in June of 1667.

The raid on Chatham was undoubtedly the nadir of England's defence of its coasts during the seventeenth and eighteenth centuries, but in all wars there were moments when trade seemed threatened, when insecure shipping lanes seemed likely to continue and therefore when coal prices remained high. The weekly records of London coal prices kept by John Houghton in the 1690s, and the daily accounts kept by a merchant during the following decade, show just how sensitive the market was to such episodes. During the 1690s, the French preyed with much success on English and Dutch trade, taking some 4,000 English ships of all sorts from 1689 to 1697.[31] Under such generally dangerous circumstances London coal prices rose and stayed high, rarely dipping below 25s. per chaldron from the beginning of Houghton's price series in 1692 and averaging 32s. over 305 entries.[32] In mid-May of 1694, Houghton reported a jump in the market to 39s. per chaldron, but between 1 and 8 June they fell quickly from 39 to 28s., a decline of almost 30 per cent over just a few days. This must have been a response to news of the progress of the coal fleet, which on the eighth was reported in the official organ *The London Gazette* to be en route from Yarmouth. The following day this 'great fleet' of 400 colliers from the Tyne arrived in the Thames.[33] By the fifteenth prices had bottomed out at 27 shillings, the lowest price the capital would enjoy for two years. Another such episode occurred in April

---

[30] Purchase recorded on 27 April in *Pepys, 1667*, 187. For convoys and the progress of the fleet, *The London Gazette*, n. 3277–80, 1–19 April 1667; *CSPD 1667*, 37–40. Nef, *Rise*, II, 82 recognizes that there were 'violent fluctuations' in prices during wartime, but he interprets Pepys' price of 23s. as the result of special contacts and insider information. This is possible, but it seems more likely that it reflects the level of the general market at a moment when there were reasons to expect continued healthy shipments.

[31] Rodgers, *Command of the Ocean*, 158.

[32] This and the remainder of the paragraph uses John Houghton, *Collection for Improvement of Agriculture and Trade* n. 1 (1692) – n. 305 (1698).

[33] *The London Gazette*, n. 2982, 7–10 June 1694.

of 1697. The previous winter prices had remained high, between 30 and 42s., but in late March they suddenly surged to 60s., presumably a response to problems securing the arrival of the first spring shipment. Whatever the cause of the surge, between 9 and 16 April prices halved as another 'great fleet' of 400 ships arrived in the Thames, probably a day or two after leaving Yarmouth on the twelfth.[34] With the coming of peace that September prices moderated, averaging only 26s. for the remaining seventeen months of Houghton's price series.

Not only were prices acutely sensitive to the security of the collier fleet, but merchants, officers, and, in all likelihood, the broader public were aware that this was the case. This was, after all, the basic assumption behind Anne's assertion that convoys should moderate prices in November 1703. But it is remarkable just how well expert observers understood the relationships between the security of the trade and the London market. On 23 February 1704, a few weeks after Anne mentioned how effectively she had already 'taken care' for convoys, two coastal shippers, Arnold Cox and Abraham Jaggard, complained that upon expectation of the coal fleet prices had fallen to 35–6 shillings and would have declined still farther but that the convoys appointed for the fleet refused to sail. This, they claimed, had driven prices up to 44s., and they would 'inevitably' advance further to £3 (60s.) the chaldron.[35] The private account book of an anonymous London merchant, which recorded daily transaction prices for fuel from 1703 to 1705, shows that Cox and Jaggard were not only accurate in their claims about current prices, but remarkably prescient about the future trend.[36] It shows that during the first three weeks of February 1704 coal retailed at 38–40s., a price which presumably includes delivery charges and if so is nearly equivalent to the merchants' claims. During the final week of February prices did indeed increase by about 20 per cent. Furthermore, after a March decline back to 38s. by early April coal retailed for £3 per chaldron, exactly as Jaggard and Cox had predicted. Their claims were evidently persuasive to London's aldermen and to the Privy Council, both of which devoted more resources and more attention during the following months to coal convoys.[37]

Such merchants understood how the market worked in large part because they and their fellow traders at Billingsgate were the ones who set prices, responded to shortfalls, and speculated on conditions of supply and demand. This expert familiarity with price movements, however, was shared among a much broader body of the urban populace who understood in general terms

[34] *The London Gazette*, n. 3279, 12–15 April 1694; n. 3280, 15–19 April 1694.
[35] Jour. 53, p. 741, 744.   [36] TNA C114/60/3.
[37] Rep. 108, pp. 288, 304, 501, 542, 546; TNA PC 2/80, 70, 111, 116–7, 124, 171, 178, 422–4.

that insecurity led directly to expensive fuel. Indeed, causing false rumours along these lines was one of the dark arts used by fuel retailers to raise prices artificially, at least according to a 1653 pamphlet. There a 'chandler' brags to his fellow 'woodmonger' that

I cause it to be noised and rumoured about that in regard our fleet is now come into harbours, those insolent Dutch pickeroons with some men of war, lie upon or near Sunderland and the north coasts, that a collier cannot stir out, and this affrights the meaner sort of people that they cry out, Alas! Would these wars were once done with the Hollander, else we poor must starve. And is not this a cunning sleight?[38]

For the pamphlet's readers this kind of market manipulation was especially threatening because experience showed how such 'noise', whether ultimately caused by real 'men of war' or by traders' 'cunning', led London's fuel prices to remain high until the arrival of another 'great fleet'.

IV. PROTECTION AND IMPRESSMENT

Despite the widely recognized importance of state power to safeguard coastal trade, when contemporaries used the word 'protection' in regard to coal shipping they most frequently referred not to protection *by* the state, but to protection *from* the state. One basic challenge of naval warfare in Europe throughout the early modern period was that while it was intended to win overseas markets and expand maritime trade, yet in practice naval warfare demanded embargoes and impressment which temporarily crippled that very sector. To man a fleet during the reign of William and Mary, as was discussed in Chapter 8, required around 20,000 men immediately upon the outbreak of war and over 40,000 within a few years.[39] Meeting such goals often required an embargo on shipping as available seamen were pressed into royal service. The implications of this for trade in general were clearly serious and London's coal supply, which required a very large number of ships manned by a very large number of mariners, was identified as a trade that would be particularly and importantly damaged. As Captain Duppa wrote to Nicholas early in 1626,

The estate of these times concerning seamen doeth much trouble me, and if the king should set out a great fleet this summer, as I fear he will have occasion to make offensive or defensive war ... then will it come to pass perforce, that all navigation must cease, whereby their seamen may help to furnish the king's ships, etc., all to little. This being granted, what shall the city of London do for fuel; if there be no seamen, there can be no sailing, and so shall the city be unfurnished.[40]

---

[38] Well-willer to the prosperity of this famous Common-wealth, *The Two Grand Ingrossers of Coles: viz. the Wood-Monger, and the Chandler* (1653), 7–8.
[39] Baugh, 'The Eighteenth Century Navy'; Rodger, *Command of the Ocean*, 206.
[40] Duppa to Nicholas, 27 March 1626 TNA SP 16/23/87.

Duppa then proceeded to describe his project for providing guns to private coal ships so as to protect this vital trade while minimizing costs to the crown. A different approach to convoying the coal fleets, as we have seen, was adopted throughout the seventeenth and eighteenth centuries, but the problem Duppa described remained: how could the coal trade act as the 'nursery of seamen' without losing its workforce to the navy during a war? How could the navy both serve and be served by the coal trade?

For some the answer to these questions was simply that such balancing acts were impossible, and the state must therefore choose whether scarce seamen were better used on board colliers or ships of war. This attitude was especially prominent in those professionally concerned with manning the fleet, whose devotion to military necessity caused them to see any hindrance to manning as an annoyance. Some such officers were capable of nuanced positions, such as Sir William Coventry who argued in 1665 that high coal prices would mean high pay on colliers, which in turn would allow them to be resupplied after their men had been impressed into royal service.[41] Others would develop this logic further in an effort to resolve the paradox surrounding the coal trade's ambivalent relationship to state power.[42] But a more commonly expressed attitude, in particular among those more immediately concerned to supply London than to man the navy, was to deny this problem, or at the least to avoid it.

It was therefore often asserted that convoys and protections from the press were complementary policies, each necessary to supply London with fuel. Indeed in practice they were normally done together, despite the inherent difficulty of reconciling such ambivalent aims. During both the Second and the Third Dutch Wars, for example, the crown provided both protection of the navy and protection from the navy. In the spring of 1665 Charles II announced by royal proclamation that the coal fleet would receive convoys and that seamen employed therein would escape the press, and the Privy Council compiled lists of dozens of ships' masters and hundreds of men who would be exempt from naval service.[43] Parliament agreed with this priority, William Prynn reporting a committee's finding in 1667 that the principle causes of the high prices of coals were lack of 'sufficient convoy' and also recommending protection for four able seamen for ships of over 100 tons and six if over 200 tons.[44] Similarly in May 1672, the Privy Council received recommendations from the City of London as to 'the most effectual way of

---

[41] Coventry to Earl of Falmouth, 24 May 1665, Bod. Carte MS 34, f. 229.
[42] 'Proposals whereby the City of London may be served during the present War with France with Coales at or about 20s p. Chaldron', BL Add MS 28,079, f. 74-v; Crosfield, 'England's Glory Revived', Bromley, ed. *Manning of the Royal Navy*, 2–8.
[43] *CSPD 1664–5*, 333; TNA PC 2/58, 111–2, 116, 124–5, 127, 160.
[44] JHC VIII, 676.

having London sufficiently supplied with coals at reasonable rates during the present war'. Protection from impressment was at the top of London's wish list, and further methods of securing and financing convoys were also agreed.[45]

Under William and then Anne, the City of London again requested protections in addition to convoys, a goal finally codified in parliamentary statutes.[46] In 1695, an act establishing duties on coal included the stipulation that colliers' masters could nominate two protected seamen per fifty tons burden.[47] Nine years later, the 'Act for the increase of seamen, and better encouragement of navigation, and security of the coal-trade' allowed one able seamen per hundred tons in addition to the ship's master, his mate, and carpenter, all to be free from impressment.[48] Both acts demonstrate the connections between London's coal supply and the fiscal-military state's capacity to wage war, in the former case to raise taxes and in the latter to man the fleet. During both conflicts protections from the press were advocated and established even as colliers were also receiving more systematic and more effective convoys than ever before. Into the eighteenth century, as the navy's global reach expanded still further, the coastal coal trade retained and even improved its protected status. In 1739, the Lord Mayor successfully lobbied for the Admiralty to protect the colliers despite the Navy's current need for thousands of men.[49] By September and October of that year colliers were privileged above other trades, and during that winter protections were extended to include not only the men aboard coastal ships but also those employed unloading and transporting coal in the Thames.[50]

V. CONCLUSION: THE STATE AND THE LONDON COAL TRADE

By the eighteenth century, then, the British state had a century-long heritage of both deploying and restraining its resources in order to ensure that London's coal consumption continued unabated. The crown never took over the coal trade but its power was essential, and was universally seen to be essential, in protecting and maintaining the capital's fuel supply. Such protections were emergency procedures restricted to wartime, but wars

---

[45] TNA PC 2/63, 232–3, 240; Rep. 77, f. 139v-140, 148, 151-v, 159v, 175, 185v, 211-212v; *HMC Le Fleming*, 91–3.

[46] HL/PO/JO/10/6/44, 97-v; *JHC XIV*, 310.

[47] John Raithby, *The Statutes Relating to the Admiralty, Navy, Shipping, and Navigation of the United Kingdom* (1823), 69.

[48] *Ibid.*, 92–6.

[49] Daniel Baugh, *British Naval Administration in the Age of Walpole* (Princeton, 1965), 171–3; TNA ADM 3/43, [Admiralty Minutes] 20–6 September 1739.

[50] TNA ADM 3/43, 26 September, 2, 31 October 1739, 3 January, 1, 22 February 1739/40; TNA ADM 3/44, 3, 13 January, 1 February 1739/40.

should not be seen as special cases during the seventeenth and eighteenth centuries. Making war was a leading, perhaps the primary, reason for the existence of a powerful central state, and wars were familiar, frequent, and expected throughout the period. For the state to devote scarce resources during wartime to the coal trade is not, therefore, an aberration that can be bracketed off from normal practice, but rather a clear demonstration of the importance of London's fuel supply to the state and, inversely, of the state to the coal trade.

Contemporaries were convinced that the failure of the colliers would be disastrous from several perspectives. As this and previous chapters have described, it was widely feared that if London lacked fuel households would suffer, many of its leading industries would cease, and its people would quickly become ungovernable. Its protection was thus not only part of economic and social policy, but a serious strategic consideration for a state frequently at war. Coal was not the only commodity threatened by warfare, of course. But it was uniquely bulky relative to its value, and it was also uniquely easy to locate colliers who had only one source of supply, one dominant market, and only route in between them. A Dutch, Flemish, or French privateer could simply lie anywhere off of England's eastern coast during the spring, summer, or autumn in order to find innumerable colliers passing by. More valuable cargoes arriving from the Caribbean or Indian Ocean were much harder to find, and coasters moving similarly mundane kinds of cargo would not have been so dependably found, month after month, plying one predictable route. An undefended coal fleet would have seemed very inviting prey indeed to enemy privateers.

Such considerations show that the debate regarding coal's place in England's industrial revolution is sometimes quite divorced from the real actions and experiences of early modern Londoners and their governors. It might have been possible (though there are good reasons to doubt this) for foreign wood to have substituted for domestic coal with minimal injury to England's economic development.[51] If so, inventive Englishmen could still have found efficient new methods of production that were not energy-intensive and could still, even without coal, have launched the first industrial revolution. But achieving this would have required a great deal more than merely paying for more expensive fuel; it would have required England to have no reason ever to fear any ships in the Baltic and North Seas. It would thus have had to retain good relations with the Netherlands, Denmark, Sweden, and, in time, Russia, all simultaneously and perpetually, or else to have controlled those seas directly. Either strategy would

---

[51] Clark and Jacks, 'Coal and the Industrial Revolution'; Mokyr, *Enlightened Economy*, 102; McCloskey, *Bourgeois Dignity*, 187.

have entailed very significant costs. The fact that this alternative was never considered suggests how unrealistic it would have seemed to early modern Londoners. Wood from Finland would have been both expensive and vulnerable.

Part of coal's utility, by contrast, was that it was both relatively cheap and relatively secure. This chapter has stressed the strategic significance of the coastal coal trade and the consequent lengths that the state would go to protect it. Previous chapters have argued that this and other kinds of state intervention were driven by the widespread belief that fuel scarcities could lead to social disorder, while a healthy fuel trade would lead to commercial, industrial, and fiscal improvement. These have been analyzed separately, but they all reinforced each other: state power both assisted and benefitted from London's coal consumption, while the London market depended on improving commercial networks as well as assistance from a strong but responsive state. Coal became, during the seventeenth and eighteenth centuries, ever more embedded in London's social relationships and its status as the economic and political heart of an expansive kingdom. It was not, therefore, something that London seemed able to do without. Those troubled by London's polluted atmosphere, therefore, needed to find ways for London, and for themselves, to live with coal. This even applied to the early modern period's most eloquent enemy of coal smoke, John Evelyn.

Part IV

Accommodations

– I I –

Evelyn's place: Fumifugium *and the royal retreat from urban smoke*

I. A CONVERSATION ON THE THAMES

One autumn day in 1661 an unusual ship sailed down the Thames. It was a yacht, which John Evelyn explained in his diary was a new kind of 'pleasure boat', perhaps the very one that had been presented to King Charles II by the Dutch the previous year on the occasion of the restoration of his kingship.[1] The ship carried the king himself, 'divers noble persons and lords', and Evelyn, a man of humbler social status, comfortably wealthy and impressively cultivated but not a 'noble person' and not a member of the restored king's inner circle. It was therefore unusual for Evelyn to spend so much time with his monarch, eating both breakfast and dinner with him, conversing while the king raced his brother's boat on the forty-mile round trip from Greenwich to Gravesend and back again. Charles had met Evelyn before, probably in Paris during the 1640s when Evelyn had close connections to leading Royalist exiles, and certainly in London several times after his return in April of 1660.[2] But, so far as the record indicates, they had never spoken as long, nor had the king ever asked Evelyn for as much, as during their return leg back up the Thames from Gravesend.

Evelyn's diary therefore records the conversation in some detail. It began with a short book that Evelyn had given the king only a few weeks before.

He was pleased to discourse to me about my book inveighing against the nuisance of the smoke of London, and proposing expedients how, by removing those particulars I mentioned, it might be reformed; commanding me to prepare a Bill against the next session of Parliament, being, as he said, resolved to have something done in it.

---

[1] E. S. De Beer, ed. *The Diary of John Evelyn*, 6 vols., (Oxford, 1955), III, 296. De Beer's note states that this ship was the yacht *Mary*, though this does not seem so certain. Evelyn's diary calls the king's boat 'new', and the previous month Pepys's diary distinguished between two 'new' yachts and two 'Dutch ones'. *Pepys 1661*, 177, 179.

[2] Evelyn married the daughter of Sir Richard Browne, royal representative in Paris throughout the 1640s and 1650s, and he was in close contact with leading royalist exiles like Arundel, Hyde, Nicholas, Earle, Cosin, and the Earl and Countess of Newcastle. Darley, *John Evelyn*, ch. 4–6.

The book was *Fumifugium: or The Inconveniencie of the Aer and Smoak of London Dissipated*, the most extensive, sophisticated, and ambitious analysis of urban air pollution produced anywhere during the early modern period. Evelyn had presented a manuscript of it to Charles two weeks previously, when the king had given his approval and ordered it published.[3] During their second conversation Charles went further, expressing support for a parliamentary bill. The conversation then shifted, only slightly, to 'the improvement of gardens and buildings, now very rare in England comparatively to other countries'. Such complaints were favourite topics of Evelyn's and were central to *Fumifugium*'s vision of a restored and beautified London. But then, whether because Charles suddenly realized the breadth of Evelyn's virtuosity or because this had been the king's real agenda for their conversation all along, Charles changed topics entirely. The previous day there had been a major diplomatic row between the Spanish and the French embassies, and Charles needed an able and discreet pen to craft an official version of events. For the next three days Evelyn did almost nothing else, until at last he left court exhausted.

The progress of this conversation mirrors in many ways the progress of Evelyn's attempts to reform London's smoke: *Fumifugium* achieved royal promises to 'have something done', and yet, in the end, there were always more pressing calls on the king's scarce money and attention. Or perhaps Charles always knew that Evelyn's project suffered from crucial flaws that rendered it impossible, both technically and politically. In any case, despite what the king had 'resolved' in October of 1661, London's smoke was never, in fact, 'dissipated'. Decades later Evelyn drafted a letter to Pepys in which he mentions 'the old smoky pamphlet'.[4] Evelyn himself had not changed his mind about London's air – he noted during the hard frost of 1684 that the smoke was so severe that he could hardly breathe – but his letter to Pepys suggests he was resigned to the work's failure.[5] During the final years of his long life he returned once more to *Fumifugium*, inserting a discussion of fuel and smoke to a new edition of his forestry treatise *Sylva*. There Evelyn suggested Timothy Nourse's recently-published plan for fuelling London with wood was worthy, even 'not impossible', but admitted that his own contribution was gone with the wind.[6]

Some historians have agreed with Evelyn's own pessimistic take on *Fumifugium*'s impact, finding it an interesting, perhaps laudable, early statement of environmental concern, but one that was also sadly isolated from

[3] Evelyn, *Diary*, III, 295–6.
[4] De la Bédoyère, ed. *Particular Friends*, 182.
[5] Evelyn, *Diary*, IV, 363.
[6] 'My Fumifugium is long since vanished in aura [the breeze]'. John Evelyn, *Sylva* (4th ed. 1706), 265; Nourse, *Fuel of London*.

his contemporaries. Evelyn, from this perspective, may well offer an insightful take on early modern London, but he was not really part of any larger movements or processes. He was an outlier, a man ahead of his time.[7] Mark Jenner, by contrast, has provided what might be seen as the view from the yacht. In 1661, he points out, restoration and renewal were in the air. As a royalist and one of the founders of the Royal Society, Evelyn was interested both in the stability and honour of the new regime and its capacity to patronize the arts of peace and improvement. His text's mixture of optimism and flattery is not uncharacteristic of the very early Restoration in general. He was not a modern environmentalist born too soon but rather a man fully integrated in his own times, writing from and to the concerns of his period.[8] If the crown, House of Lords, and Church of England could be so wonderfully restored (and Evelyn thought 'such a restoration was never seen' in history since the ending of the Babylonian captivity) then reducing London's coal smoke should hardly be impossible.[9]

It is certainly true that *Fumifugium* was a product of a particular historical moment when radical changes were worth proposing and one's talents were worth advertising to a new regime. Jenner is quite right to stress all of this, but this chapter aims to place Evelyn and *Fumifugium* not merely within the immediate context of Restoration, but also within a trajectory of politicized concern for London smoke that spanned his lifetime. *Fumifugium*, it will argue, was not an early environmental manifesto but rather the culmination of existing traditions which approached London's air as a problem to be reformed and improved. The 1660s, however, were also the beginning of the end of the English crown's attempts to effect such improvement. Charles II showed real concern with urban smoke, but during his reign the practical results of this concern became ever more constricted. By the years around 1700 English monarchs continued to dislike London's smoky atmosphere but they finally abandoned their predecessors' attempts to reform it. Rather than reduce pollution they chose to remove themselves from it.

II. EVELYN'S PLACE: *FUMIFUGIUM* AND THE POLITICS OF LONDON SMOKE

Evelyn opens his text with a direct address to the 'The King's Most Sacred Majesty' in which he positions the work as a response to an experience of London's material reality.

---

[7] Darley, *John Evelyn*, 176, 339 n. 18; Thorsheim, *Inventing Pollution*, 5, 17; Radkau, *Nature and Power*, 143.
[8] Jenner, 'The Politics of London Air'.   [9] Evelyn, *Diary*, III, 246.

Sir, It was one day, as I was talking in Your Majesty's palace at Whitehall (where I have sometimes the honour to refresh my self with the sight of your illustrious presence, which is the joy of your peoples' hearts) that a presumptuous smoke, issuing from one or two tunnels near Northumberland House, and not far from Scotland yard, did so invade the court, that all the rooms, galleries, and places about it were filled and infested with it, and that to such a degree, as men could hardly discern one another for the cloud, and none could support, without manifest inconveniency.[10]

In this opening passage, Evelyn introduces many of the themes that will be central to the rest of the text. First, smoke is a problem that has been experienced through the senses and the bodies of Londoners in general, and the king and his court in particular. Indeed, it is through its penetration of the space of the court itself that urban smoke is most offensive; it inconveniences the bodies of all in the royal palace, but it also renders impossible the process of 'discern[ing] one another' that was one of the court's primary purposes. Smoke was observed and experienced with and through the senses, a material reality that posed immediate and dire political consequences.

In a following address to the reader Evelyn then expands the sphere of those affected to include all 'the people of this vast city'. They suffer in two distinct but related ways. First, they are all 'pursued and haunted by that infernal smoke', as their 'health and felicity' are damaged through its influence on their bodies. Second, these many individual harms collectively form an assault on the public good and thereby on the city's and the nation's honour. It is here that the interests of 'the reader' and of the king unite, as Charles II is 'a prince of so magnanimous and public a spirit', that whatever may be done for either 'health or ornament' is now possible. 'We have a prince who is resolved to be a father to his country, and a parliament whose decrees and resentments [i.e. grievances] take their impression from his majesty's great genius, which studies only the public good. It is from them, therefore, that we augur our future happiness.' In both of these opening addresses, then, Evelyn is at pains to present smoke as a current problem without relevant historical depth, a harm that the current benevolent government may remove to its own honour and glory.[11]

Despite his rhetorical efforts to remove his complaint from history, both his diagnosis of the problem and his proposed solutions drew on precedents stretching back into the early decades of the seventeenth century. In the first of the text's main three sections, Evelyn described the nature of the air and its medical importance. As the examination of early modern medical and natural philosophical writings in Chapter 6 stressed, Evelyn was far from unusual in arguing that air was important to health. As one of the six

---

[10] *Writings of John Evelyn*, 129.   [11] *Ibid.*, 131–2.

non-naturals in the classical medical tradition, its general significance was universally acknowledged. Evelyn entered into this tradition through both classical and modern sources. He drew on a substantial range of ancient texts, including obvious authorities on the intersections of medicine and the environment like Hippocrates, Galen, Pliny the Elder, and Vitruvius but also poets and philosophers like Virgil, Plato, Lucretius, and Cicero, as well as moderns like Avicenna and Paracelsus and his contemporaries like Kenelm Digby and Arnold Boate. He had clearly discussed air's and smoke's importance with many of his learned contemporaries, including an unnamed 'most learned physician'. He had further unnamed sources for the air's role in the death of Thomas Parr, whose autopsy by Harvey was not yet published.[12] He addressed the idea, to be advanced in print the following year by John Graunt, that the 'corrupt air' of cities shields them from plague.[13] Finally, Evelyn's sources also included common experience and *vox populi*, 'common notices', 'that which was by many observed', and 'how frequently ... we hear men say'.[14] Here, as throughout his substantial published oeuvre, Evelyn's work is characterized by a syncretic approach that refused to distinguish the methods of the natural philosopher from those of the humanist or the political advisor.[15]

While he happily drew on this wide variety of authorities, Evelyn's central agenda was to improve urban air in politically significant ways, and in this his key influences were more limited and more recent. The political relevance of London's air built, silently, on the claims and initiatives of the personal rule of Charles I.[16] Evelyn was not keen to advertise this, and since he was only a child when Charles I came to the throne he may not even have appreciated the novelty of that regime's politicization of the London environment. Evelyn's own access to this was likely primarily through the Earl of Arundel and/or other royalist exiles that had executed the anti-smoke policies of the 1630s.[17] Whatever precisely he knew about the measures of the 1620s and 1630s and, however, he learned about them, *Fumifugium* shared several basic assumptions with the initiatives of the personal rule.

[12] Evelyn knew the Earl of Arundel, who brought Parr to London, and through him and his circle could certainly have met Harvey or heard about the Parr autopsy. Darley, *Evelyn*; Craig Ashley Hanson, *The English Virtuoso: Art, Medicine, and Antiquarianism in the Age of Empiricism* (Chicago, 2009), ch. 1–2.

[13] *Writings of John Evelyn*, 142. Cf. Graunt, *Natural and political*, 68–70.

[14] *Writings of John Evelyn*, 134, 139, 141.

[15] For discussions of these themes see Hanson, *English Virtuoso*, ch. 2; Michael Hunter, 'John Evelyn in the 1650s: A Virtuoso in Quest of a Role', in Therese O'Malley and Joachim Wolscke-Bulmahn, eds. *John Evelyn's 'Elysium Britannicum' and European Gardening* (Washington, 1998), 79–106.

[16] For details of this see Chapter 4 and Cavert, 'Environmental Policy'.

[17] Though Evelyn's diary has little to say about it, his visits to the English resident in Paris, Richard Browne, would have allowed for substantial contact with Queen Henrietta Maria's

Most fundamentally Evelyn accepted the explicit connection between environment and political ideology, between the beauty and health of London and the prestige and honour of the regime. Just as the 1624 Breweries Bill claimed that smoke 'doth greatly diminish not only the pleasure and delight but the health and soundness' of London and Westminster, Evelyn's address to the king emphasized the 'hazard to your health' and to the 'luster and beauty ... splendor and perfection' of the royal court.[18] For both Evelyn and his predecessors, smoky air was an insult to what should have been spaces of royal display. He also followed precedent in refusing to acknowledge any tension between this and the interests of the generality of the population, arguing instead that smoke assaulted both the dignity of the crown and the common good of the entire populous. Evelyn's invocation of the collective interest of 'this vast city' thus parallels Charles I's 1634 proclamation in which smoke annoyed all inhabitants of great cities.[19] In both cases smoke mattered primarily because it sullied royal honour in ways that also provided room for a complementary rhetoric of the public good.

Evelyn understood the causes, as well as the meanings, of air pollution in terms that shared much with Charles I's government. The Privy Council concerned itself with a few specifically identified coal burners, primarily but not exclusively beer brewers. They were arrested, sued, and fined as individuals rather than as members of a trade. Nor did the king or his officers argue that coal burning in general was a problem. Quite the contrary, the council worked hard to regulate the coal trade with an eye towards maintaining high quality and low prices, and also to increased royal taxes. Evelyn took a very similar line, arguing that the 'the health and felicity of so many' Londoners was prejudiced merely by 'the sordid, and accursed avarice of some few particular persons'.[20] In section two Evelyn leaves no doubt as to who these persons were:

such trades, as are manifest nuisances to this city ... such as in their works and furnaces use great quantities of sea coal, the sole and only cause of those prodigious clouds of smoke, which so universally and so fatally infest the air ... Such we named to be brewers, dyers, soap and salt-boilers, lime burners, and the like.[21]

---

court, including Endymion Porter, as well as with Caroline Privy Councillor Francis Cottington and the Council's clerk Edward Nicholas. The writers Abraham Cowley, William Davenant, and John Denham were also there, whose descriptions of London smoke are discussed in the next chapter. Darley, *Evelyn*, ch. 4,6. Brimblecombe's suggestion that Laud was the foremost influence on Evelyn is misguided. Brimblecombe, *The Big Smoke*, 40.

[18] HL/PO/JO/10/4/1; *Writings of John Evelyn*, 129.

[19] Larkin, *Stuart Royal Proclamations*, 426–7. Evelyn describes Londoners as 'ad eundem fumum degentes', living in the same smoke, a reference to Aristotle's *Politics* (1252b). Though not pursued by Evelyn, the implications of this reference are that urban political organization proceeds from the common experience of the urban environment.

[20] *Writings of John Evelyn*, 131.   [21] *Ibid.*, 147.

Whereas the complaints against industrial smoke during the 1620s and 1630s tended simply to ignore the possibility that widespread consumption of coal in households might also have been a significant cause of dirty air, Evelyn explicitly denied this possibility. It was 'not from the culinary fires, which for being weak and less often fed below, is with such ease dispelled and scattered above, as it is hardly at all discernible'. Only the emissions of 'private trades' mattered, 'one of whose spiracles [vents] alone, does manifestly infect the air more than all the chimneys of London put together besides'.[22] This was a gross exaggeration; a brewery near to Whitehall might have consumed about 400 tons of coal annually, but Westminster's hearths and 'culinary fires' burned something like 100 times that amount.[23] Whether Evelyn realized this is a separate, unresolvable question.[24] He does, however, address such a sceptical response. The previous quotation continues:

And that this is not the least hyperbole, let the best of judges decide it, which I take to be our senses. Whilst there are belching it forth their sooty jaws, the City of London resembles the face rather of Mount Ætna, the Court of Vulcan, Stromboli, or the Suburbs of Hell, than an Assembly of ration creatures and the imperial seat of our incomparable monarch.[25]

This language is idiosyncratic; neither Charles I nor his council would have described smoke through this sort of prose. But Evelyn's rhetoric is a response to a problem that had been recognized for decades, namely whether universal coal use was consistent with a clean and beautiful city. In *Fumifugium* Evelyn makes explicit, through an onslaught of his most baroque rhetoric, a position that was defended, but only implicitly, during the personal rule.

These similar assessments of smoke's cause led to similar prescriptions for improvement, though here again Evelyn's vision was more detailed and more clearly articulated than anything produced earlier. If the problem was smoke from a few specific manufacturing houses, then for both Evelyn and the Caroline regime the obvious response was to remove them. During the 1630s leading brewers were instructed to remove to 'remote places' that would not damage the court.[26] Evelyn knew the perfect remote place: 'five

[22] *Ibid.*, 137–8.

[23] Cavert, 'Industrial Coal Consumption'. Londoners burned something close to 1 ton of coal per capita, and Westminster's population exceeding 40,000 by mid-century. Merritt, *Social World*, 262.

[24] Cf. Hiltner, *What Else is Pastoral*, 107, which makes unsupportable claims about what 'we' know regarding the relative importance of industrial and domestic fuel consumption.

[25] *Writings of John Evelyn*, 138. Evelyn's volcanic language has recently been studied by Toby Travis, '"Belching Forth Their Sooty Jaws": John Evelyn's Vision of a "Volcanic" City', *London Journal* 39 (March 2014), 1–20.

[26] TNA PC 2/43, 239 and Chapter 4 above. Parliament also considered a bill in 1657 that would have moved lime and brick kilns to five miles beyond London. John Rutt, ed. *Diary of Thomas Burton* (1828), 221; *JHC VII*, 532, 554.

or six miles distant from London below the River of Thames ... or at the least so far as to stand behind that Promontory jetting out and securing Greenwich from the pestilential air of Plumstead Marshes'.[27] London's industrialists would thus have been relocated to the future sight of the royal arsenal at Woolwich. Greenwich Palace, whose redesign Evelyn was soon to discuss with the king, would be protected, as too would Evelyn's own house and garden just to the west at Deptford.[28] The inconvenience to brewers would be trivial, Evelyn argued, but the benefits to the city substantial. Besides the reduction in smoke, large riverside warehouses could be converted into 'noble houses for use and pleasure'. The removal of industry, then, was part of a plan, quite literally, to gentrify London. The rhetoric of civic and metropolitan honour was central to Evelyn, as it had been for the breweries bill, but with a crucial difference. That bill described the need to preserve a clean capital for the royal family and the 'nobles and other most eminent persons' living there.[29] Evelyn agreed that London should be a space fit for noble and eminent persons, but his vision was one of transformation rather than preservation.

This stress on transformation and improvement also drew substantially on other pre-Restoration currents. Removing industry from London was only part of Evelyn's plan, to be complemented by a positive programme for managing the nature of urban air. This would be achieved by a series of 'square plots of fields of twenty, thirty, and forty acres or more', filled with aromatic 'fragrant and odiferous' flowering shrubs, flower beds, herbs, and 'blossom-bearing grain'. 'By which means', Evelyn wrote,

the air and winds perpetually fanned from so many circling and encompassing hedges, fragrant shrubs, trees, and flowers ... not only all that did approach the region, which is properly designed to be flowery, but even the whole city, would be sensible of the sweet and ravishing varieties of the perfumes.

These were not, then, merely to be pleasant little gardens for temporary diversion and retreat, but rather permanent and substantial instruments of 'ornament, profit, and security'. Though he listed several varieties of flowers and herbs that would be most suited to his purposes, the scale of Evelyn's project was not that of a garden bed but rather of a 'field'.[30] In addition to flowers and herbs they would include beans, peas, hops, and grains 'marketable at London' as well as sheep and cattle. To facilitate such plantings Evelyn hoped that further expansion of the 'tenements, poor and nasty cottages near the city' would be prohibited as they had been before the Civil War.[31] This prohibition of further sprawl was not framed as a measure against

[27] *Writings of John Evelyn*, 148.   [28] Evelyn, *Diary*, III, 313.
[29] PA HL/PO/JO/10/4/1.   [30] *Writings of John Evelyn*, 154–5.
[31] Barnes, 'Prerogative and Environmental Control'.

growth or development, and Evelyn would surely have vigorously denied Mendeville's later assertion that a 'flourishing' city could not be 'cleanly'.[32] On the contrary, Evelyn's proposal claimed to unite 'ornament, profit, and security'. Or, as he claimed in the tract's penultimate sentence, such gardens and fields would benefit London for 'health, profit, and beauty'.[33]

Evelyn therefore did not aim to conserve an existing cityscape, but to create something entirely new; his goal was not protection but improvement. Indeed he used this keyword of the mid-seventeenth-century for the heading of part III 'an offer at the improvement and melioration of the air of London, by way of plantation', and used it three further times in the section's 850 words. The improvement of London air should therefore be seen alongside the many other visions of increased 'health, profit, and beauty' that were characteristic of the men who sought to realize Francis Bacon's utopian vision of a natural philosophy grounded in empiricism and oriented towards utility. As Charles Webster's magisterial research as well as subsequent work has shown, Evelyn was a key member of these circles.[34] He, like other members of the 'Hartlib circle' and later founders of the Royal Society, advocated a program of experiment and research whose ultimate goal was to restore the Edenic knowledge of and (therefore) control over nature.[35]

But if the ideal of restoring a lost paradise of knowledge was one influence on Evelyn and his colleagues, so were the transformative powers associated with alchemy. As several recent commentators have argued, the alchemical notion of transmutation was central to the thought of early modern natural philosophy and medicine in general, and to several of Evelyn's long-time collaborators in particular.[36] Contemporary political arithmetic and economics, as well as philosophy and medicine, were understood as processes of transformation. Evelyn, whom Hartlib described as a 'good chymist', shared with his Hartlibian colleagues a belief that air could be altered by the emissions of plants, and that it, in turn, could then alter the bodies and hence minds of those who breathe it.[37] Evelyn also shared with his contemporaries the

---

[32] Mandeville, *Fable of the Bees*, 22. Discussed further in Chapter 8.

[33] *Writings of John Evelyn*, 156.

[34] Charles Webster, *The Great Instauration: Science, Medicine, and Reform 1620–1660* (New York, 1975); Mark Greengrass et al., *Samuel Hartlib and the Universal Reformation*.

[35] This strand in Evelyn's career is stressed in Steven Pincus, 'John Evelyn: Revolutionary', in Frances Harris and Michael Hunter, eds. *John Evelyn and His Milieu* (2003), 185–220. For improvement more broadly Paul Warde, 'The Idea of Improvement, c.1520–1700', in *Custom, Improvement, and the Landscape*, Richard Hoyle, ed. (Farnham, 2011), 127–48; Paul Slack, *The Invention of Improvement: Information and Material Progress in Seventeenth-Century England* (Oxford, 2015).

[36] See the discussion in Chapter 6, as well as McCormick, *William Petty*; Carl Wennerlind, *Casualties of Credit: The English Financial Revolution, 1620–1720* (Cambridge, MA, 2011).

[37] HP 28/2/66B-67A, 28/2/71B; John Beale to Evelyn, 30 September 1659, 67/22/3B-4A; *Writings of John Evelyn*, 141.

belief that such alterations would be wrought by improvements to nature that would be no less profitable than healthy. Just as plantations might render the wastes of Ireland, America, or England at once healthier and more productive, so Evelyn advocated

improving those plantations ... in the moist, depressed, and marshy ground about the town, to the culture and production of such things, as upon every gentle emission through the air, should so perfume the adjacent places with their breaths, as if, by a certain charm or innocent magic, they were transferred to that part of Arabia styled the happy.[38]

London's wastes would thereby be transformed into agents of transformation. They would become carefully managed natural spaces, changing the urban environmental and, thereby, improving the health, happiness, prosperity, and behaviour of the capital's inhabitants. *Fumifugium*, in sum, might be seen as at once the culmination of the Caroline politicization of London air and a typical mid-century celebration of the virtues of improvement, a Restoration attempt to re-establish order through innovation and transformation.

III. 'TO HAVE SOMETHING DONE': THE RECEPTION OF *FUMIFUGIUM*

The foregoing has assessed *Fumifugium* as a text and a set of ideas, but it should also be seen as a political tool, a tactic to achieve real results. There is a strong argument that, from this practical perspective, *Fumifugium* was (and was always likely to be) a failure. Enforcing a prohibition on new buildings would always have been difficult politically and administratively, creating new gardens at royal expense would have required significant support from an inadequately funded crown, prohibiting urban industry would surely have encountered significant opposition, and it was simply untrue that non-industrial coal consumption was trivial. Evelyn's proposals certainly were too ambitious to be realized and were based on some unwarranted assumptions that would have caused significant problems even if his proposals had, miraculously, come to fruition. But it does not follow that they were wholly insignificant; indeed, Evelyn's ideas, as set out most extensively in *Fumifugium*, were known and did have influence. The practical results of their influence, however, were restricted in a way that must have disappointed Evelyn. Such restrictions were characteristic of the ways that, during the century following *Fumifugium*'s publication, responses to urban air pollution moved steadily from reform to adaption.

---

[38] *Writings of John Evelyn*, 130. See also 154.

Understanding the nature of *Fumifugium*'s failure to create a new London requires attention to its reception, and this begins with its mode of publication and circulation. The only study of *Fumifugium*'s publishers notes that they, like Evelyn, were actively pursuing court patronage during this period. While this is taken to mean that Evelyn's text is therefore unimportant (as if any book, regardless of its content, might make a good gift for a king), it actually reveals something crucial about how Evelyn's proposal and his method of publication worked together.[39] Evelyn was certainly not opposed to publication, and indeed published often both before and after the Restoration, producing original as well as translated works. When he sought to teach English householders the virtues of classical architecture, or to show landowners how to reforest their estates, commercial publication offered a welcome opportunity to maximize his audience.[40] But when he counselled the king on a sensitive matter of public policy there was less of an incentive to focus on publication. Both Evelyn's diary and the tract's title page state that the work was published by command of the king himself, framing the publication less as a commercial endeavour or an attempt to sway public opinion than as an officially authorized record of royal support for Evelyn's project.[41]

Fumifugium, therefore, was not and was never intended to be a best seller. But it was circulated, both in manuscript and in print, within a limited circle. Its first and most important audience, of course, was King Charles II himself. Evelyn seems likely to have composed his text during the opening months of 1661, probably not long before the date of 1 May given at the end of the address to the reader.[42] This was only a week after the king's coronation ceremony and after Evelyn presented him in person with a panegyric, enjoying 'a magnificent feast' afterwards with the leading figures of the court.[43]

---

[39] Peter Denton, 'Puffs of Smoke'.
[40] John Evelyn, *Parallel of the Antient Architecture* (1664); *Sylva* (1664).
[41] Denton notes that the royal imprimatur was removed at some stage during *Fumifugium*'s printing, but his discussion is inconclusive. Mark Jenner has recently highlighted the stakes of publication for a regime highly suspicious of the public sphere, showing how a proposal for rebuilding London after the Great Fire earned its author imprisonment not because of its content, but because its discussion was excessively public. Mark S. R. Jenner, 'Print, Publics, and the Rebuilding of London: The Presumptuous Proposal of Valentine Knight', unpublished paper presented at the Institute for Historical Research, British History in the Seventeenth Century Seminar, 30 January 2014.
[42] The address to the king refers to a statement by the king's sister that is probably datable to 21 December 1660. Though the address to the reader is dated 1 May, it discusses in the present tense the parliament which did not sit until 8 May, suggesting perhaps that Evelyn did write the address on the date stated, when parliament's sitting was imminent. *Writings of John Evelyn*, 129 n. 3, 132.
[43] Evelyn recorded meeting Lord Chancellor Clarendon, Ormonde, and other unnamed 'noble men' and 'great persons'. Evelyn, *Diary*, III, 284.

A few weeks later Evelyn spoke with the king again, this time about the Royal Society, and there were several other trips to court during the summer of 1661, during one of which Charles expressed some kind of 'resentment' towards the nuisance of urban smoke.[44] When he 'presented' *Fumifugium* to Charles on 13 September it is clear that Evelyn had enjoyed access and might therefore have hoped for influence. The king 'was pleased I should publish it by special command, being much pleased with it'.[45] It was published accordingly, but the only subsequent reference to it in his diary reveals that Evelyn was more concerned with its efficacy as a policy document than its success in the bookshops.

This reference occurs in the following January, when Evelyn recorded receiving from Sir Peter Ball 'a draft of an act against the nuisance of the smoke of London, to be reformed by removing several trades which are the cause of it, and endanger the health of the King and his people'. Ball had been attorney to the Queen Mother, Henrietta Maria, since before the civil war, and was thus a personal link between Evelyn's project and the complaints of King Charles I and his wife during the 1630s. Evelyn could have known him through his contacts with the exiled court during the 1640s and 1650s, but a more immediate association is through Ball's son, William, a prominent astronomer and fellow member of the Royal Society.[46] Despite his service to the queen Ball carried little influence at court. He was, however, a leading lawyer and former MP, so it would seem likely that Evelyn approached him to render *Fumifugium*'s learned and courtly prose into parliamentary language. He seems to have done so, but the diary's 'It was to have been offered to the Parliament, as his Majesty commanded' is the last we hear of *Fumifugium* as an active project.[47]

While it neither transformed public opinion nor led to sweeping legislation, *Fumifugium* was not without influence. A month after first receiving *Fumifugium*, and ten days after he talked through it with Evelyn in his yacht, Charles is recorded to have declared at a meeting of the Privy Council 'that it was his express will and pleasure that hence forward no other fuel should be burned in the Council Chamber but charcoal only'. The order was recorded by Sir Richard Browne, the council's newly reinstated clerk and Evelyn's father-in-law.[48] If Charles could not yet have a capital free of industry and surrounded by flowerbeds, he could at least keep parts of his own palace free of that mineral coal which, Evelyn had

---

[44] *Ibid.*, III, 288. Also 285, 287, 290, 293. 'Resentment' in BL Add MS 78,298, f. 113-v, see note 61 below.
[45] Evelyn, *Diary*, III, 297.
[46] *Ibid.*, 286; Joseph Gross, 'Ball, William'; Wilfred Prest, 'Ball, Sir Peter', *ODNB*.
[47] *Diary of John Evelyn* III, 310.
[48] TNA PC 2/55, 402; TNA LS 13/170, 112.

argued, had such pernicious effects on clothes, pictures, and wall hangings.[49] Charcoal or wood, rather than sea coal, were the usual fuel preferred for the front rooms of houses and the meeting rooms of parish vestries, so there was nothing surprising in the content of Charles's order. Its timing, however, strongly suggests that he had taken Evelyn's ideas – whether he received them by reading *Fumifugium* or in conversation with its author – to heart.

The most notable echo of *Fumifugium* came somewhat later, in the aftermath of the Great Fire of London. In early September of 1666 most of London within the ancient walls burned, including St Paul's Cathedral, the Royal Exchange, scores of churches, company halls, public structures, and over 12,000 houses.[50] Despite his wartime duties overseeing care for the sick and wounded and prisoners, Evelyn presented Charles with a 'a plot for a new City, with a discourse on it' within a week of the fire, even as the ruins still smoldered.[51] The discourse was *Londinium Redivivum*, which again urged many of the same measures proposed in *Fumifugium*.[52] Again Evelyn stressed London's potential to be 'improved', the complementarity of 'commerce and intercourse, cheerfulness and state', as well as the desirability of classical regularity both in building and in urban design.[53] These goals were shared with other proposals, including Wren's famous plan as well as several others.[54] As in *Fumifugium*, however, Evelyn's vision also included the management of the air through a new social and industrial geography meant to reform 'the hellish smoke of the town'. Large warehouses were therefore to be removed to the Southwark side of the river, while 'the necessary evils, the brewhouses, bake houses, dyers, salt, soap, and sugar-boilers, chandlers, hat-makers, slaughter houses, and some sort of fishmongers, etc.' should be removed into the eastern or northern suburbs.[55] Such improvements, again as in *Fumifugium*, would soon render the capital 'fitter for commerce, apter for government, sweeter for health, more glorious for beauty'.[56] And once again Evelyn recorded a discussion of 'near a full hour', this time with the king, queen, and duke of York.[57] That same day, the 13th of September, the king issued a 'declaration' endorsing many aspects of Evelyn's proposal. Most notably,

---

[49] *Writings*, 138.
[50] Stephen Porter, *The Great Fire of London* (Stroud, Glouc., 1996), 70.
[51] *Diary of John Evelyn*, III 462–3. For Evelyn's wartime work in the days following the fire, *ibid.*, 457–8, and Darley, *Evelyn*, 192–203.
[52] *Londinium Redivivum or London Restored not to its pristine, but to far greater Beauty Commodiousness and Magnificence*, in *Writings*, 335–45.
[53] *Writings*, 337, 339.   [54] *Ibid.*, 341–2; Porter, *Great Fire*, ch. 4.
[55] *Writings*, 339, 341.   [56] *Ibid.*, 345.
[57] *Diary of John Evelyn*, III 463.

That there shall be a fair key or wharf on all the river side, that no house shall be erected within so many foot of the river as shall be within a few days declared ... nor shall there be in those buildings which shall be erected next the river, which we desire may be fair structures, for the ornament of the city, any houses to be inhabited by brewers, or dyers, or sugar bakers, which trades by their continual smokes contributed very much to the unhealthiness of the adjacent places; but We require the Lord Major and Aldermen of London upon a full consideration and weighing all conveniences and inconveniences that can be foreseen, to propose such a place as may be fit for all those trades which are carried on by smoke to inhabit together, or at least several places for the several quarters of the town for those occupations ... it being Our purpose that they who exercise those necessary professions, shall be in all respects as well provided for and encouraged as ever they have been, and undergo as little prejudice as may be by being less inconvenient to their neighbours.[58]

This, it seems very likely, was composed with Evelyn's words still ringing in the king's ears. The 'necessary professions' of trades 'carried on by smoke' are to be sequestered into new areas where they will be 'less inconvenient to their neighbours'. This was a small part of Evelyn's proposal in *London Redivivum*, but it was the centrepiece of *Fumifugium*. It therefore seems quite likely that some part Evelyn's hour-long conversation with the king, his wife, and his brother returned to this earlier project or to its basic proposals. If so, then Evelyn's ideas strongly pushed the king, at least during the immediate aftermath of the fire, towards a model of urban renewal that sought to create clean air through the same principles of order, regularity, and circulation that would also create an honourable city and prosperous, healthy, and loyal subjects.

This influence on those in power, however, was short-lived, as the immediate needs of a city still suffering from both plague and a naval war trumped longer-term planning. Parliament sought above all 'some speedy way of rebuilding the City' and the Rebuilding Act of 1667 established building codes but dropped the industrial zoning envisaged by Evelyn and the king.[59] One final echo of the hope that *Fumifugium* might provide a blueprint for London's renewal came from the pen of Evelyn's correspondent John Beale. A brilliant scholar and member of the Royal Society, Beale was, like Evelyn, particularly interested in the moral, spiritual, and medical aspects of improvement. Indeed, in a 1659 letter to Evelyn, Beale advocated using plants to correct the air, noting in particular that such work might provide 'a sweet and easy remedy against the corrosive smoke of their sea coal' and may therefore have been an important influence on *Fumifugium*'s proposals.[60] In

---

[58] *His Majesties Declaration to his City of London, Upon Occasion of the Calamity by the Lamentable Fire* (1666), 7. Draft in TNA SP 29/171/94.

[59] Caroline Robbins, ed. *The Diary of John Millward* (Cambridge, 1938), 9. On the rebuilding see Porter, *Great Fire*, ch. 5.

[60] HP 67/22/3B-4A, printed in Greengrass *et al.*, *Hartlib and Universal Reformation*, 357–64.

February of 1667, just as parliament was passing the Rebuilding Act, Beale wrote to Evelyn of the many plans to rebuild London. His own preferences derived much from the *Fumifugium* which he himself influenced: 'noble gardens to yield a pure air, fit to take the place of lungs', and brewers to be banished 'up and down the Thames'. To underline the moral powers of such gardens, Beale suggested they would be defended from brewers by a 'flaming sword' of a few miles, thus likening smoky industry with original sin and the renewed London with the Garden of Eden.[61]

Beale's letter shows that, though *Fumifugium* was primarily intended to explain a set of policy suggestions to the king and his advisors, it was also a work of humanistic natural philosophy and as such was also circulated among friends and colleagues. On 13 September 1661, the day that Evelyn left his home in Deptford for the following day's presentation of the treatise at court, Evelyn sent a copy of *Fumifugium* to Robert Boyle. He called the pamphlet a 'trifle', and apologized for its indignant tone and the presumption of its dedication to the king. He further protested, 'Not that I believe what I have written should produce the desired effects; but to indulge my passion, and in hopes of obtaining a partial Reformation; if at least his Majesty pursue the resentment which lately expressed against this nuisance since this pamphlet was prepared.'[62] There is no direct evidence for the extent, if any, of its influence on Boyle, largely because air, its composition, and its relationship to health was so central to Boyle's work, but in at least one place he did consider that 'fires burning in our chimneys' emitted salts that contributed to make air 'so exceedingly compounded a body'.[63] In addition to Boyle, a drafted letter of 1688 shows that Samuel Pepys asked his long-time friend and colleague for a copy of 'the old smoky pamphlet', with which he was clearly familiar, though it is not clear if one was actually sent or not.[64] Royal Society fellow, MP, coal magnate, and friend of Evelyn Sir John Lowther read and took notes on *Fumifugium*.[65] Copies are to be found in the libraries of some other fellow natural philosophers, including Kenelm Digby and John Woodward.[66] An anonymous manuscript ballad lampooning the research of the Royal Society devoted considerable space to *Fumifugium*, treating it in four of the poem's twenty-eight stanzas, more than is devoted to any other philosopher or experiment.[67] Lines

---

[61] BL Add MS 78,312, f. 43.   [62] BL Add MS 78,298, f. 113-v.

[63] 'The General History of Air,' in Michael Hunter and Edward B. Davis, eds. *The Works of Robert Boyle* (2000), Vol. 12, 30–1.

[64] De la Bédoyère, ed. *Particular Friends*, 182.

[65] Sir John Lowther's general notebook, *c.*1676–80, Cumbria Record Office, D LONS/W1/32.

[66] *Bibliotheca Digbeiana, sive, Catalogus librorum in variis linguis editorum quos post Kenelmum Digbeium eruditiss* (1680), 126; *A Catalogue of the Library, Antiquities, & c. of the Late Learned Dr. Woodward* (1728), 141.

[67] Dorothy Stimson, ed. 'The Ballad of Gresham College', *Isis* 18 (July 1933), 103–17, esp. 115–6.

like '[l]et none at Fumifuge be scoffing, who heard at Church our Sunday's coughing' show an easy familiarity with the text and assume the same from the ballad's audience.[68] Yet despite these instances, Evelyn himself regretted in the early eighteenth century that his treatise had 'vanished', and in 1772 a new edition was justified it on the grounds that it was a 'very scarce tract'.[69] *Fumifugium*'s influence as a book, therefore, was indeed limited, yet it remains significant as a record of the views and projects that Evelyn described in person to some of England's leading natural philosophers, as well as its king and his closest advisors.

IV. REMOVING: THE ROYAL RETREAT FROM ENVIRONMENTAL REGULATION

Evelyn's influence on his king, it has been suggested, was significant, yet Charles II did not need Evelyn to convince him that London's smoke was excessive and undesirable. His father's campaign against Westminster brewers during the 1630s was justified, in part, through concern for the fragile health of young prince Charles and his siblings.[70] Authors like John Denham and William Davenant, who had mocked urban smoke in plays and poetry, had been among royal exiles and enjoyed royal favour at the Restoration.[71] The surgeon who faithfully attended young Charles during the civil war, and was later granted a place at court, described in print smoke's damage to scrofula patients.[72] The first instance of Charles's council taking any action against coal smoke in London, however, was prompted not by the king nor by his advisors, but by subjects. In March of 1664 a petition was submitted to the Privy Council from Sir James Austin, an inhabitant of the area of Southwark west of London Bridge, complaining that a new glass factory was being constructed which would 'by its continual raising of smoke be a great damage and annoyance as well to passengers, and those citizens of London who retire for the benefit of the air and their health in the summer time to their garden houses there'. This was answered by a counter-petition from the factory's builders who protested that a prohibition would ruin themselves and the many 'poor

---

[68] *Ibid.*, 115. Stimson's discussion of Joseph Glanvill's candidacy for authorship, as well as a copy of the poem in Henry Power's papers, suggest that the satire was intended for members, or well-informed fellow travellers, of the Royal Society.

[69] John Evelyn, *Fumifugium: Or, the Inconvenience of the Aer, and Smoake of London Dissipated* (1772), iii.

[70] TNA PC 2/43, 238–40.

[71] Denham, Davenant, and literary representations of urban smoke are examined in the following chapter.

[72] Wiseman, *Severall Chirurgicall Treatises* (1676), 255. For Wiseman's career, John Kirkup, 'Wiseman, Richard', *ODNB*.

families' whom they would employ, and that the claim of nuisance was invalid because the area already contained two other glass factories. The council used the seldom-enforced orders against new buildings to find in the petitioners favour, and the Surveyor General Denham was therefore ordered to prevent further building.[73]

A few months later the council considered a quite different case, one in which the king and his court were the aggrieved party. A brewer named John Breedon was called before the council and informed that 'his brewhouse situate so near unto his majesty's palace of Whitehall was found to be a great nuisance and prejudicial to the health of Their Majesties. And being thereupon required to find out some fit place more remote, and thither to transplant himself.'[74] This brewhouse, 'so near' to Whitehall on its north side, must have been the same one that Evelyn records in the opening sentence of *Fumifugium* as the source of the 'presumptuous smoke' that invaded the court. Breedon's house was on the current site of Northumberland Avenue, only 100 metres or so from Scotland Yard and just 300 from the Banqueting House itself.[75] Breedon's case was in some ways a return to the initiatives of the 1630s. Breedon, like his predecessors, was ordered to find a fitter place to conduct his trade, more 'remote' from court. But whereas the brewers of the 1630s resisted in the face of fines and arrest, Breedon immediately offered 'ready and dutiful compliance' even as he explained that the king's order would ruin his estate. Charles too showed a willingness to compromise not evident during the 1630s. 'Taking in good part' his deference, the king ordered Surveyor Denham to assist Breedon in finding a new location, and the Lord Treasurer and Chancellor of the Exchequer to 'repair and indemnify the prejudice and loss he may sustain by reason of this his remove'.[76]

In the end, nothing came of these orders, and the significance of the Breedon case lies in what it reveals about Charles II's departures from his father's methods. First, while Breedon's brewhouse was a long-standing annoyance to the court (having been included in suppressions of the 1630s), after 1664 Charles and his government preferred to try to limit the further spread of smoky industry rather than remove it from places where it was already established.[77] This can be seen in its sympathetic attitude to a petition in 1665 from

---

[73] TNA PC 2/57, 43, 63, 74.   [74] TNA PC 2/57, 188.

[75] The brewhouse was in Hartshorne Lane, probably quite near the present site of the Playhouse Theatre. For the area's general history G. H. Gater and E. P. Wheeler, eds. 'Northumberland Street', *Survey of London: Volume 18: St Martin-in-the-Fields II: The Strand*, British History Online, www.british-history.ac.uk/report.aspx?compid=68268.

[76] TNA PC 2/57, 188, 196, 214.

[77] On 1 August 1637 a brewhouse in Hartshorne Lane was cited by Comptroller of the Household Henry Vane (the elder) as 'an annoyance to the king's house'. GL MS 5445/16. Its existence is shown on a map of 1621, WAC Acc. 1815.

inhabitants of Lambeth who opposed a large new glass house. The council at first prohibited its builders from proceeding, but the complaint was dropped once it emerged as the property of the Duke of Buckingham, who had recently received a royal patent to develop a domestic glass industry.[78] Complaints regarding another new brewhouse in the new developments north of Piccadilly were scrutinized (but rejected) in 1672, while in 1676 the Earl of Norwich was granted permission to expand Arundel House gardens on condition that he build no brewhouses or other buildings.[79] In 1675 the council restricted water supplies to some Westminster brewhouses located away from the river, but the year before it granted a prominent brewer expanded water rights.[80] Finally, during Charles II's reign some lawsuits continued to be brought in the king's name against smoky trades, but there is no evidence that these proceeded from the king or his council, nor that these were part of a concerted campaign as such suits had been under his father.[81] Altogether, then, the measures pursued by the restored king do show a perception that coal smoke was undesirable and objectionable, but measures against it were limited, local, sporadic, and rearguard.

This may be in large part because Charles was not interested, as his father had been, in developing Westminster into an appropriately regal capital. The exception to this came late in 1664, just a few months after the Breedon case, when Charles was planning to rebuild Whitehall in the neoclassical style of the Banqueting House.[82] But thereafter he lavished his architectural ambitions on projects outside of London. 'It was at Windsor', according to Kevin Sharpe, 'that Charles showcased Restoration kingship.'[83] Simon Thurley has stressed different projects, a new palace at Greenwich as 'the ceremonial gateway to his kingdom' and, at the end of the reign, a great new house at Winchester.[84] In the wake of the tumult of the Exclusion Crisis of

---

[78] TNA PC 2/58, 44–5, 59, 70; *CSPD 1663–4*, 186–7.

[79] TNA PC 2/63, 166, 171; F. H. W. Sheppard, ed., *Survey of London: Volumes 31 and 32: St James Westminster, Part 2* (1963), 118–9. The Arundel House grant reads, 'And that no wharfes, brewhouses, dye houses, or any other buildings whatsoever be erected thereupon, with other provisos and clauses, as are usual in grants of like nature', TNA PC 2/65, 287.

[80] TNA PC 2/64, 384, 389; 2/64, 280, 289; 2/66, 158.

[81] Bartholomew Shower, *The Second Part of the Reports of Cases and Special Arguments, Argued and Adjudged in the Court of King's Bench* (1720), 327. These are also discussed in Chapter 5.

[82] Simon Thurley, *The Whitehall Palace Plan of 1670* (1998), 6; Kerry Downes, 'Wren and Whitehall in 1664', *The Burlington Magazine* Vol. 113, n. 815 (February 1971), 89–93. There also seem to have been some plans for a new Whitehall during the summer of 1661, the period during which the king met with Evelyn and pronounced his own agreement with them. Thurley, *Whitehall*, 5; Evelyn to Boyle, 13 September 1661, BL Add MS 78,298, f. 113-v.

[83] Kevin Sharpe, *Rebranding Rule: The Restoration and Revolution Monarchy, 1660–1714* (New Haven, 2013), 119–21.

[84] Simon Thurley, 'A Country Seat fit for a King', in Eveline Cruickshanks, ed. *The Stuart Courts*, (Thrupp, 2000), 214–39, quote at 226.

1679–81 it made sense for Charles to avoid London, which had been so central in the 'great rebellion' against his father, in pursuit of his own Versailles. While it was probably the capital's politics that was most responsible for pushing the king away from the capital, one of the effects of this was that the city's environment would not be a priority for Charles II in the ways that Evelyn advised or as it had been for Charles I.

Charles II's successors continued this retreat from regulating the urban environment. The only whiff of governmental regulation of smoky air during James II's short reign is a relation decades later in a volume describing itself as 'reports, lies, and stories' from the years before the Glorious Revolution of 1688–9. James's notoriously high-handed Lord Chancellor George Jeffreys is said to have summoned a Westminster brewer, appropriately named Mr England, and 'rattled him severely because the smoke of the brewhouse offends his lordship and all the company that comes to his lordship's house' in King's Street, Whitehall. England protested and refused to remove, so Jeffreys threatened to ruin him. But England, 'some say', was a prominent dissenter and was crucial to James's plans to abolish the Test and Penal Acts and was therefore Jeffreys's 'match'.[85]

While John England was indeed a prominent Westminster brewer in King's Street, there is no evidence that he played such a role for James, nor any other evidence of the confrontation with Jeffreys.[86] The story has interest, however, because of the positions ascribed to the Chancellor and the brewer. Jeffreys announces that the smoke is a nuisance which damages the 'king's business', a consideration that ought to trump all others. Other sites in Westminster were available which would be more fit for such a trade. England counters with his legal rights: the brewhouse is ancient and is his property – indeed is 'his right'. Even if the confrontation itself is an invention, its ingredients were the familiar conflicting claims of property right versus public/royal interest that had been at stake in Jones's case and Charles I's campaign against the Westminster brewers. These opposing legal principles retained their potential to create conflict, even if there is no real evidence that any royal policy under James II actually pursued such an agenda.

After 1689 the king's position regarding urban smoke was fairly clear: William III was an asthmatic and therefore had limited tolerance for

---

[85] *Revolution Politicks: Being a Compleat Collection of all the Reports, Lyes, and Stories Which were the Fore-runners of the Great Revolution in 1688* (1733), part IV, 18.

[86] For England as the king's brewer after the revolution, Guy Miege *The New State Of England Under Their Majesties K. William and Q. Mary* (1693), 390. Scott Sowerby's authoritative research on James's dissenting allies has found no role for John England (email communication, March 2013). This story could, perhaps conflate England, William and Mary's brewer, with Michael Arnold, the royal brewer and MP under James, who was in fact an ally of the king.

London's urban environment, but his response was to avoid, rather than to reform, his capital. This can be explained in large part by his priorities lying elsewhere, above all with the demands of warfare. Perhaps such reform would have appeared excessively difficult in any case, as London's continual growth made Westminster a very different place in the 1690s than it had been in the 1630s. In any case William chose Hampton Court, over twenty miles upriver from Westminster, and Kensington House, then beyond the westernmost expansion of the metropolis, as his primary residences.[87] William's successor, Anne, followed her brother-in-law in preferring Kensington Palace as her primary residence, and her uncle Charles II in making extensive use of Windsor.[88] The last Stuart monarch chose, as her predecessors had done since the Restoration, to escape rather than reform their capital.

Anne's reign did, however, witness one final flicker of interest in policing industrial coal smoke in Westminster. In December of 1706 a bill was introduced in the House of Commons to ban 'every glasshouse, brewhouse, melting house, dyeing house, or other workhouse consuming a great[er] quantity of sea coal than is usual for a dwelling' from being built within one mile of the Banqueting House in the heart of Whitehall.[89] Though the bill followed earlier royal claims by stressing Westminster's status as a seat of the monarch, it also invoked the city's role as the site of parliament, administration, archives, the courts, and as the home of England's landed elite more broadly. For all of these reasons 'the air of the said City ought to be preserved as clear, wholesome, and free from annoyances as is possible'. The bill was introduced by a Whig, Sir Henry Dutton Colt, who served as MP for Westminster, and by a Tory, William Lowndes, an important treasury official who lived near the west front of Westminster Abbey and was thus both an interested party in local environmental policing and a likely source for the bill's reference to Westminster's role as home to royal records and officers.[90]

If passed this would have been the strongest governmental initiative to limit coal smoke of the early modern period, but it was less ambitious in scope than the other failed parliamentary bill regarding urban coal smoke. The reach of Charles I's Breweries Bill of 1624 would have included much

---

[87] Béat Louis de Muralt, *The Customs and Character of the English and French Nations* (1728), 76–7. Also Abel Boyer, *The History of King William the Third* (1702–3), 84; Tony Claydon, *William III and the Godly Revolution* (Cambridge, 1996), 72–3.

[88] Gregg, *Queen Anne*, 48, 51, 76, 136.

[89] BL Stowe MSS 597 105v–106.

[90] The bill was read twice but failed a vote to send it to committee, after which it disappears from the record. *JHC* XV, 207, 220, 230, 238; E. Cruickshanks, S. Handley and D. W. Hayton, eds. *The House of Commons, 1690–1715* (Cambridge, 2002), III, 654–60; IV, 674–82, 996–9; V, 150–1.

of the City of London as well as Westminster, and would have required existing brewhouses to abandon fossil fuels. Colt and Lowndes' bill only embraced a one-mile radius from the centre of Whitehall, a distance that would have included all of Whitehall, St James's, Westminster, and Lambeth palaces, plus the residential districts around St James Square, Covent Garden, the Strand, the Temple, and the developed areas of the city of Westminster, but that would have excluded the City of London, Southwark, the outlying fringes of Westminster, and the entire northern and eastern suburbs.[91] Within this radius, moreover, only new manufacturing houses were prohibited, as structures erected before 1705 were permitted to remain. It was thus not an attempt to improve London air as the 1624 bill and Charles I's subsequent actions were, but rather, like most of Charles II's measures against breweries and glass houses, a rearguard action intended to prevent Westminster air from becoming any smokier than it already was. On the other hand, while the 1624 bill confined its attention to breweries, the 1706 bill recognized the diversifying industrial landscape of the capital. Glasshouses, in particular, were increasingly visible sources of smoke. One final difference between the 1706 measure and that of 1624 is that while Charles's bill reflected a personal commitment to reform Westminster's environment, there is no evidence that the 1706 bill reflected any similar royal policy. It does not seem to have been pushed by the court, nor is there any reason to suspect that it was connected to any broader ministerial agendas that could compare with Charles I's politics of space in the 1630s.

V. CONCLUSION: THE VIEW FROM 1772

When *Fumifugium* was reprinted in 1772 its editor emphasized that the intervening century had brought 'some alterations that appear worthy of notice'. The first of these was the growth of smoky industry: while Evelyn had stressed brewers, dyers, and soap boilers,

since his time we have a great increase of glass houses, foundries, and sugar bakers to add to the place catalogue, at the head of which must be placed the fire engines of the waterworks at London Bridge and York Buildings, which (whilst they are working) leave the astonished spectator at a loss to determine whether they do not tend to poison and destroy more of the inhabitants by their smoke and stench that they are able to supply with their water.[92]

---

[91] This radius may be roughly traced on a modern map by drawing lines connecting Blackfriars, Holborn, Oxford Circus, Hyde Park Corner, Pimlico, and Southwark rail stations.

[92] Evelyn, *Fumifugium* (1772), iii–iv. Geoffrey Keynes, *John Evelyn: A Study in Bibliophily and a Bibliography of his Writings* (Cambridge, 1937), 89, attributed this introduction to the antiquary Samuel Pegge.

These new steam pumps, in addition to expanded glass, sugar, and metal industries, rendered London even more industrial than it had been at the Restoration. The effect of this, the editor claimed, was that smoke had become even worse. While Evelyn complained that trees in urban gardens bore no fruit, 'the complaint at this time would be ... that they would not bear even leaves.[93]

Perhaps the most dramatic change between 1661 and 1772, however, was the political rather than the environmental context. For while Evelyn was 'unfortunate in recommending a work of such consequence to so negligent and dissipated a Patron [as Charles II]: The Editor is encouraged by a more promising appearance of success'. This success, he concludes, is possible because of the 'present public-spirited and active magistrates'.[94] It is to them, therefore, that he 'submit[s]' 'with deference' the 'hints' in Evelyn's book. For all that he shared with Evelyn, therefore, including a perception that smoke was dangerous and ugly and that it was caused by industry and therefore might be reformed by good policy, he did not share the perception that the monarchy was a likely agency of such transformation. The 1772 editor does not mention the king at all, even though only a few years before a proposal for a new royal palace argued that George III required a new house free from industrial smoke.[95] Nor does he mention parliament, despite the omnipotence that the legislature than attained by the eighteenth century and despite the importance afforded to it in Evelyn's own proposal. By 1772, it seems, policing the London environment had become a local affair, one of some national significance perhaps but not driven or directed by the centre. Westminster remained the capital, but its 'public-spirited and active magistrates' would govern it in ways not fundamentally different from any other city or town. The idea that London would be a showcase for the power and honour of the royal (or national) government was no longer even worth addressing. This seems the culmination of the long trend by which the monarchs of England, and then Britain, chose to remove themselves from urban smoke rather than remove it from themselves. As the following chapters describe, so too did their subjects.

---

[93] *Fumifugium* (1772), iv.   [94] *Ibid.*, iv, viii.
[95] John Gwynn, *London and Westminster Improved, Illustrated by Plans* (1765), 11.

$-12-$

Representations: coal smoke as urban life

1. MOTTEUX'S JEST

In autumn of 1696 London theatre-goers flocked to a new comedy, *Love's a Jest*, written by the Huguenot exile Peter Anthony Motteux. The witty protagonists court, woo, trick, avoid, rebuff, and succumb to each other in ways that tried to offer audiences welcome escape from the hard times of the 1690s. Indeed, England's monetary crisis is the immediate occasion for several characters' arrival in the country. In the first scene Samuel Gaymood announces that he has arrived at his brother's home in Hertfordshire to avoid his creditors in London. The brother, Sir Thomas Gaymood, has no use for London life but begs to hear of the capital's news and novelties. 'I hate it as a smoky kitchen', he says, 'but good things come out of it sometimes, though 'tis bad living in it'. Their friend Railmore refuses to provide any news, listing instead the many absurd, immoral, and hypocritical habits of citizens and courtiers. Immediately following this other characters enter and the story begins. The action of *Love's a Jest* is in part an investigation of the familiar proposition that London really is the exclusive home of vice. Of course bumpkins are not virtuous either, so the play constantly raises and then undermines the association between urbane life and sin. This agenda is announced at the play's very opening when Gaymood orders his servant to sing him a song, a song in which smoky air is a central part of what makes London distinctive:

> Slaves to London, I'll deceive you,
> For the Country now I leave you;
> Who can drink, and not be mad,
> Wine so dear, and yet so bad?
> So much noise, and air so smoky,
> That to stun, this to choke ye;
> Men so selfish, false, and rude;
> Nymphs so young, and yet so lewd.
> If we play, we're sure of losing;
> If we love, our doom we're choosing;
> At the play-house tedious sport,

> Cant i'th city, cringe at court.
> Dirty streets, and dirtier bullies,
> Jolting coaches, whores and cullies;
> Knaves and coxcombs ev'ry where,
> Who that's wise would tarry here?
> Quiet, harmless country pleasure
> Shall at home engross my leisure:
> Farewell, London, I'll repair
> To my native country air.'[1]

Here is an entirely typical list of urban ills, in which smoky air, noise, and dirty streets gesture towards a series of moral and practical problems. The immigrant, born in the purity of his 'native country air', is choked by coal smoke even as he is subjected to dangers and frustrations everywhere. Smoky London is a city where temptations abound and all lead to trouble; the 'love' of young 'nymphs' leads to the 'doom' of disease, 'play' leads inevitably to loss, and even that particularly urban temptation the theatre is, in fact, 'tedious'. Coal smoke thus joins the material and moral problems of city life. The only 'wise' response, therefore, is to return home to the country. Motteux's final couplets, however, upend this virtuous conclusion:

> And leave all thy plagues behind me;
> But at home my wife will find me:
> Oh! ye Gods 'tis ten times worse,
> London is the milder curse.

This misogynistic joke nicely obliterates the pious denunciation of urban vice, revealing the narrator to be, along with his listeners, yet another 'slave to London'.

Motteux's song and its role in introducing an ambivalent and playful exploration of urban immorality is a revealing example of the way that coal smoke became an increasingly common and legible symbol of urban life during the century or so after 1650. Smoke provided a useful sign of London and Londoners because it was particular to the metropolis and was immediately sensible to visitors and residents alike. It therefore came to stand in for a range of attitudes and practices often considered specifically urban. Some, like Sir Thomas Gaymood, denounced both London and its air confidently and apparently without reservation. But even Sir Thomas admitted that the great 'smoky kitchen' had its uses. The audience of *Love's Jest* may well have laughed in agreement with its suggestion that, compared to rural simplicity and boredom, urban life, even with its dirty environment, was indeed the milder curse. Through such representations London smoke

---

[1] Peter Anthony Motteux, *Love's a Jest. A Comedy: Acted at the New Theatre in Little-Lincoln's Inn-Fields* (1696), 2. Its commercial success is discussed in Motteux's preface to the reader and in Montague Summers, ed. *Roscius Anglicanus by John Downes* (no pub. info.), 44, 253.

was addressed as a problem, but one that to be properly understood needed to be set in context of urban life as a whole. Such contextualization in some cases fatally undermined the moral dichotomy between the virtuous country and sinful city. Many authors, to be sure, did denounce smoke as earnestly as they denounced the rest of the town's vices. Others toyed with this convention, mocked it or questioned it. Coal smoke's meanings, therefore, were not simple or straightforward, a flexibility which made it all the more useful as a sign. However it was deployed and for whatever creative purposes, throughout the seventeenth and eighteenth centuries London coal smoke became ever more entrenched as a synecdoche of urban space and the lifestyles it contained. London, through such uses, became nearly synonymous with its smoky air.

II. 'SELFISH, FALSE, AND RUDE': STAGING SMOKE AS URBAN LUXURY

There was no more vital medium through which to represent early modern urban life than the theatre. Beginning with the city comedy tradition of the late Elizabethan stage and culminating in the witty and rakish town comedies of the Restoration and Augustan periods, London life and manners were endlessly modelled, mocked, reproduced, and parodied before audiences of substantial social depth. While some of these plays have acquired places in the classical Anglophone repertoire most are usually found derivative and frivolous, light entertainment comparable to television sitcoms. But the many hundreds of plays performed before audiences in early modern London, whether of enduring literary quality or not, did crucial cultural work, not least by helping to define the nature and meaning of urban manners and social relations. Within this theatre of urban life the idea of coal smoke played a crucial role. It served as a metaphor for all that was particular and prominent about the metropolis, and as such was at once easily understood and yet also flexible and multivalent. It could be used to invoke almost any aspect of Britain's capital. But 'Slaves to London' was typical in using smoky air to stress all that was ugly and regrettable about the city. Under London's 'air so smoky', men were 'selfish, false, and rude', women lewd, and all the pleasures of the fashionable city led only to tedium and doom.

Coal smoke began its career as a sign of urban life early in the seventeenth century. Much Jacobean city comedy turned on the distinction (real or aspirational) between the mercantile citizens and the leisured, landed inhabitants of the capital. In George Chapman, Ben Jonson, and John Marston's *Eastward Ho* (1605) coal smoke is invoked to define this social boundary by a character notable primarily for her desire to transcend it. Gertrude, the daughter of a rich goldsmith, wants to marry Sir Petronel Flash, whom she believes to be a gentleman, and thereby to become a lady and her parents' social superior. Escaping her family, their middling status, and her native environment combine in her plea to her betrothed: 'Sweet knight, as soon as ever we are married, take me to thy mercy out of this miserable City! Presently carry me out of the scent of Newcastle coal, and the hearing of Bow-bell.' The smell of coal smoke, for her, defines London as effectively as the famous bells of St Mary-le-Bow. To leave the city's atmosphere is also to trade the mercantile world for the chivalrous values which she dreams will be upheld by her 'sweet knight'.[2]

Coal smoke invoked the unity between city space, civic society, and the base mercantile values and manners against which landed gentlemen struggled to distinguish themselves. This can also be seen at work in a play by Henry Glapthorne, *The Lady Mother* (c.1635).[3] In it a young Londoner named Crackby is frustrated in his attempts to debauch a country maid, which leads him to complain that the country has 'metamorphised' him for the worse. He then reveals, however, that what he values about the city are its absurdly high prices and pretentions, while rural honesty provides too little scope to 'dissemble'. These moral inversions are gestured towards in his complaint that the fields and country air are unhealthy to him compared to 'the wholesome smell of sea coal'.[4] For Gertrude, coal smoke represented a civic past to be cast off and transcended, while Crackby is lost away from the city where his sharp practices are successful and appreciated. In both cases, however, smoke was a useful shorthand for the civic society and culture which were the plays' subjects.

Among the authors for whom urban smoke was most consistently important was the poet, dramatist, and theatre manager William Davenant, whose career encompassed both the Caroline court of the 1630s and the re-establishment of commercial drama after the Restoration. In *The Wits* a young man protests surprise at meeting the old knight Sir Morglay Thwack 'mongst so much smoke, diseases, law and noise'. It is immediately revealed that Sir Morglay is not out of place in the immoral city, as he plans to make his fortune by cheating citizens of as much money as possible.[5] *The Wits* was immediately successful and remained so decades later. King Charles himself seems to have discussed details of the text with the Master of Revels, and the play was performed at court before the king and queen in January,

[2] George Chapman, Ben Jonson, and John Marston, *Eastward Ho*, in James Knowles, ed. *The Roaring Girl and Other City Comedies* (Oxford, 2001), 76–7. See also the later adaptation by Nahum Tate, *Cuckolds-haven* (1685), 12.
[3] For dating and attribution see Julie Sanders, 'Glapthorne, Henry', *ODNB*.
[4] Henry Glapthorne, *The Lady Mother* (Oxford, 1958), 9.
[5] *The Works of Sir William Davenant* (1673, reissued New York, 1968), II, 173.

1634 – just a few months after three of Westminster's leading brewers were ordered by the Privy Council not to brew with smoky coal when the royal family were at home. The following year another play of Davenant's, *Newes from Plimouth*, performed at the Globe on Bankside, described how a typical, grasping Londoner would acquire for himself another's ancient country estate. The citizen would leave his 'smoky habitation' to insinuate himself into landed society, culminating at his death when 'their sinful sea coal dust mingled with the ashes of your warlike ancestors'.[6] In his dramatic work of the 1630s, as well as his poetry, Davenant used smoke to invoke mercantile, bourgeois London, a city of trade and greed.[7] He did so, furthermore, while closely associated with many of the leading figures at court – including the royal couple themselves – who were responsible for developing and enforcing the campaign of the 1630s against industrial smoke in and around Westminster.[8]

Davenant survived military service in the royalist cause and time in the Tower to remain involved in theatre, even during the rule of Oliver Cromwell when commercial playhouses were suppressed. In 1656 he created an entertainment that has since assumed a central role in English dramatic history, a staged performance of 'declamations and music, after the manner of the ancients'. This, the *First Dayes Entertainment at Rutland House*, included speeches in which personifications of London and Paris contested superiority, each describing in detail the shortcomings of the other. Paris informed London that, 'the plentiful exercise of your chimneys makes up that canopy of smoke which covers your city'. Smoke, Paris repeated more than once, was an essential component to London's overall lack of decorum and beauty, joining narrow streets and irregular architecture to produce a hopelessly jumbled urban space in which 'every private man hath authority, for his own profit, to smoke up a magistrate'.[9] The performance then concluded with a song, said to have been written by the leading composer Charles Coleman, which proclaimed 'London is smothered with sulph'rous fires; Still she wears a black hood and cloak, of sea coal smoke, As if she mourn'd for brewers and dyers.'[10] Davenant's career reached its pinnacle after the Restoration, and his work, including those associating London society and manners with its smoky environment, remained popular. *The*

[6] Davenant, *Works* II, (2nd page), 2.
[7] See also reference to 'the mists of sea coal smoke' in a poem written to Queen Henrietta Maria. A. M. Gibbs, ed. *Sir William Davenant: The Shorter Poems, and the Songs from the Plays and Masques* (Oxford, 1972), 47.
[8] For his associations with the king, queen, the earl of Dorset, Endymion Porter, and Inigo Jones, see Mary Edmond, *Rare Sir William Davenant: Poet Laureate, Playwright, Civil War General, Restoration Theatre Manager* (Manchester, 1987), ch. 4–5. For their role policing smoke see Chapter 4 and Cavert, 'Environmental Policy'.
[9] Davenant, *Works*, I, 353, 352.   [10] *Ibid.*, 358; Edmond, *Rare*, 126.

Wits was revived and played before King Charles II and other members of the royal family in August of 1661, just a few weeks before Evelyn discussed his *Fumifugium* with the king.[11] Samuel Pepys recorded himself to be 'too much in love with plays' to attend to business after seeing *The Wits* for the second time in three days, and in 1664 he read London's and Paris's speeches 'with great mirth'.[12]

The leading comedic playwright after Davenant's death in 1668 was probably Thomas Shadwell, and Shadwell's 1672 work *Epsom Wells* contained perhaps the period's most influential comedic denunciation of London's social, moral, and environmental corruption. In it Clodpate is a country justice, notable above all for his strong preference for the virtuous, healthy, and self-consciously English life of his country estate rather than the habits and environment of England's 'Sodom', 'that damned town of London'.[13] The play's action is outside of London yet it constantly looks towards the capital. Most of the characters are Londoners – either citizens or polite inhabitants of the 'town' – down to the spa for a visit, and the play continually stages a tension between the town-in-the-country and the country itself. Clodpate clarifies this tension through his ridiculously drawn distinction between London and its rural opposite; he is an anti-urbanite who constructs a wholly virtuous and healthful country through his denunciation of a wholly sinful and dangerous town. His town is one in which dirtiness and dishonesty are inextricably connected and repeatedly associated. 'That place of sin and sea coal', thunders Clodpate, breeds gout and venereal disease, 'pride, popery, folly, lust, prodigality, cheating knaves, and jilting whores'. Its 'beastly pleasures' are no better than its vices:

to sit up drunk till three a clock in the morning, rise at twelve, follow damned French fashions, get dressed to go to a damned play, choke yourselves afterwards with dust in Hyde Park or with sea coal in the town, flatter and fawn in the drawing room, keep your wench and turn away your wife.[14]

London's smoke, for Clodpate, is emblematic of a world turned monstrously upside-down, a place that exchanges marital chastity for whoredom, turns day to night, and prefers French to English customs.

This moral censoriousness, however, is ridiculous rather than damning. Clodpate's hatred for London derives not from a clear perception of the town as it really is but rather from near ignorance of the capital itself and from a failure to evaluate it reasonably. His anti-metropolitan prejudice is

---

[11] Edmond, *Rare*, 165–6; *Diary of John Evelyn* III, 295–7.
[12] Edmond, *Rare*, 50, 126, citing Pepys' diary for 7 February 1664.
[13] *Epsom Wells* in *The Complete Works of Thomas Shadwell*, Montague Summers, ed. (1968 ed.), II, 110.
[14] *Ibid.*, 111.

so excessive that in the play's third act a scheming woman, mistress Jilt, immediately wins his interest simply by declaiming against 'that villainous lewd town'.[15] She hires a fiddler to perform a song that begins 'Oh, how I abhor, The tumult and smoke of the town', before expanding upon the vain follies which it contains. This alone is sufficient evidence for Clodpate of her good sense and he proceeds quickly towards marriage, only discovering Jilt to be 'a Londoner, and consequently a strumpet' at the play's conclusion.[16]

Clodpate's position, of course, may not be the play's and clearly cannot be Shadwell's. Not only does the entire plot mock the country justice's pretensions but the witty Lucia rebuffs his courtship by informing him that she resolves to live in London because 'people do really live no where else; they breathe and move and have a kind of insipid, dull being; but there is no life but in London'.[17] Clodpate's image of smoky sin is countered with the claim that, even if it may fairly be called a 'stinking town', London's wit and sociability are life itself, a position to be examined in more detail at the end of this chapter. Here the point is that in Shadwell's enormously popular comedy – seen several times by King Charles II and regularly performed for decades after its first staging in 1672 – the Clodpate character firmly established the comedic trope of London as both dirty and dishonest, a place of 'sin and sea coal'.

After *Epsom Wells* playwrights could expect audiences to be familiar with coal smoke as a physical manifestation of urban sin, even if they chose to reframe and subvert that metaphor for their own purposes. The anonymous *Woman Turned Bully*, performed less than three years after *Epsom Wells* and also in the Dorset Garden theatre, staged a kind of moderate female version of Clodpate. Mrs Goodfield begins the play certain, according to her son, of London's debauchery: 'She believes this town spoils all young men that come to it; but for women, she's confident the very air of London meets 'em, and debauches 'em at Highgate.'[18] Like Clodpate her opinions are based on very limited, and dated, personal experience, but unlike Clodpate she soon arrives in the capital. She first finds her senses assaulted by noises and 'ugly smells', and, like Clodpate, contemplates the relationship between sin and sea coal. But unlike Clodpate she is open to the possibility that they may be dissociated, perhaps because her Derbyshire is more industrial than Clodpate's Sussex:

Your smells, here in London, are most wickedly bad: I have been told, the occasion is your sea coal; but I rather think 'tis the sins of your naughty city that makes it stink

[15] *Ibid.*, 138. [16] *Ibid.*, 177.
[17] *Ibid.*, 121. [18] Anon., *The Woman turn'd Bully* (1675), 5.

above ground. Had not you Londoners better come down to Darbyshire and fetch of our coal from the pits, which is (though I say it) a delicate fuel?[19]

Having discarded the association (which she has 'been told' by unnamed sources) between sin and smoke, Widow Goodfield then proceeds to change her views of urban morality more generally once she has acquired for herself a husband and allowed her daughter to do the same.

The association between urban smoke and urban sin continued into the 1690s when, as we have seen, Motteux's 'slaves to London' endured its 'air so smoky' and Sir Thomas Gaymood denounced it, in the Clodpate tradition, as 'a smoky kitchen'.[20] Mary Pix's *The Innocent Mistress* (1697) opens with two old friends examining whether London offers a 'dear epitome of pleasure' or merely 'noise and nonsense', and soon a woman announces her happy return from Jamaica to her 'native air ... the smoke of this dear town'. There she is finally at home because she can find those who understand what intrigues are necessary to 'deal with those deluders, men'.[21] In another contemporary play a lady rebuffs a suitor by recommending him to offer himself 'to some woman that never was within the smoke of London', since no urbanite could be wooed by mere 'sincerity'.[22] In yet another production from the 1690s, John Vanbrugh's *The Provok'd Wife*, two women imagine a world free of men. Lady Brute concludes that despite her terrible husband a world without men would be undesirable because it would be a world without sociability, display, and entertainment, all of which they understand to be grounded in the London landscape: 'adieu plays, we should be weary of seeing 'em ... Adieu Hyde Park, the dust would choke us ... Adieu St. James, walking would tire us ... Adieu London, the smoke would stifle us ... And adieu going to Church, for religion would ne'er prevail with us'.[23] And so into the eighteenth century, when the works of George Farquhar, Samuel Foote, Isaac Bickerstaffe, David Garrick, and others similarly used coal smoke to gesture towards city practices, habits, and morality.[24]

---

[19] *Ibid.*, 27.

[20] Motteux, *Love's a Jest*, 2. Gaymood gestures towards Clodpate too in his partiality for urban news, borrowing from Clodpate's love of gazzettes.

[21] Mary Pix, *The Innocent Mistress a Comedy* (1697), 1, 8.

[22] Thomas Dilke, *The Lover's Luck* (1696), 11; see also Anonymous [Robert Dodsley], *The Footman: An Opera.* (1732), 40.

[23] John Vanbrugh. *The Provok'd Wife a Comedy: As it is Acted at the New Theatre in Little Lincolns-Inn-Fields.* (1698), 38. See also (John Vanbrugh) Colley Cibber, *The Provok'd Husband, or, A Journey to London* (1728), 65.

[24] Shirley Strum Kenny, ed. *The Works of George Farquhar* (Oxford, 1988), I, 264 (*Sir Harry Wildair*); Paula R. Backscheider and Douglas Howard, eds. *The Plays of Samuel Foote* (New York, 1983), I, 'Epilogue' (*The Author*); II, 18 (*The Mayor of Garret*); II, 2 (*The Lyar*); David Garrick, *The Country Girl. A Comedy. (Altered from Wycherley)* (1766), 27; Peter A. Tasch, ed. *The Plays of Isaac Bickerstaffe* (New York, 1981), I, 68 (*Love in a Village*); II, 60 (*Lionel and Clarissa*). Also James Carlile, *The Fortune Hunters, or, Two Fools Well Met*

Often examined in explicitly gendered and sexualized terms, such plays repeated, manipulated, and sometimes undermined increasingly famil-iar associations between smoky London and a specifically urban moral-ity. Genre is crucial to interpreting how plays used urban air, as comedies could easily make use of smoke as a sign for urban self-interest and deceit in ways not available to tragedy or other morally elevated productions. Smoke was therefore a useful metaphor for Shadwell but not for Dryden, for Vanbrugh but not Congreve.[25] But if genre provided a comic space for smoke to represent urbanity, the physical setting of London itself was also crucial. Playhouses were situated within dense neighbourhoods where coal smoke was long-established as an aspect of daily life. Davenant's *News from Plymouth* was performed amidst the brewhouses and dyers of bankside, his *First Dayes Entertainment at Rutland House* across Aldersgate street from the Earl of Bridgewater's annoying neighbours, and Dorset Gardens nestled within a dense area between Bridewell and Whitefriars from which the Fourth Earl of Dorset was annoyed by brewers during the 1630s. The smoke that town comedies used to signify urban life was therefore at once a theatrical and textual convention, and yet also very much a real aspect of urban space.

III. 'QUIET, HARMLESS, COUNTRY PLEASURE': SMOKE AND THE POETICS OF RETREAT

Genre shaped how urban smoke offered a useful metaphor for poetry no less than it did for drama, but in a nearly opposite way. While theatrical references to smoke usually evoked urban duplicity in the service of comedy, poetic smoke signified urban immorality in much more ethical and didactic contexts. If town comedies used smoke to help create spaces in which vice and avarice, or alternately sociability and culture defined social relations, much pastoral poetry of the seventeenth and eighteenth century drew simi-lar portraits in order to celebrate the superior merits of retreat.

A recent study of Renaissance pastoral literature in England has argued that environmental change was central to the genre, but as something so fleeting and mutable that it could only be gestured towards rather than rep-resented.[26] For many poets, however, it was the city rather than the country

(1689), 67; Thomas Baker, *Hampstead Heath. A Comedy. As it was Acted at the Theatre Royal in Drury Lane* (1706), 20; Susannah Centlivre, *The Platonick Lady* (1707), 60; John Breval, *The Play is the Plot* (1718), 32; William Popple, *The Double Deceit, or A Cure for Jealousy* (1736), 62.

[25] Shadwell and Dryden's approaches to London are contrasted in Robert W. McHenry, 'Dryden and the "'Metropolis of Great Britain"' in W. Gerald Marshall, ed. *The Restoration Mind* (1997), 177–92.

[26] Hiltner, *What Else is Pastoral?*, 21.

that was inescapably mutable, and so the country became the place to which one must fly in order to contemplate, or it was the place in which one meditates with equanimity on a well-moderated rural life. London, of course, produced huge amounts of poetry throughout the period, and yet it became a commonplace that urban business and care, figured by smoky air, rendered the city unfit for true inspiration:

> Alas! Sir, London is no place for verse;
> Ingenious harmless thoughts, polite and terse:
> Our age admits not, we are wrapped in smoke;
> And sin, and business, which the muses choke.
> Those things in which true poesie takes pleasure,
> We here do want; tranquility and leisure.[27]

By the eighteenth century this had become commonplace: in the smoky city poetry was just impossible. Almost a century later a poem in the *London Magazine* repeated the claim that London life, with its business, hurry, and dirty environment, is a bar to true poetry: 'While you, my Florio, breathe untainted air, And to the Muse-inviting shades repair; Immers'd in smoke, stun'd with perpetual noise, In vain I strive to tune my harsh, hoarse voice.'[28] The obvious irony here is that these poets, living in the city, choose to affect an anti-urban stance. They are Londoners writing poems about the impossibility of poetry in London.

This is not, however, mere hypocrisy, but rather an attempt to achieve a certain kind of retreat through a dismissal of a certain kind of urbanity, both of which are preconditions for renewed virtue. The city, in this tradition, is repeatedly described so as to be rejected. A key foundational work here was John Denham's 'Cooper's Hill', first written on the eve of civil war in 1642.[29] Looking down on London, physically and morally, from 'Parnassus', his estate in Surrey, the Royalist Denham denounces his rebellious contemporaries from his moral and topographical high ground:

> ... raised above the tumult and the crowd
> I see the City in a thicker cloud
> Of business, than of smoke; where men like ants
> Toil to prevent imaginary wants;
> Yet all in vain, increasing with their store,

[27] Alexander Brome, *Songs and Other Poems* (1664), 185.

[28] *London Magazine* XIII (1743), 43. See also Mary Wortley Montagu, 'Now with fresh vigour Morn her Light Displays' (1729), in Robert Halsband and Isobel Grundy, eds. *Lady Mary Wortley Montagu Essays and Peoms and Simplicity, A Comedy* (Oxford, 1977), 252.

[29] Other contemporary poets also associated the smoky city with work and care. See Sir Richard Fanshawe, 'An Ode, upon occasion of His Majesties Proclamation in the Year 1630. Commanding the Gentry to reside upon their Estate in the Country' in H. J. C. Grierson and G. Bullough, eds. *The Oxford Book of Seventeenth Century Verse* (Oxford, 1938), 451; Francis Kinnaston, 'To Cynthia', in *Leoline and Sydanis* (1642), 137–8.

> Their vast desires, but make their wants the more.
> As food to unsound bodies, though it please
> The appetite, feeds only the disease;
> Where with like haste, though several ways they run:
> Some to undo, and some to be undone.[30]

This first version of 'Cooper's Hill' showed the possibilities of smoke as a metaphor for urban life, allowing the poet to elide London's unique smoke 'cloud' with its moral decadence, fruitless busyness, immoderate greed, diseased bodies, and – what follows from these – its rebellious politics. The immediate political context of 1642 is obviously crucial here, but Denham's more general claim that urban smoke figured insatiable and unreasonable desire remained relevant throughout the period.

This association of smoke with work and care, greed and excess was a central feature of poetry drawing on the tradition of Horace's 29th Ode of Book III. In Dryden's translation this begged the poet's friend to:

> Come away,
> Make haste, and leave thy business and thy care,
> No mortal interest can be worth thy stay...
> The smoke, and wealth and noise of Rome,
> And all the busy pageantry
> That wise men scorn, and fools adore:
> Come, give thy soul a loose, and taste the pleasures of the poor.

Horace then describes the moral and mental benefits of retreating into a simple and humble rural life. 'Happy the man, and happy he alone, He, who can call today his own', and such happiness is soon found to consist of calm acceptance of ill fortune.[31] Horace thus associates urban smoke with noise and wealth, as well as the vanity of urban politics and the vagaries of fortune, against which are ranged good, simple food and wine, moderation, and equanimity. Repetitions and elaborations on these themes, as shown in Maren-Sofie Røstvig's classic study *The Happy Man*, were a crucial strand of neoclassical literature between the Renaissance and Romanticism. Røstvig judged much of this literature to be mere 'abuse' of a noble tradition, 'virtual begging letters addressed to a patron' rather than true Horatian 'sober advice'.[32] But the point here is not to distinguish great from plodding literature, but to realize how fully the smokiness of London became embedded in

---

[30] Sir John Denham, *Cooper's Hill. A Poeme* (1642), 2–3. Denham revised his text repeatedly in response to the current political climate, removing the denunciation of London for the edition of 1655 when he had returned to Cromwellian England. W. H. Kelliher, 'Sir John Denham', *ODNB*.

[31] Translation of Horace, Ode 29. Paul Hammond, ed. *The Poems of John Dryden; Volume II 1682–1685* (1995), 370–1, 374.

[32] Maren-Sofie Røstvig, *The Happy Man: Studies in the Metamorphoses of a Classical Ideal. Volume II 1700–1760* (Oslo, 1971 ed.), 11.

206 The Smoke of London

sophisticated philosophical positions and literary conventions that associated it with an unvirtuous, and yet ultimately tolerable, urban world.

During the seventeenth century, Røstvig argued, the 'classical motif of the happiness of rural retirement had been virtually monopolised by Royalists and Anglicans'. No one, perhaps, was more fully a member of these circles nor as popular with readers as Abraham Cowley. In 'The Garden', a short work addressed to John Evelyn and so published both in Cowley's collected works and in later editions of Evelyn's *Sylva*, Cowley argued that the garden was God's first gift to mankind, even 'before a wife'. The city, by contrast, was first made by Cain, which reinforced the vanity and destructiveness of worldly ambitions.[33] Cowley's garden is thus a haven against excessive and inappropriate desires, perfectly tuned to delight our true natures. Against this stands the opposite environment, a place repugnant both to reason and to the senses, a place of gender inversion and unnatural creations:

> Who, that has reason, and his smell,
> Would not among roses and jasmin dwell,
> Rather than all his spirits choke
> With exhalations of dirt and smoke?
> And all th'uncleanness which does drown
> In pestilential clouds a pop'lous town?
> The earth it self breaths better perfumes here,
> Than all the female men or women there,
> Nor without cause about them bear.[34]

Cowley's 'populous town' is fundamentally disordered, a place of female men who perfume themselves while living under pestilential clouds, dirt, and smoke. Like in 'Cooper's Hill', London's smokiness is both a metaphor for and a physical creation of its fundamental moral flaws. For Denham these are primarily greed and disloyalty, for Cowley unreasoning ambition and pride. For Denham these are placed explicitly in London, for Cowley in a 'populous town' that unambiguously points to the metropolis where he was born and bred.

Throughout the later-seventeenth and eighteenth centuries these themes were widely adopted and adapted, appropriated or subverted. One of Edward Young's satires, published with *The Love of Fame* in 1728, argued that women were silly in specifically urban ways, associations quite familiar from the contemporary stage. But Young's poetry couples the denunciation of urban folly with a didactic celebration of the superior pleasures of retirement. 'Britannia's daughters', Young informs, are too fond of a series of unmistakably metropolitan temptations:

[33] *The Works of Mr. Abraham Cowley* (1668), 116; 'The Garden', in John Evelyn, *Sylva*, (4th ed., 1706).
[34] *Works of Cowley*, 117.

Thro' every sign of vanity they run;
Assemblies, parks, coarse feasts in city halls,
Lectures, and trials, plays, committees, balls,
Wells, Bedlams, executions, Smithfield scenes...
Taverns, exchanges, Bridewells, drawing-rooms
Installments, pillories, coronations, tombs.

Later in the same satire such exhaustive lists can more efficiently be gestured towards by summarizing the spaces and environments in which they flourish:

fresh air
(An odd effect!) gives vapours to the fair;
Green fields, and shady groves, and crystal springs,
And larks, and nightingales, are odious things;
But smoke, and dust, and noise, and crowds, delight,
And to be pressed to death transports her quite.

In the face of such dirty and depraved urban pleasures, Young extols the virtues of the country:

Is stormy life preferred to the serene?
Or is the public to the private scene?
Retired, we tread a smooth, and open way;
Thro' briars, and brambles in the world we stray.[35]

Here Horatian retirement becomes feminine retreat, with the smoky city carrying all the weight Cowley and Denham have assigned to it but also signifying a public-ness entirely inappropriate for women.[36]

Young, a clergyman, merges the classical motif of retreat with a Christian denunciation of 'the world', and this fusion of classical and Christian modes of anti-urban retirement was attractive for others as well.[37] For Røstvig no one could have been farther from the Royalist tradition of Abraham Cowley than the pious dissenting minister Isaac Watts.[38] In a remarkable elegy on the death of a friend, Watts imagined the peaceable retirement that they would have enjoyed at his new house in Stoke Newington, just a few miles north of the city. There they would have conversed of 'heavenly things', until sometimes their thoughts would

[35] Edward Young, *Love of Fame, The Universal Passion. In Seven Characteristical Satires* (4th ed., 1741), 84, 99.
[36] For the garden as the pre-eminent space of feminine retreat during the eighteenth century, Stephen Bending, *Green Retreats: Women, Gardens and Eighteenth-century Culture* (Cambridge, 2013).
[37] For the development of Christian renunciation of 'the world', see Peter Brown, *Through the Eye of a Needle: Wealth, the Fall of Rome, and the Making of Christianity in the West, 350–550* (Princeton, 2012).
[38] Røstvig, *Happy Man*, 15, 102–8.

> lower their lofty flight,
> Sink by degrees, and take a pleasing sight,
> A large round prospect of the spreading plain,
> The wealthy river, and his winding train,
> The smoky city, and the busy men.[39]

Thoughts are only 'lowered', however, in order to reject the 'insolent and proud' vanity of ambition and wealth, positioning the contemplative observers at a distance from such urban weakness. These vices, again, are specifically urban, and yet also intrinsically human, as Watts then proceeds to consider that 'man is a restless thing: still vain and wild'. The city, visible and legible from the suburban villa through its smokiness, is thus the embodiment of human sinfulness against which the Christian must be vigilant. The smoky city, one might conclude from Watts, was entirely consistent with a depraved and weak human nature. Watts' position is thus very close in some ways to Mandeville, who similarly saw urban dirt as the result of grasping human nature. Mandeville's radicalism was in his celebration of this nature, while Watts was much more conventional in his counsel to struggle against it.[40]

While Watts represented, for Røstvig, the appropriation of the *beatus ille* tradition into dissenting piety, Stephen Duck was the most eminent of a series of lower-class authors whose celebrations of rural retirement incorporated the 'wistfulness' of the poor.[41] Despite his personal familiarity with the realities of rural labour, Duck suffered, in Raymond Williams's phrase, 'social absorption' by his royal patrons and thereafter celebrated the moral superiority of the country in conventional terms quite similar to those of Cowley or Young.[42] 'Two young ladies leaving the country', for example, were advised to shun the 'smoke, coaches, fops, and carmen' of the city in favour of a 'blissful grove'.[43] Similarly, his adaptation of a translation by Cowley described – rather ridiculously – the modest and moderate desires of 'the swain of Bethnal Green'. Despite living on the periphery of London's eastern suburbs, the swain is presented as wholly alien to all the temptations and vices of the modern city, its court, rich foods, stock market, and imperial trade. He 'nor ever prov'd the smoky Town, But breath'd a purer air'.[44]

For Røstvig what was notable about Duck's work, and about other contemporary middle- and lower-class poets, was the 'alarmingly precise' way

---

[39] Isaac Watts, *Horæ lyricæ. Poems, Chiefly of the Lyric Kind* (1715 ed.), 295.

[40] For Mandeville see above, Chapter 8.

[41] Røstvig, *Happy Man*, 157, and 11, 152–61.

[42] Raymond Williams, *The Country and the City* (1973), 89.

[43] 'On Two Young Ladies leaving the Country' in *Poems on Several Occasions By the Reverend Mr. Stephen Duck.* (1753 ed.), 111–2.

[44] *Ibid.*, 221–3.

in which they expressed their desires for material happiness in 'a veritable orgy of pleasant day-dreaming' quite at odds with the detached spirit of their classical models.[45] Perhaps so, but despite this divergence these authors' approach to the urban environment's moral significance was utterly conventional. A 'Parson's Wish', for example, surprises Røstvig by requesting 'an income easy tax'd and clear, Around two hundred pounds a year', but his dream that this comfortable life would be 'exempt from city-smoke and strife' is more orthodox.[46] Much of this is not great poetry, but its very mediocrity illustrates how widely London's smoke had become accepted as a metaphor for vain, worldly greed and ambition, how accessible and applicable this tradition appeared.

This metaphor was a central aspect of a neoclassical genre, but the genre was capacious enough to provide room for divergent or even subversive uses. Rigorous moralists like Watts figured smoke as the world, with rural retreat therefore entailing the physical benefits of a healthy environment and the denunciation of greed, ambition, urban fashion, hurry, and mutability. Others recognized the difficulty of this project and mocked it. MP Soame Jenyns, for example, wrote a poem against worldly ambition that questioned the possibility of escaping from worldly ambition:

> Why shou'd we then to London run,
> And quit our chearful country fun
> For bus'ness, dirt, and smoke?
> Can we, by changing place, and air,
> Our selves get rid of, or our care;
> In troth tis all a joke.[47]

Others showed no great concern to renounce the city, presenting its pleasures as inevitably associated with its troubles, both of which ought to be taken in moderation. The poet Gilbert West, according to Samuel Johnson, inscribed a poem extolling this model of alternating rural retreat and urban pleasure onto his country villa itself:

> Not wrapt in smoky London's sulphurous clouds,
> And not far distant, stands my rural cot:
> Neither obnoxious to intruding crowds,
> Nor for the good and friendly too remote.
> And when too much repose brings on the spleen,
> Or the gay city's idle pleasures cloy;
> Swift as my changing wish, I change the scene;
> And now the country, now the town enjoy.[48]

---

[45] Røstvig, *Happy Man*, II, 159.

[46] 'The Parson's Wish', in *The London Magazine*, Volume 11, October, 1742, 506–7. See also 'The Annual Recess' in *The London Magazine*, Volume 3, December 1734, 661.

[47] Soame Jenyns, *Poems, By* * * * * (1752), 135.

[48] Gilbert West, 'Imitation of Ausonius, "Ad Villam"' in Samuel Johnson, *Works of the English Poets* (1790), LVII 324.

Such poetry, then, figured London's smoky environment as a physical embodiment of a variety of urban perils. It derived from, or gestured towards, or could signify: greed and ambition, busyness and care, immoderate passions of both body and mind, fashion and mutability, inconstancy and the passage of time, inverted gender roles and sexual excess, political disloyalty and love of the crowd. Against this, or rather *through* such constructions, poets imagined a rural anti-type characterized by Christian and philosophical moderation, subdued passions, divine benevolence visibly manifested in the natural creation, political virtue, stable social relationships, and healthy bodies. This was a very powerful ideology during the seventeenth and eighteenth centuries, and yet it should not be taken to derive from a straightforward denial of the city. On the contrary, such poetry objected too much to urban pleasures that were increasingly seen as irresistible.

IV. 'THE MILDER CURSE': CHOOSING LONDON

It is an obvious observation, though still a crucial one, that most of the authors of the preceding paeans to rural simplicity can be accused of hypocritically denouncing an urban life that they themselves had no intention of avoiding.[49] Many authors, moreover, discovered the benefits of retreat only after suffering political reverses, and then forgot them again when their fortunes improved. In 1691 the former Archbishop of Canterbury William Sancroft, who twenty-five years before had responded to the Earl of Bridgewater's complaint regarding smoky industry in the Barbican, at last embraced the benefits of 'sweet air, and quiet ... much to be preferred to the smoke and noise of London'. Sancroft wrote these words from enforced retirement, just twelve months after being deprived of his office and leaving London for his native Suffolk. His career had spanned almost fifty years, most of it in London, and included a prominent role in appropriating coal taxes for the rebuilding of St Paul's Cathedral.[50] His embrace of smokeless rural air, therefore, might be read as another 'compensatory myth', comparable to a twenty-first-century politician who discovers a desire to spend more time with their grandchildren precisely when their position becomes untenable.[51] While we do not need to doubt that such figures genuinely did appreciate peace and health, it is still clear that the rejection of worldly ambition and the smoky city in which it flourished was an enforced position, something imposed, in Sancroft's case, upon a man who had spent a lifetime helping London to function as an ecclesiastical and political capital.

---

[49] Thomas, *Man and the Natural World*, 251.
[50] William Sancroft, *Familiar Letters of Dr. William Sancroft, Late Lord Archbishop of Canterbury* (1757), 10; R. A. P. J. Beddard, 'Sancroft, William' *ODNB*.
[51] See also Thomas, *Man and the Natural World*, 252, drawing heavily on Røstvig, *Happy Man*.

For many others, moreover, the language of rural retirement described in the poetry analysed above, as well as in the many prose sources which this chapter has not discussed, was deployed to celebrate a very temporary and very circumscribed retirement from the city. Gilbert West's villa in Kent celebrated the suburban dream of access both to city and country, *hortus et* (rather than *in*) urbe. In 1710 the tremendously popular periodical *The Tatler* described a kind of rural retirement, introduced with a quotation from Horace, that was explicitly temporary, even fleeting. The narrator walked from London into a suspiciously flowery and rural countryside, 'which formed the pleasantest scene in the world to one who had passed a whole winter in noise and smoke'.[52] He finds himself to be, like (Londoner) John Milton's Satan, 'one who long in populous city pent, Where houses thick and sewers annoy the air, Forth issuing on a summer's morn, to breathe Among the pleasant villages and farms'.[53] This is a remarkable appropriation of the moment when Satan first encountered Eve in *Paradise Lost*. The hidden Satan gazes on Eve, innocent and vulnerable as she tends to her 'delicious' garden; 'much he the place admired, the person more'. For Milton the passage from hell and the company of demons into Eden and the company of Eve is best explained through the experience of a city-dweller taking a walk in the country. *The Tatler* quotes the simile without acknowledging the referent; it positions the narrator as being like Satan stalking Eve in the garden, and yet proceeds as if this ramble amongst the flowers is innocent and pleasant. The narrator enjoys his philosophical walk, returning home musing on the beauty of nature and the 'bounty of providence' rather than less elevated city business.

Whatever the tensions in the *Tatler*'s choice of allusions, the crucial point here is that this exploration of the classical motifs of retirement and retreat begins and ends in the capital itself. Both in *The Tatler* and in numerous other texts, smoky London was described as something from which one might escape, but to which one would also return. For those writers urging withdrawal, the city was associated so powerfully with human passions that retreat becomes all the more desirable and beautiful because it is so nearly impossible. As one author put it towards the end of the eighteenth century:

methinks I hear the good man say, though my business requires me to live among the sons of industry and the daughters of dissipation, though the greatest part of my time is spent in the labourious avocations of an active tradesman, or the narrow

---

[52] The actual environs of London, while undoubtedly greener than the city, were also home to intensively managed farms and market gardens as well as workshops, factories, and garbage heaps. Charles Jenner, 'The Poet', turns on this tension between the literary convention of the beautiful countryside and the sordid realities of London's outskirts. *Town Eclogues* (1772).

[53] *The Tatler*, no. 218. 31 August 1710; John D. Jump, ed. *The Complete English Poems of John Milton* (New York, 1964), 274 (*Paradise Lost* IX, 445–51).

confines of a retail shop, yet I must confess myself pleased with the expectation, and amused with the tranquil prospect of ere long retiring from the tumultuous abodes of the smoky and crowded town, into the peaceful dwelling of a country retreat.[54]

Smoky London thus becomes the world, a place which the Christian would hope to escape, but from which his less philosophical neighbours neither could nor would.

The possibility that London's inhabitants would choose the smoky city's industry and dissipation over the countryside was fundamental to poetry as well as to comedic theatre, though in opposing ways. If would-be Horaces counselled their readers to shun smoke and noise, comedies played on the question of whether this was actually good advice. In *Epsom Wells* the bumpkin Clodpate denounces London's 'sin and sea coal', but the play itself rewards the rakish urbanites, rather than Clodpate, with good marriages. The victim of Clodpate's courtship, Lucia, speaks for many comedic characters when she informs him that 'there is no life but in London'.[55]

In some ways contemporary anti-urban poetry makes the identical point, associating the city with busyness, care, work, public life, and sociability. If it is true that 'almost everything which anybody *does* in the countryside is taboo' in seventeenth-century rural landscape poetry, one explanation for this may be that such countrysides are imagined in opposition to a city in which everyone is constantly doing.[56] For both poetry and drama, then, smoky London was often figured as the place of action, of industry both in the sense of industriousness and in the sense of production. Much writing about urban smoke was therefore always also about these visions of urban life itself. In literature the only reasonable response to the dirty city was to come away into the country, and, as the next chapter examines, lived experience and social practice drew heavily on this ideal of retreat from urban smoke and all that it represented.

---

[54] 'On Solitude and Retirement', *The London Magazine* 49 (March 1780), 112.
[55] *Works of Shadwell*, II, 121.
[56] James Turner, *The Politics of Landscape* (Oxford, 1979), 165, quoted in Thomas, *Man and the Natural World*, 251.

$-13-$

Movements: avoiding the smoky city

I. BLUESTOCKING AIRS

The women represented in the poetry and drama analysed in the last chapter were inordinately fond of the smoky city, with its silly fashions and the superficial excitements of crowds, business, and the theatre. Real women, among them some of the best-educated and most intellectually ambitious figures of the mid-eighteenth century, were well aware of such gendered tropes and even described their own relationship with London through them. Bluestocking Elizabeth Carter wrote in 1779 to her friend, the writer and leading *salon* hostess Elizabeth Montagu, asking her to take care of her health and to rest 'after all your fatigues, in the shades of Sandleford'. This referred to Montagu's luxurious country house, built from the profits of an enormous estate which included several coal mines around Newcastle. Such wealth allowed Montagu to act as England's leading patron for female intellectuals. It was appropriate, then, for Carter's letter to Montagu to echo Horace's address to Maecenas, in which the hot Roman summer prompts a retreat from the smoke, wealth, and noise of Rome (*fumum et opes strepitumque Romae*). 'The shades of Sandleford', Carter continued, 'as a shelter from the dog-star, which must rage with great fury among crowds and smoke and brick-houses.'[1] In a letter of the following year Carter returned to the same associations, suggesting that the busy building activity around Montagu's new mansion in Portman Square was a mortal threat to her health. This specific annoyance then merges into the familiar association between London's general smoke and noise and the 'business' which is their cause. Against this danger Carter 'rejoices' once Montagu has escaped into pastoral retreat: 'fresh air and tranquility, to basking on the lawn, or sitting under the shade of your groves at Sandleford'.[2] In such letters, Carter was

---

[1] Montagu Pennington, ed. *Letters from Mrs. Elizabeth Carter, to Mrs. Montagu, Between the Years 1755 and 1800. Chiefly Upon Literary and Moral Subjects* (1817) III, 105. A footnote by Pennington, Carter's godson and student, quotes the line from Horace.
[2] *Ibid.*, 137

capable of depicting London, in the well-established pastoral mode, as dirty and dangerous, smoky and busy, as the city from which the poet counsels retreat.

Carter's letters, and many others like them, were both more playful and more serious than poetic representations of the smoky metropolis. They were more playful because, like some poetry and much comedy, they used this familiar language selectively, alternately endorsing and subverting its implications. In the world of the pastoral the smoke of London was only something to shun and to escape, but as Stephen Bending has recently argued such conventions were highly problematic for women who wished to retain sociability, to achieve retreat without becoming hermits.[3] The smoke of London, and the urban intellectual, moral, and emotional community which it contained, could therefore also be presented as an object of desire by an author stuck in a quiet, boring, and lonely country. Carter wrote to Montagu in 1765, for example, that 'in the brightest splendor of summer suns, and amidst all the beauty of smiling prospects' of her beloved home on the Kentish coast, 'I look forward with joy to the dark days of January and the smoke of London.'[4] This is a conscious inversion of the normal language, deriving in part from the problematic fact that London was not always actually a place of busyness, hurry, and sociability. On the contrary, the seasonal rhythm of elite residence in the country during the summer and town during the winter meant that for Carter 'the smoke of London' could signal either the pleasures of society or the pain of 'solitude'.[5] Montagu, similarly, wrote to the Duchess of Portland that the country would certainly be preferable to the 'sin and sea coal' of London, but only because the 'amiable society as Mrs. Delany and Miss Hamilton' brought the intellectual and social advantages of the town into the country.[6]

Nor was the distinction between the healthy country and the dangerous city any more stable than the ideal of rural retreat. Carter wrote in 1764 that the 'stifling' smoke of London was indeed a 'positive evil', but that the weather in the country was not always preferable. In Kent, she announced, they were 'alternately scorching and freezing, by a hot sun, and a cold north-east wind'.[7] The air at Sandleford, which Carter had praised in

---

[3] Bending, *Green Retreats*.

[4] Pennington, ed., *Letters from Elizabeth Carter*, I, 270.

[5] Carter to Susanna Highmore, 8 June 1748, Gwen Hampshire, ed., *Elizabeth Carter, 1717–1806: An Edition of Some Unpublished Letters* (Cranbury, NJ, 2005), 136; *A Series of Letters Between Mrs. Elizabeth Carter and Miss Catherine Talbot from the Year 1740 to 1770. To Which Are Added, Letters from Mrs. Elizabeth Carter to Mrs. Vesey, Between the Years 1763 and 1767*. 4 vols. (1809), IV, 66, 364.

[6] Montagu to the Duchess of Portland, 6 December 1783 *HMC Longleat*, I, 351; See also *Letters Between Carter and Talbot*, I, 251.

[7] Pennington, ed., *Letters from Elizabeth Carter*, I, 223.

letters written during the summer, was found one October to be too damp compared to the preferable 'warm air' of her house near Berkeley Square. In such correspondence authors like Montagu and Carter were capable of drawing on the conventional, didactic language of the pastoral in order to express highly ambivalent, even flatly contradictory, attitudes towards urban life. London's smoke, and all that it represented, was therefore for Carter alternately an object of longing and an unpleasant annoyance, either a healthier or a more dangerous alternative to the country.

These varying and playful uses of the urban atmosphere as metaphor were highly literary, and yet were also entirely serious and entirely real. Carter and Montagu were not building castles in the air or idly contemplating rural happiness amidst urban care. On the contrary, the pastoral ideal of retreat, however ambivalent and unstable, was a central model through which these Bluestocking women, along with many other Londoners from a range of social classes, understood their real movements into and out of the metropolis. The classicizing ideal of temporary, annual, and socially limited retreat from the dirty city influenced behaviour as well as poetry, real estate transactions as well as landscape painting. Representations of urban smoke, therefore, were not free-floating cultural productions; it was through and against such representations that real people formulated, questioned, and understood their behaviours, their movements into, out of, and around the city.

This final chapter describes such movements because they, along with the languages and ideas used to describe and understand them, were the most common and important responses to air pollution in early modern London. Whereas many historians have found modern reactions to pollution to be defined only by the creation of environmental movements, this focus on everyday behaviour suggests as alternative way to understand the place of urban smoke in early modern society. We have seen throughout the previous chapters that there were strong reasons for disliking smoke – aesthetic, political, and medical – and that very many people did so. But we have also seen how their responses to London's smoke were partial approaches to the problem, limited by political culture and political contingency, legal procedures, scientific models of knowledge, and the distinctions inherent in an aggressively hierarchical society. No response, therefore, was able to dislodge the powerful and positive associations that coal consumption came to carry, nor to achieve the kinds of political victories associated with late-twentieth-century environmentalism.

And yet, the absence of modern environmental politics is not the absence of environmental concern. If early modern Londoners never successfully reformed their smoky air, they did develop ways of adapting to it. The most important of these tactics was to find some practical version of the pastoral ideal, to shake off what *The Spectator* called 'the smoke and gallantries ...

sin and sea coal' of the town for the country and its 'breezes, shades, flowers, meadows, and purling streams'.[8] Such absolute distinctions between
city and country have been shown to be falsifications of social realities, but
achieving some version of this ideal, however attenuated and limited, nevertheless did influence the movements of real people in and around London in
the seventeenth and eighteenth centuries.[9]

II. 'SMOKE AND BRICKHOUSES': ELITES AND THE ENVIRONMENT OF THE WEST END

Beginning in the middle decades of the seventeenth century, London's West
End became something new in England: an extensive residential district
from which industry was largely excluded and whose inhabitants were
largely social elites. As we have seen in Chapter 6, privately negotiated leases
between landlord and tenant were the primary means through which these
new developments excluded smoky, noxious, or loud industries. They did
so because the aristocrats and gentry who increasingly came to spend significant parts of the year in the West End wanted spaces fit for polite urban
living, clean and airy and as free as possible from the fumes, smokes, and
people that they considered nuisances.

   Good air, therefore, was an important part of what made the West End a
unique district, a set of neighbourhoods in which elites from across England
could meet, observe, emulate, quarrel with, and marry each other. Such air
was generally seen as both natural and man-made. Nature provided for
westerly winds so that, as William Petty observed, 'blowing near ¾ of the
year from the west, the dwellings of the West End are so much the more free
from the fumes, steams, and stinks of the whole easterly pile; which where
seacoal is burnt is a great matter'.[10] One of Charles II's surgeons agreed that
western suburbs enjoyed air that was both naturally superior and also free
of dangerous coal smoke.[11] By the early eighteenth century publications like
John Macky's *Journey Through England* (1714) and *The Foreigner's Guide*
(1729) informed readers which West End squares enjoyed 'open free air',
'clear air', and 'the best air about London'.[12]

---

[8] Donald F. Bond, ed. *The Spectator* (Oxford, 1965), IV, 390.
[9] The classic discussion of the falsehoods and obfuscations within the city/country distinction
  is Williams, *The Country and the City* (1973); useful recent developments include Bending,
  Green Retreats; Elizabeth McKellar, *Landscapes of London: The City, The Country and the
  Suburbs, 1660–1840* (New Haven, 2013).
[10] Hull, ed. *Economic Writings*, 41.
[11] Wiseman, *Severall Chirurgicall Treatises* (1676), 255.
[12] John Macky, *A Journey through England, In Familiar Letters from a Gentleman here to
  his Friend Abroad* (1714), 121; *The Foreigner's Guide*, 122, 32, referring to Bloomsbury,
  Cavendish and St James Squares, respectively.

By the eighteenth century the distinction between the central City and the West End's 'town' could be expressed in mutually reinforcing environmental and social terms. Moneyed citizens trying to escape urban space, therefore, were sometimes equated with social climbing. So in 1755 Edward Moore's periodical *The World* mocked what it saw as the pervasive disease of social emulation, the processes of competitive consumption that meant that 'every tradesman is a merchant, every merchant is a gentleman, and every gentleman one of the noblesse'. For the past fifty years, it claimed, commercial wealth had blurred social boundaries in ways that defied what should have been a precisely mappable urban social and cultural geography: 'Hair has curled as genteelly on one side of Temple Bar as on the other, and hoops have grown to as prodigious a magnitude in the foggy air of Cheapside as in the purer regions of Grosvenor Square and Hill Street.'[13] Catherine Talbot stressed the opposite point about her lodgings in the Deanery of St Paul's Cathedral in which she was surrounded by smoke and considered herself in the antipodes of the 'beau monde'.[14] For such authors smoke meant the commercial, bustling, bourgeois city, while the landed leisured town was similarly defined through its 'purer' air.[15]

Moving westward therefore meant avoiding smoke and achieving the fashionable town, but as Petty realized this was inevitably a temporary solution because the target of the urban frontier steadily moved outwards.[16] As the western suburbs grew, the desire of many elites to live on its periphery led to certain kinds of problems. The heavy smoke of brick burning, which has been cited by Keith Thomas and Emily Cockayne as typical examples of urban industry, was in fact a complaint specific to those living at the very edge of the urban zone.[17] The extremely wealthy people whose townhouses were annoyed by brick kilns would not have endured that nuisance had they not lived on the border of urbanization, near to the open fields which provided

---

[13] Adam Fitz-Adam (Edward Moore), *The World*, Number 125, 22 May 1755, (Dublin, 1755–7), III, 105, 107.

[14] Catherine Talbot to Elizabeth Carter, 8 June 1751, 14 March 1752, and 23 September 1759, *Letters Between Carter and Talbot*, II, 33, 68, 297; Margaret Heathcote to Catherine Talbot, 18 December 1754, BLARS L30/21/4/3. It is noteworthy that despite Talbot's claim that the city and the beau monde were different worlds, Heathcote's own family were among the most spectacular examples of the kinds of social mobility satirized in *The World*. Her husband's grandfather was Sir Gilbert Heathcote, one of the early modern period's richest London merchants, and her father was Philip Yorke, First Earl of Hardwicke, a prominent lawyer and leading ally of Whig governments.

[15] Hannah Greig, *Beau Monde: The Fashionable Society in Georgian London* (Oxford, 2013), 10–11.

[16] Such expansion was not quite constant, however. For a survey of eighteenth-century cycles of boom and bust in London construction, see White, *London in the Eighteenth Century*, ch. 1–2.

[17] Thomas, *Man and the Natural World*, 245; Cockayne, *Hubbub*, 207–8.

the earth and the undeveloped plots that the bricks would help develop. The Duke of Chandos, for example, complained in 1741 that he was 'every night poisoned with the brick kilns and the other abominate smells which infect these parts'.[18] 'These parts' were not the London metropolis in general, and were certainly not the heavily urbanized City, but rather the still-unfinished Cavendish Square, immediately north of which lay unbuilt suburbia. A few years earlier the inhabitants of the area just east of Berkeley Square, who included a Duke, two future Earls, and an MP among others, brought a suit complaining that the smoke of nearby brick-burning would 'make the air very unwholesome and spoil the houses and furniture of plaintiffs who were near neighbours'.[19] As in Chandos's case, this brick burning was not, as Thomas implies, 'carried on in the middle of the city'.[20] Rather, the defendants, including brickmaker Stephen Whitaker, carpenters Francis Hillyard and Edward Cock, and plasterer William Perrit, were engaged in creating Mayfair.[21] Similarly, when Elizabeth Carter warned Elizabeth Montagu of urban 'crowds and smoke and brick-houses', Montagu was then arranging a move from her house on Hill Street, near the edge of West End, to a new construction on Portman Square which, like Chandos's great house, faced open country to its north.[22]

Complaints such as these are not, therefore, evidence of industry in the city, but rather of one kind of industry that always flourished precisely where the city met the country, precisely where the elite preferred to build their townhouses. The archive of their correspondence and legal proceedings contains their complaints about smoky brick kilns both because brick kilns did in fact emit large amounts of smoke and because England's elite had settled in the West End in large measure to escape such nuisance. In many ways this was an impossible goal; houses on the fringe of the city were sure to be engulfed by further urban growth, and that growth itself depended on the coal-fuelled production of bricks, lime, and tiles. The Duke of Buckingham was hardly exaggerating when, during the reign of Anne, he inscribed his new house at St James's Park – on the edge of western development – with the motto *rus in urbe* (the country in the city). But by the time Buckingham House became a royal residence at the end of the century it

---

[18] HEHL, Chandos MSS, ST 57, Vol. 55, 262.
[19] Duke of Grafton v. Hilliard (1736), TNA C 11/2289/87. The plaintiffs included Charles FitzRoy Second Duke of Grafton; Hendrick [Henry] van Ouwerkerk, First Earl of Grantham; James Brudenell, future Fifth Earl of Cardigan; and William Townshend, MP for Great Yarmouth.
[20] Thomas, *Man and the Natural World*, 245.
[21] TNA C 11/2289/87. Their building activity can be traced in *The Survey of London* 36 (Covent Garden), 32 (St. James Westminster), and 40 (Grosvenor Estate).
[22] *Letters Between Carter and Talbot*, III, 105.

was surrounded by further urban sprawl, the motto an anachronistic fantasy.[23] Escaping urban pollution entirely, then, was hardly more possible than escaping the poor.[24] Nor was it a useful solution for London's systemic environmental problems, as the environmental protections afforded by the West End's landlords were, by definition, exclusive and exclusionary. They were not, however, altogether in vain, as the eighteenth-century West End supported more and larger gardens than were possible in the crowded areas of the city or the poorer suburbs to the south and east.[25] For the privileged few, therefore, the 'purer regions' of the West End did offer a place where clean air was more available than anywhere else in the early modern metropolis.[26]

III. 'FRESH AIR AND TRANQUILITY': MIDDLING SORTS, TEMPORARY MOVEMENTS

Despite the airy squares of Mayfair, for the vast majority of Londoners the primary means of escaping smoky air was not the West End but the cottages and fields of the suburbs. While Britain's elite families structured their lives around the annual rhythm of townhouse and country estate, the capital's permanent residents pursued different arrangements of both time and space. The capital's rich or comfortable merchants, artisans, as well as a broad swathe of its working poor – in short, those whom we might include within a capacious understanding of the urban middling sort – moved in and out of London in pursuit of pleasure, relaxation, variety, and health.[27] They generally did so more frequently than the landed residents of the town, and for shorter amounts of time. Their weekend excursions, therefore, were necessarily not to far-flung rural estates but to the capital's nearest suburban retreats. These areas are now generally within Zone 2 on the Underground system and might plausibly be classified as parts of central London. During the early modern period they were distinctive spaces, were clearly outside of the metropolis and yet accessible by

---

[23] Macky, *Journey*, 125–6.
[24] For the Westminster poor see Malcolm Smuts, 'The Court and its Neighborhood: Royal Policy and Urban Growth in the Early Stuart West End', *Journal of British Studies* 30:2 (1991), 117–49; Jeremy Boulton, 'The Poor Among the Rich: Paupers and the Parish in the West End, 1600–1724' in Paul Griffiths and Mark S. R. Jenner, eds. *Londinopolis: Essays in the Cultural and Social History of Early Modern London* (Manchester, 2000), 197–226.
[25] Dan Cruickshank and Neil Burton, *Life in the Georgian City* (1990), 190–203.
[26] Moore, *The World*, III, 107; John Fransham, *The World in Miniature: or, The Entertaining Traveller* (1740), II, 131.
[27] General studies of the urban middling sorts include Peter Earle, *The Making of the English Middle Class: Business, Society and Family Life in London, 1660–1730* (Berkeley, 1989); Margaret Hunt, *The Middling Sort: Commerce, Gender, and the Family in England, 1680–1780* (Berkeley, 1996). Their uses of housing and urban space are the focus of McKellar, *Landscapes*; and Peter Guillery, *The Small House in Eighteenth Century London* (2004).

foot, carriage, or short boat trip for those wishing, in John Stow's phrase, 'to recreate and refresh their dulled spirits in the sweet and wholesome air'.[28]

Drawing clear lines between the kinds of retreat achieved in suburban cottages and country retreats on the one hand and West End townhouses on the other is, in some cases, difficult and perhaps arbitrary. This is in part because the richest had no need to choose between these styles of habitation. The First Earl Spencer, for example, owned an ornate townhouse on Green Park, a suburban manor at Wimbledon, and a magnificent country pile at Althorp, and his family used each of them.[29] Moreover, some literary descriptions of retreat increasingly celebrated moderate distance as the golden mean between immersion in urban sin and sea coal and rural seclusion. Gilbert West's poem inscribed on his own suburban villa, quoted in the previous chapter, stressed access to both city and country, not the choice between them, as the ideal: 'swift as my changing wish, I change the scene; And now the country, now the town enjoy'.[30]

Whether a particular 'scene' should be classified as 'country', 'town', or suburban can therefore be unclear. But to many observers there were cases that did seem clear, in particular those villages outside of London full of little else but the cottages of Londoners. In his *Tour Thro' the Whole Island of Great Britain* Defoe focused on the grand houses that bordered the Thames for several miles upstream from London. These, he argued, were individually beautiful and collectively 'truly great', especially as they were not estates but

> Gentlemen's mere summer houses, or citizens' country houses; whither they retire from the hurries of business and from getting money, to draw their breath in a clear air, and to divert themselves and families in the hot weather; and that they are shut up, and as it were stripped of their inhabitants in the winter, who return to smoke and dirt, sin and sea coal (as it was coarsely expressed) in the busy city.[31]

Such citizens' retreats were not only found to the west. Eastern Hackney, according to the *Foreigner's Guide* (1729), contained scores of houses

---

[28] Charles Lethbridge Kingsford, ed. *A Survey of London by John Stow* (Oxford, 1908) I, 127. Discussed in Laura Williams, '"To Recreate and Refresh Their Dulled Spirites in the Sweet and Wholesome Ayre": Green Space and the Growth of the City', in Julia Merritt, ed. *Imagining Early Modern London: Perceptions and Protrayals of the City from Stow to Strype, 1598–1720* (Cambridge, 2001), 185–213. It should be noted that Stow is here not describing a transition from rural to urban, but rather from 'fields' to suburban facilities like 'Garden houses and small cottages … garden plots, tenter yards and bowling allies'. Cf. Hiltner, *What Else is Pastoral?*, 63–4.

[29] Accounts, Spencer papers, BL Add MS 75,765.

[30] Samuel Johnson, *Works of the English Poets* (1790), LVII, 324; also Gilbert West to Philip Dodderidge, 26 April 1751, Thomas Stedman, ed., *Letters to and From the Rev. Philip Doddridge, D.D.* (1790), 451.

[31] Daniel Defoe, *Tour Thro' the Whole Island of Great Britain* (1968), I, 167, 169.

owned by city merchants, especially dissenters.[32] The volumes of the *Victoria County History* for Middlesex record endless names of Londoners, both courtiers and professionals, who held lands and houses all around London's outskirts, and doubtless many treated these as weekend retreats.[33]

While Defoe suggested that citizens and gentlemen used their 'country houses' in similar ways, other observers noticed differences. *The Foreigner's Guide* found that both gentlemen and 'tradesmen' had houses 'in the villages about London', but that the latter also sometimes had only 'private lodgings', to which they repaired 'on a Saturday in the afternoon to take the pleasure of the country air, and return on the Monday following to business again in the town'.[34] Pehr Kalm noticed a similar schedule in the 'large brick houses' of Fulham and Chelsea, where 'gentlemen and others ... now and then, especially on Saturday afternoons, went to take the fresh air and to have the advantage of tasting the pleasures of a country life'.[35]

That such weekend getaways were generally employed by the urban middling sort, but understood or represented through the aristocratic language of country retreat, provided an opportunity to mock this as yet another example of social emulation and the plague of fashionable consumption. One of Charles Jenner's *Town Eclogues* began describing Sundays as the day when 'cits take their weekly meal of air; Whilst, eastwards of St. Pauls, the well-dressed spark, Runs two long miles to saunter in the park'.[36] This observation, however, leads into a long mockery of social climbing and emulation, in which the middling sort are said to ape the styles of suburban retreat that are more appropriate to their social betters.

The mid-eighteenth-century periodical *The Connoisseur* was especially interested in pursuing this line, suggesting several times that there was something ridiculous in *cits* – citizens of the city, the urban middling sorts – spending their money and time in gardens which were presented as the opposite of the urban world from which all their prosperity came. This contradiction was a fruitful object of satire, as in the poem describing 'Sir Thrifty's' search for a 'country seat' three or four miles from the city where he could pursue 'exercise and country air', while his wife furiously schemed to impress all her friends. Thus, the retreat becomes yet another object of feminized urban business, consumption, and social emulation rather than the Horatian/Christian negation of such worldly cares.[37] *The Connoisseur's* cits play the ludicrous game of aiming at such models of retreat without ever

---

[32] *Foreigner's Guide*, 138.

[33] *A History of the County of Middlesex* 13 Volumes, (1969–2009).

[34] *Foreigner's Guide*, 128   [35] Kalm *Account*, 35.

[36] Jenner, *Town Eclogues*, 8.

[37] George Colman, *The Connoisseur. By Mr. Town, Critic and Censor-General*, (1757–60), 135, 26 August 1756, IV, 233–8, 'country air' at 235.

even attempting to discard their bourgeois values or even to escape their urban environment. Their 'weekly excursions into the villages about town' were confined to suburban Wandsworth and Hampton, and seldom even 'out of the sight of the London smoke'.[38] From one would-be country seat in Kennington Common one cit beheld not a landscaped estate but the decaying bodies of executed criminals and 'a distant view of St. Paul's Cupola enveloped in a cloud of smoke'. As the essay's opening epigram claimed, "Tis not the country, you must own, 'Tis only London out of town!'[39] In such satire coal smoke functioned, as it long had on the stage, as a symbol of bourgeois wealth and values. But it achieved its bite in part through the perception that citizens eagerly sought out suburban clean air in self-negating ways, that they used their money to run away from the smoky environment which had made them rich and so remained their proper home.[40]

For Colman the 'cit' was a substantial merchant or tradesman, wealthy enough to sell a bit of stock and so buy a suburban house whenever his wife convinced him to do so. But he also mocked another form of suburban leisure, the pleasure gardens that surrounded the metropolis, in particular on its west and north. For Colman it was ridiculous that while the polite culture of the capital was emulated in country towns across England, in London itself, 'where we are almost smothered in smoke and dust', inhabitants seemed to desire nothing so much as a taste of the country.[41] Despite such mockery, for others there was no need to apologize for this aspect of pleasure gardens' appeal. When the popular tenor Thomas Lowe leased the long-established gardens at Marylebone in 1763, he chose to announce his suburban musical venue as a rural retreat:

> Now the summer advances, and pleasure removes
> From the smoke of the town, to the fields and the groves
> Permit me to hope, that your favours again,
> May smile, as before, on this once happy plain.[42]

---

[38] *Ibid.*, number 79, 31 July 1755, III, 58–9.

[39] *Ibid.*, Number 33, 12 September 1754, I, 255–62, quotes at 258 and 255.

[40] See also *ibid.*, 93, 6 November 1755, III, 165. Such cits also had their defenders: 'There is no popular subject of satire, on which the modern commonplaces of wit and ridicule have been exhausted with more success, than on the subject of a mere cockney affecting the pleasures of the country. The dusty house close to the road side, the half acre of garden, the canal no bigger than a wash-hand-basin, etc. have all been marked out with much humour and justice; but after all, it is not unnatural for a tradesman, who is continually pent up in the close streets and alleys of a populous city to wish for fresh air, or to attempt to indulge a leisure hour in some rural occupation.' *The Annual Register, or a View of the History, Politicks, and Literature of the Year 1761* (1762), 208.

[41] *The Connoisseur*, 23, 4 July 1754, I, 180.

[42] *The Gentleman's Magazine* (May 1763), XXXIII, 252. For Lowe and the garden's history, William Wroth and Arthur Edgar Wroth, *The London Pleasure Gardens of the Eighteenth Century* (1896), 93–110, esp. 101.

A few years later the poem *The Art of Living in London* agreed that 'fragrant air' was one of the primary reasons to visit commercial establishments like Bagnigge Wells (south of the present King's Cross station) or Cromwell's Gardens (near Gloucester Road station in Kensington): 'Here, leaving city smoke and noise behind, at ease indulge the wand'rings of the mind.'[43] Clean air was among the many attractions of these commercial gardens, and the anti-urban, pastoral language of ease and groves offered an attractive formula through which to mask what was in reality an uneasy combination of relaxation, commercial entertainment, and sexual possibility.[44]

Colman's project of denying the classical language of retreat to cits was all the more pressing because the polite spaces of pleasure gardens were equally open to citizens and leisured men and women. Much less attention is paid, both by Colman and by other commentators, to the more modest tradesmen, artisans, and their families who also sought out suburban sociability and pleasure. But there is evidence that the search for clean air encompassed many fairly far down the social scale. A satirical description of a typical day's rhythms described 'handicraft tradesmen marching to and from Islington and Chelsea with their offspring in their arms, followed by their wives'.[45] A popular geographical text distinguished places like the Inns of Court, Somerset House gardens, and the Charterhouse, which were accessible only by 'persons of gentle appearance', from 'Stepney Fields, Moor Fields, Islington, Red Lion Fields, and St James Park; as also Chelsea, and Kensington' where no such restriction is mentioned.[46] This list travels from east to west, and might be extended to include the open fields and gardens surrounding Southwark to the south. London and its suburbs were ringed with a series of fields and open spaces that were accessible to anyone able to walk. Even by the middle of the eighteenth century, when the capital contained about three-quarters of a million people, no one was much more than a mile away from open space. Considering this, 'the vast multitudes of people poured forth by this city into the fields around it on summer evenings' described in 1751 must have included very many of quite modest circumstances.[47]

For most people simple drinks like beer, wine, and the 'home-brewed ale' advertised by the George and Vulture pub in Tottenham were probably quite enough to make for an enjoyable day out.[48] Many of the innumerable

[43] [William Cooke] James Smith of Tewkesbury, *The Art of Living in London* (1785), 21.
[44] Wroth and Wroth, *London Pleasure Gardens*; Ogborn, *Spaces of Modernity*, ch. 4.
[45] *Hell upon Earth: or The Town in an Uproar* (1729), 8–9.
[46] *The Present State of Great Britain, and Ireland ... Begun by Mr. Miege* (1745), 103.
[47] Corbyn Morris, *Observations on the Past Growth and Present State of the City of London* (1751), 24.
[48] Jacob Larwood and John Camden Hotten, *The History of Sign-Boards, From the Earliest Times to the Present Day* (1866), 291.

alehouses and inns surrounding the metropolis offered large gardens and greens, space to walk, talk, play, see and be seen. The George and Vulture (located about five miles north of the City), even described itself through the pastoral language that people like Colman thought was only appropriate for country estates.

> If lur'd to roam in summer hours
> Your thoughts incline tow'rd Tottenham bow'rs
> Here end your airing tour, and rest
> Where Cole invites each friendly guest...
> Who leaving city smoke, delight
> To range where various scenes invite:
> The spacious gardens, verdant field
> Pleasures beyond expression yield.[49]

Talk of bowers, airing tours, and verdant fields may seem a bit much for what was in fact a tavern in central Tottenham. But this hyperbole underlines a crucial point, that the language of pastoral was available for appropriation by the middling sorts in part because they, like the rich of the West End, perceived their urban environment as something to escape, if only for a weekend afternoon.

IV. 'A POSITIVE EVIL': MEDICINE, HEALTH, AND AVOIDING THE CITY

Despite the poetic convention celebrating the 'pleasures beyond expression' to be found in rural retreats, it is clear that for many the decision to avoid London was motivated as much by the perceived dangers of urban air as by the perceived benefits of the country. The rhetoric of air and exercise appears a cruel euphemism when we consider the serious medical implications of urban air pollution and the real impact they had on many lives. For many, retreat from London's environment was not so much a question of feeling as good as possible, but rather of maintaining a basic minimum of health, or even of clinging to life itself. Elizabeth Carter described herself, quite conventionally, in a letter of the summer of 1748, as impaired in both health and spirits by the London smoke. 'I am not so happy as to be running wild in the nettle groves of Enfield [Middlesex], but am panting for breath in the smoke of London.' Her breath, however, did not suffer excessively. 'This is sad confinement for my active genius, however I read, write, sing, play, hop, and amuse myself as well as I can: and every afternoon walk as if I was bewitched, to keep myself in health.'[50] When she wrote this Carter was

---

[49] *London Advertiser and Literary Gazette* 1 April 1751. Issue 25.
[50] Elizabeth Carter to Catherine Talbot, 5 August 1748, *Letters Between Carter and Talbot*, I, 287.

thirty years old and had another fifty eight years before her. But for those of less robust health the capital's atmosphere was not endured so easily.

Physicians, as was discussed in Chapter 6, frequently claimed that coal smoke threatened health and was particularly dangerous for those suffering from respiratory trouble. The successful physician Gideon Harvey explained during the 1660s why consumption was more common in London:

> By means of those sulfurous coal smokes the lungs are as it were stifled and extremely oppressed, whereby they are forced to inspire and expire the air with difficulty in comparison to the facility of inspiring and expiring the air in the country, as people immediately perceive upon their change of air; which difficulty, oppression, and stopping must needs at length waste the lungs, and weaken them in their function.[51]

Harvey was far from alone, as the perception that London's smoky air was particularly dangerous to sufferers of consumption or asthma was widespread. The physician Richard Morton's influential study of consumption was published during the reign of his former patient (and asthmatic) William III, and it stressed the negative power of coal smoke many times, both in general terms and during case histories.[52] Other prominent physicians agreed, suggesting not only that London's smoke caused or worsened consumption and asthma, but also that this meant that removal to the country was a crucial treatment for such patients.[53]

It is easily demonstrated that in learned treatises these and other physicians asserted that coal smoke contributed to ill health and that healthy country air contributed to effective therapy. The evidence of practice is less clear, but there are good reasons to conclude that such assertions were not restricted to the pages of learned tomes, but were also frequently communicated to patients. Richard Morton's case histories, for example, describe men whose treatment included change of air, such as 'Mr. Foster, apothecary of London', whose severe lung trouble was effectively managed by leaving the 'smoky air of London' for the 'open air at Highgate', in addition to taking frequent doses of laudanum.[54] One Mr Luff, however, returned from the country 'into our air, that is filled with the smoke of coals' only a few weeks after a severe bout of consumption, 'much sooner than was fit'. This was too much for poor Luff, who 'departed this life' within a few weeks.[55] Morton must have been typical rather than exceptional in

---

[51] Gideon Harvey, *Morbus Anglicus: or, The Anatomy of Consumptions.* (1666), 166.

[52] Morton, *Phthisiologia*, 75, 96, 152, 214, 235, 237. For Morton's career and influence Stephen Wright, 'Morton, Richard', *ODNB*.

[53] John Pechey, *The Store-house of Physical Practice Being a General Treatise of the Causes and Signs of All Diseases Afflicting Human Bodies* (1695), 128, 156, 458; Pitcairn, *The Whole Works*, 126–7; Arbuthnot, *An Essay*, 208.

[54] Morton, *Phthisiologia*, 245–6. Highgate is an elevated area five miles north of the city of London.

[55] *Ibid.*, 236–7.

recommending travel into the country to convalesce or to prevent further damage. Indeed, in the 1770s the prominent Quaker physician John Fothergill argued that physicians needed to be much more careful regarding precisely which countryside they recommended to their patients. While the experience of asthmatics varied, for consumptives 'the air of all large cities is found by experience to be particularly injurious', and the convalescence in healthy suburban villages was therefore widespread. But to which of these many villages the patient repaired should not, Fothergill argued, be a matter of indifference.

> The town itself is covered almost continually with an atmosphere of smoke, embodied with other exhalations, so as to form a cloud, more or less dense, which is visible at a great distance. This vast body of smoke is seen to extend for several miles beyond the limits of the city and its suburbs, and is driven by the winds that prevail in the several seasons, according to their direction. In the summer season, for instance, whilst the southerly winds prevail, this dense body is driven to the north and northeast parts of the environs, and covers the herbage, the trees, and everything, both living and dead, with black penetrating soot. In the winter and spring, while the northerly winds prevail, the opposite villages on the southwest and west side of London receive this thick atmosphere.[56]

The physician should therefore not only advise his patient to leave this 'vast body of smoke', but to leave for a destination carefully chosen according to its local characteristics and seasonal weather patterns.

Such medical advice may have been easy to accept, as it seems to have agreed with widely shared common sense regarding the superiority of country airs to urban smoke. This, as the previous chapter described, was a commonplace assumption of pastoral poetry and town comedies throughout much of the seventeenth and eighteenth centuries. The unhealthiness of coal smoke, moreover, was the basic premise of the political regulation, nuisance suits, and private arrangements described in Chapters 4 and 5. In a few exceptional cases it is possible to examine how specific people translated these general perceptions into particular courses of action, and thereby to understand some of the human costs of London's unhealthy atmosphere. The best-documented such case is perhaps John Locke, whose early medical studies made him particularly well qualified to diagnose the effects of urban air on his own health. These, he was quite certain, were devastating. Immediately after returning to England from exile early in 1689 Locke reported to a Dutch correspondent that he began to suffer from his old 'lung trouble', to cough and to 'pant'. This was caused, he stated unequivocally, by 'the malign smoke of this city'.[57] As spring approached Locke reported an

---

[56] 'Further Remarks on the Treatment of Consumptions, &c. by John Fothergill, M.D. F.R.S.', *Medical Observations and Inquiries. By a Society of Physicians in London* (1776), V, 362–3.
[57] Locke to Egbertus Veen, 8 and 12 March, 1689. De Beer, ed. *Correspondence*, III, 577, 583.

improvement, but in subsequent years he was only able to make periodic visits to the capital during the comparatively clear summer months.[58] He spent the rest of his time as a permanent guest with Francis and Damaris Masham at their house in Oates, Essex, about twenty-five miles north-east of London. It seems quite likely that this semi-retirement limited Locke's involvement in government during the fifteen years between the Glorious Revolution and his death in 1704. It is impossible to prove that Locke would have played a more prominent role in politics had he been able to breathe more easily, but in any case it is revealing that Locke's earliest biographers found it easy to believe that his political career had in fact been hindered by the effects of London smoke on his 'asthmatic disorder'.[59]

Locke's was not the only career impacted by an inability to tolerate London's atmosphere. In some cases this was a minor impediment. In 1772, the Maryland-born merchant (and future American consul in London) Joshua Johnson reported that he had been compelled to leave town because of illness brought on by 'the disagreeable damps and smoke of London'. He soon recovered, however, and returned to busily pursuing his commercial interests in the capital.[60] Around the same time a provincial MP, Frederick Cornewall, hoped that the ministry could do without him in parliament, as he found 'country air' much better for his health than 'the smoke of London'.[61] Johnson, who recovered quickly, and Cornewall, who showed little desire to come to London in any case, each experienced the capital's unhealthy air as a fairly minor inconvenience.

Opposing such examples are figures whose ambitions were quite incompatible with London's effects on their health. Locke's former pupil the 3rd Earl of Shaftesbury shared with his mentor philosophical ambition, Whig politics, and weak lungs. In 1707 he wrote of 'being excluded as much as I am from the public service by my ill health which will not suffer me in the winter and chief season of business to live in or near our capital city where coal is burnt, so that I am half banished society and civil life'.[62] He lived in Chelsea, beyond the westernmost spread of the metropolis, and yet even there the easterly winter winds described above by Fothergill prompted

---

[58] Isabella Duke to Locke 5 April 1689; Locke to Philippus van Limborch 12 April 1689, *Ibid.,* 595, 599.

[59] Jean Le Clerc, *The Life and Character of Mr. John Locke* (1706), 10, 17; *Memoirs of the Life and Character of Mr. John Locke* (1742), 8; Thomas Birch, *The Heads of Illustrious Persons of Great Britain* (1743), 140.

[60] Jacob M. Price, ed. *Joshua Johnson's Letterbook 1771–1774: Letters From a Merchant in London to his Partners in Maryland* (1979), 55.

[61] Lewis Namier, 'Cornewall, Frederick (1706–88)' in History of Parliament, Common 1754–90. Accessed at www.historyofparliamentonline.org/research/members/members-1754-1790.

[62] Rex A. Barrell, *Anthony Ashley Cooper Earl of Shaftesbury (1671–1713) and 'Le Refuge Français' – Correspondence* (Lewiston, MA, 1989), 54.

immediate flight into the country.[63] There were several factors, political and philosophical, constraining Shaftesbury's political activity during the reign of Anne, but the capital's 'cloud of smoke that I can neither be in it nor near it' was undoubtedly important amongst them.[64] In other cases inability to suffer urban smoke was offered as a very plausible cover for movements that may well have had other, more politically sensitive explanations. The prominent ophthalmologist John Woolhouse, for example, claimed to have emigrated to Paris because the 'sea coal air of London' had given him a consumption. Perhaps, but he lived another forty-five years after leaving London, and there were accusations of Jacobitism.[65] Before the Glorious Revolution, Princess Anne absented herself from the court and the French special envoy struggled to discern whether the cause was opposition to her Catholic father or an actual fear that 'the air of London' would harm her pregnancy.[66] In each of these cases London's smoky air either really did, or was plausibly claimed to, impact the behaviours and careers of politically prominent Londoners.

Those whose lives were most profoundly impacted by the urban environment, however, were probably those who cannot be documented in this way. There were some elderly men who, like the antiquary William Stukeley, wished to retire 'out of the influence of the London smoke', since old age, like infancy, made one more susceptible to unhealthy air.[67] How many more elderly Londoners removed themselves from homes in the capital when they found they could no longer bear the urban atmosphere cannot be known, but is likely to have been substantial. Nor is it possible to judge with any precision what role smoke played in the widespread practice of sending infants to rural wet nurses. The unhealthy air of London, however, played a prominent part in the reforming efforts of the mid-eighteenth century, and demographers from John Graunt in 1662 to Thomas Short in 1767 assumed that children were especially vulnerable to the 'smokes, stinks, and close air' of the city.[68] As Arbuthnot argued, the 'air of cities is unfriendly

---

[63] Anthony Ashley Cooper, Earl of Shaftesbury, *Letters from the Right Honourable the late Earl of Shaftesbury, to Robert Molesworth, Esq.* (1721), 6, 12. 18, 36.

[64] *Ibid.*, 6. Also TNA PRO 30/24/21/189. For his career in general, Lawrence Klein, 'Cooper, Anthony Ashley, Third Earl of Shaftesbury' *ODNB*.

[65] Letter to Hans Sloane, 11 June 1724, BL Sloane MS 4047, f. 191; Anita McConnell, 'Woolhouse, John Thomas', *ODNB*

[66] TNA PRO 31/3/170. Anne had suffered a recent miscarriage and this pregnancy terminated with a stillbirth.

[67] *The Family Memoirs of the Rev. William Stukeley, M.D.* (Durham, 1887), III, 20; W. J. Hardy, ed. *Middlesex County Records. Calendar of Sessions Books 1689–1709*, British History Online, www.british-history.ac.uk/report.aspx?compid=66121; William Penn, *A collection of the works of William Penn* (1726), I, 148; John Hill, *The Old Man's Guide to Health and Longer Life* (Dublin, 1760), 20.

[68] Graunt, *Natural and Political Observations*, 45. Discussed in Chapter 6 above. Also Jonas Hanway, *Serious Considerations on The Salutary Design of an Act of Parliament for a Regular, Uniform Register of the Parish Poor* (1762), 17.

to infants and children' because they lacked the 'habit' necessary to make dirty, 'artificial' air tolerable.[69] Most Londoners did not read such texts but they behaved as if they did, and the belief that young children in particular required good air appears to have been widespread.[70]

By the mid-eighteenth century it was not uncommon to observe that London enjoyed rather clean and healthy air, with a few fairly important exceptions. 'London is not an unhealthful Place', asserted an encyclopedist in 1739, adding the proviso: 'unless for children, and such as are troubled with a difficulty of breathing'.[71] Another reference work dwelt on this ambiguity in more detail. London's air was 'sweet and good', and yet London was full of 'stinks and smokes'. It enjoyed a 'clean situation in the side of an hill', and yet:

> produces so much filth and occasions so many nasty vapors from tradesmen's furnaces that it is impossible the air should retain its necessary fitness for breathing, which is the reason that many asthmatic people can't stay in it a day, and others, who have colds, or are troubled with phlegm, bear the winter but uneasily because then the air is thickest by reason of the multitude of coal fires; but at the distance of four or five miles from London the air is so good that the physicians think it sufficient to send their sick patients to Kensington Gravel Pits, Greenwich Park, Hampstead Heath, and such like places not far from London.[72]

Benjamin Franklin registered this ambiguity too. 'The whole town', he explained to his wife, 'is one great smoky house, and every street a chimney, the air full of floating sea coal soot, and you never get a sweet breath of what is pure, without riding some miles for it into the country.'[73] He also argued, however, that for all this London's air was actually basically healthy, at least 'for grown persons whose lungs can bear the cloud of smoke which generally hovers over them'.[74]

According to such logic London's dirty air was not for everyone, but neither was it a general public health problem. For grown men and women in robust health, especially those who could occasionally enjoy a few breaths of fresh suburban air, the metropolitan environment was tolerable. But for children, for the elderly, for asthmatics, for consumptives, for those with a cold, for anyone weak or anyone not yet habituated to it, London's smoke was a serious danger. Such people could not expect to change the city, so they were obliged to stay away.

[69] Arbuthnot, *An Essay*, 208–9.

[70] Laura Gowing, *Common Bodies: Women, Touch and Power in Seventeenth-Century England* (New Haven, 2003), 199–200. This belief, it should be noted, was accurate. For the WHO's summary of air pollution's impact on children, see www.who.int/ceh/risks/cehair/en.

[71] Salmon, *Modern History*, IV, 489. Repeated in John Mottley, *A Survey of the Cities of London and Westminster, and the Borough of Southwark* 6th ed. (1754–5), II, 562.

[72] Thomas Cox, *Magna Britannia antiqua & nova* (1738), III, 52.

[73] Benjamin to Deborah Franklin, 19 February 1758, accessed at franklinpapers.org.

[74] Jared Sparks, ed. *The Works of Benjamin Franklin; Containing Several Political and Historical Tracts* (Boston, 1844), VI, 320.

V. CONCLUSION: *THE ART OF LIVING IN LONDON*

Those who did come to London needed guidance, at least according to the author of the 1768 poem *The Art of Living in London*. The reader, presumed to be a young man of some means but neither vane nor rich, is led through the practical and ethical traps that the metropolis lay for the green newcomer. The author professes, in a style derivative of Pope, that the poem's chief duty is philosophical, 'to know mankind' and so to avoid its common mistakes.[75] The reader is thus provided an armory against moral and monetary ruin, particularly the temptations of prostitution and gambling. These are the chief lures of the night, but by day the difficulty is more complex, as one should avoid both the expense of vain pride and 'the miser's narrow care'.[76] Several pages are devoted to where such a moderate young man may eat wholesome yet reasonably priced food. Indeed, the art of metropolitan living is largely the search for a golden mean between miserliness and excess, riotous debauchery and dreary boredom. What such moderation looks like in practice is only rarely described, but the poem does explicitly recommend four institutions of urban life: the tavern for good beer and good fellowship, 'a cheap resource to entertain the mind'; the stage to 'humanize the mind'; the coffee house to 'know, from the press, what schemes the world engage'; and the pleasure garden, to leave 'city smoke and noise behind, at ease indulge the wand'rings of the mind'.[77]

This is one of the few moments when the poet actually describes what one should do (rather than avoid) in London. 'If 'tis summer, and the ev'ning fair, miss not th'advantage of the fragrant air.' In suburban gardens the mind's wanderings should be indulged, as 'verdant prospects' warm the heart and entertain the eye. In such places the blessings of 'boundless nature' may be enjoyed in a specifically urban way. While 'the peasant' works *in* nature, 'the cit' only 'partakes' of nature's pleasures while *away* from labour, 'in recess from toil'. Such gardens join coffeehouses as the twin philosophical spaces of the capital, places where the mind may 'unbend' or 'wander'. While the subject of coffeehouse chatter is politics, the space of the garden is also implicitly national, political, and military.[78] In such gardens the virtuously moderate, masculine young Londoner achieves the mental steadiness necessary for an honourable and prosperous urban life. 'Leaving city smoke and noise behind' is here, as in so many of the sources considered in the previous two chapters, an essential

---

[75] Cooke, *Art of Living*, 6.   [76] *Ibid.*, 9.   [77] *Ibid.*, 36, 33, 20, 21.

[78] The gardens mentioned include Chelsea hospital, William III's palace at Kensington, and the Brompton Garden thought to have been 'where Cromwell liv'd, tyrannically great'. *Ibid.*, 19–20.

complement to success in the city. London's dirty air contributed in crucial ways to models of behaviour as well as representations which asserted the importance of movement and retreat as antidotes to the necessarily ills of urban life. In many such poems, and in many lives, the art of living in London consisted in knowing how to leave it.

Epilogue

When William Wordsworth celebrated the calm majesty of London one morning in 1802, with its 'ships, towers, domes, theatres, and temples' lying under a sky 'bright and glittering in the smokeless air', he was building upon a 200-year history that placed smoke near the centre of London, itself the centre of an increasingly global empire. The city that had come to represent constant work, movement, doing, exuberant sociability, and a dizzying whirl of human commerce was, in Wordsworth's vision, for a moment utterly still. London sleeping like a baby was 'a sight so touching in its majesty' precisely because it was so unusual. The canopy of 'smokeless air' helped to negate all that made London itself, a 'mighty heart lying still' rather than beating.[1]

This moment of poetic calm is charged with tension in part because it is so clear that it must be fleeting. The mighty heart of London, the reader knows, will beat again. If smokelessness offered Wordsworth a useful image to suggest stasis, smoke provided Lord Byron, writing twenty years later, an equally convenient metaphor for London's much more familiar hubbub. When Don Juan overlooked the Thames at his first approach to London he saw,

> A mighty mass of brick, and smoke, and shipping
> Dirty and dusky, but as wide as eye
> Could reach, with here and there a sail just skipping
> In sight; then lost amidst the forestry
> Of masts; a wilderness of steeples peeping
> On tiptoe through their sea-coal canopy;
> A huge, dun cupola, like a foolscap crown
> On a fool's head – and there is London town.[2]

Byron's London was all that Wordsworth's was not: busy, alive, awake, dirty, and smoky. Both drew on literary conventions dating back far into

---

[1] 'Composed upon Westminster Bridge, 3 September 1802', *The Works of William Wordsworth* (Ware, Herts., 1994), 269.
[2] Frederick Page, ed. *Bryon Poetical Works* (Oxford, 1970), 788 (Canto X, LXXXII). Written late 1822; Jerome McGann, 'Byron, George Gordon Noel', *ODNB*.

the seventeenth century, to *Eastward Ho's* association between coal smoke and mercantile society or to Evelyn's claim that industrial smoke blighted the public spaces and monuments of what would otherwise have been a grand city. Byron's view down onto London from above recalls Denham's 'Cooper's Hill', but it is Percy Bysshe Shelley that draws most clearly on that poem's collapse of London's morality into its materiality. 'Hell is a city much like London', he wrote in 1819, 'a populous and a smoky city. There are all sorts of people undone.'[3]

When the Romantic poets incorporated urban pollution into their critiques of modern society and morality they used well-established and long-familiar tropes.[4] The smoke of London was, for them as for several generations of poets, playwrights, and literary hacks before them, a symbol of all that made Britain's metropolis different: its size and wealth, importance and self-importance, its unequalled commerce and industry, sprawling scale, dissipation and sophistication. A celebration of nature rather than the works of man is often said to characterize the novelty of Romanticism, but their uses of London pollution were less revolutionary than derivative, or at least building on long traditions.

Wordsworth's poem is set on Westminster Bridge, a few hundred metres from the smoky breweries that bothered Charles I in the 1630s, from the brewhouse that Evelyn blamed for filling Whitehall with smoke in 1661, and from the breweries and glasshouses targeted by the 1706 parliamentary bill. He could have looked upriver to the site of the glasshouse against which the residents of Vauxhall complained to the Privy Council in 1665.[5] Looking downriver, towards the curve of the Thames, he could see the cupola of St Paul's. Completed a century before, Wren's smoke-blackened cathedral had replaced the medieval edifice whose walls were found unacceptably smoky during the reign of James I. Wordsworth's river that 'glideth at his own sweet will', invokes Edmund Spenser's soft-running 'sweet Thames', lines that were written almost two decades after Spenser's Queen informed London's brewers that their smoky coal interfered with her 'pleasure upon the water'.[6] Wordsworth's London may have been smokeless one morning in 1802, but it had rarely been so during the previous two centuries.

---

[3] 'Peter Bell the Third', Donald H. Reiman and Sharon B. Powers, eds. *Shelley's Poetry and Prose* (New York, 1977), 330. Cf. 'through several ways they run: Some to undo, and some to be undone'. Denham, *Cooper's Hill* (1642), 3. For 'Cooper's Hill', *Eastward Ho*, and Evelyn see above, Chapters 12 and 11, respectively.

[4] Cf. William Sharpe, 'London and nineteenth-century poetry', in Lawrence Manley, ed. *The Cambridge Companion to the Literature of London* (Cambridge, 2011), 126–7.

[5] See discussions in Chapters 4, 5, and 11.

[6] Edmund Spenser, *Prothalamion or a spousall verse* (1596); for Elizabeth see Ch. 4.

Byron's Don Juan joined a similarly venerable tradition of travellers who found London 'dirty and dusky'. One hundred and seventy years previously Lodewijck Huygens climbed St Paul's – not yet crowned with Wren's 'dun cupola' – and peered down into such a 'great cloud of smoke' that much of the City lay hidden.[7] A few years later his brother Christian noted both a mighty mass of shipping and a general smoke that shocked his senses.[8] Byron's Don Juan, in the foreground of his view of London from atop Shooter's Hill, could have seen Deptford from which John Evelyn had contemplated removing the 'sea coal canopy' 160 years previously. Looking to his right he would have seen Woolwich, which Evelyn considered a more appropriate home for London's smoky industries. To the east of the City lay industrial districts like Whitechapel, where Lord Mansfield had said nuisance laws against smoke and smells did not apply.[9] Shelley's similitude between Hell and London was similarly long-standing, running from Thomas Dekker in 1610s through Milton's mid-century *Paradise Lost* and *The Tatler* in 1710.[10]

A few months before Shelley sat in Florence writing of smoky London, the House of Commons considered two separate bills that examined the nature and meaning of London's air. The first, debated in May 1819, was an attempt to equalize coal duties that were still, just as they had been in the 1690s, levied primarily on London consumers. Adam Smith was cited against imposing any duty on 'the necessaries of life', the tax's contribution to the public revenue was conceded, and coal's power to stimulate or starve manufacturing industries carefully considered, all familiar positions after more than a century of similar debate. Reforming this tax, it was argued, would expand London's coal consumption, which would lead to a 'still darker cloud and denser atmosphere of smoke', admittedly a 'subject of general regret'. But 'in this improving age' a technological fix was surely imminent, and soon smoke itself would be 'applied to some useful purpose'.[11]

A few weeks later a bill was introduced that intended to establish just what technology was available to mitigate smoke emissions, and then to require its adoption on rapidly proliferating steam engines. These engines, MP Michael Angelo Taylor argued, emitted smoke that 'clouded the atmosphere' and was 'prejudicial to public health and public comfort'. Coal smoke, according to 'all lawyers', was 'a nuisance, and an indictable

[7] Lodewijck Huygens, *The English Journal 1651–1652*, A. G. H. Bachrach and R. G. Collmer, trans. and eds. (Leiden, 1982), 65, also 110, 134.
[8] C. D. Andriesse, *Huygens: The Man Behind the Principle* (Cambridge, 2011), 209–10.
[9] See Chapters 5 and 11.
[10] Thomas Dekker, *A Strange Horse-Race* (1613), sig. D3. *Complete English Poems of Milton*, 274; *The Tatler*, no. 218, 31 August 1710. Cf. Hiltner, *What Else is Pastoral?*, 111–2.
[11] Holmes Sumner, 20 May 1819, in *The Parliamentary Debates from the Year 1803 to the Present Time* (1819), XL, 569–72.

offense', and serious efforts must therefore be made to abate it where possible. Had Taylor felt the need he too, like the Romantic poets, could have cited two centuries of legal precedent declaring smoke to be nuisance, or a similar tradition of medical literature claiming that it harmed the lungs and elevated urban death rates. Perhaps he had no need, as his initial motion was supported by MP John Christian Curwen, a Westmoreland colliery owner. Curwen may have looked on such abatement technology as a way finally to reconcile coal's undeniable commercial, industrial, fiscal, and strategic benefits with its offences against 'public health and public comfort'. Perhaps Taylor's bill reminded Curwen of his friend Arthur Young, who had both celebrated escape from smoky London into the 'freshness and sweetness of the air' of the countryside and yet could also 'Thank God for the coal fires of London.'[12]

Taylor's bill, like the anti-urban language of the Romantic poets, built on the special place of London in the landscape of early modern Britain even as it looked forward to the very different world of the nineteenth century. It adopted scientific, medical, and aesthetic claims that had been voiced by many others. Like Queen Elizabeth, Charles I, Charles II, and those who pursued legal redress or governmental prohibition of specific nuisances, Taylor grounded his objection to smoke in his own experience and his own household. During an 1821 discussion of the bill Taylor described his displeasure, which he shared with his neighbour, Prime Minister Lord Liverpool, at not being to walk in their Whitehall gardens without 'being overclouded with smoke' from the waterworks across the river in Lambeth. Almost two centuries after Charles I advanced a bill against smoky breweries in the 1624 parliament and over a hundred years after the 1706 bill would have outlawed new smoky industries in Westminster, Britain's most powerful people were still considering using the tools of central government to mitigate dirty air in the heart of the nation's capital.

But Taylor's bill also diverged from seventeenth- and eighteenth-century precedent in one crucial respect. Though his own immediate complaint was the state of Westminster, he also invoked the common interest of those living throughout the United Kingdom. The technological cure he proposed 'might be generally applied in town and country, where such nuisances existed'.[13] That such nuisances did exist in town and country, across what was the world's most rapidly industrializing and urbanizing society, was made clear in the subsequent debate. The Member for Newcastle stated that similar improvements had already been made in Northumberland,

---

[12] Young, *Travels* I, 128, 503; Matilda Betham-Edwards, ed. *The Autobiography of Arthur Young, With Selections from His Correspondence* (Cambridge, 2012), 352.

[13] *The Parliamentary Debates. New Series* (1822), V, 440 (18 April 1821) and 439–41 *passim*.

but others disputed whether the many steam engines in the south-west, Midlands, and Wales should fall within its purview. Though Taylor's own experience was metropolitan, his perspective and his defence of the bill were national. Against the suggestion that the interest of Cornish mines, where nuisance complaints had not been filed, should limit the bill's applicability to capital, Taylor replied that the poor everywhere suffered from smoke but lacked the means to pursue legal redress. Smoke pollution, he argued, was emphatically not a metropolitan, nor even a specifically urban problem. 'Why should a clergyman who kept a school in a village be smoked out? Why should Manchester, Liverpool, Leeds, be annoyed by nuisances for the sake of Cornish miners?'[14] The 'nuisance' of smoky air was now everywhere.

By the early nineteenth century coal smoke and air pollution were no longer, as they had been during the early modern period, problems that were specific to Britain's uniquely sprawling and busy capital. They had become problems typical of a new age in which industrial production might be located in any place – city, town, or countryside – with the requisite combinations of capital, labour, and transport links. Coal-fired mills, pumps, and factories, all of which employed armies of labourers who lived in coal-heated homes, connected to each other by smoke-belching trains and steam ships, proliferated at first throughout and then beyond Britain. As industrialization spread during the later nineteenth century, more or less following the British pattern, smoky skies and dirty air appeared to very many as the price of entry into the modern world of prosperity, economic growth, and political power. Throughout Europe, the United States, and Japan the unbound Prometheus of technologically driven growth seemed necessarily to entail environmental costs that observers ignored, tolerated, or celebrated.

During the last century this dynamic has changed even as it has become universal. Smoke-belching chimneys are no longer necessarily the only or the dominant image of economic production, though in the unavoidable case of China air pollution remains pervasive. Other varieties of toxicity and danger, from mercury to radiation to a warming climate and changing oceans, are also the grim products of the contemporary world. Rachel Carson pushed the world to realize that pollution is very often invisible, a process of imperceptible and slow violence.[15] The causal link between growth and degradation, what Bernard Mandeville identified 300 years ago as the impossibility of a 'cleanly' environment in a 'flourishing' city, is now often seen, working on global scale, as a

---

[14] *Ibid.*, 537, and 535–8 *passim* (7 May 1821) and 654–5 (10 May 1821).
[15] Rob Nixon, *Slow Violence and the Environmentalism of the Poor* (Cambridge, MA, 2011).

Faustian bargain at the heart of modernity.[16] The need for 'sustainability' might be read as the wish to uncouple Mandeville's dyad, a search for cleanly flourishing.

No one in London during the seventeenth or eighteenth centuries, of course, anticipated the grey skies of twenty-first century Beijing, much less global warming. But over the course of the early modern period Londoners, their governors, and those who reflected on the meaning of urban life in Britain came to think about their coal-powered city in ways that provided powerful models for the subsequent period of widespread industrialization. As this book's opening chapters showed, early modern London received most of its energy from coal by about 1600, and almost certainly remained the world's largest fuel consumer throughout the early modern period. This was only possible because of the abundant and accessible coal imported from north-eastern England, the combustion of which filled London's air with pollutants that modern environmental regulators would find unacceptable. The following chapters described a series of contestations: monarchs, ministers, rich and middling inhabitants, physicians, and natural philosophers who all objected in their different ways to London's smoky air. The grounds for these objections varied but also tended to bleed into each other, as smoke was thought to harm bodies, properties, and the dignity of politically-charged spaces of display. Despite these objections, coal became ever-more embedded in London's social relationships and in the state's political economy and security strategy. By 1703 England's queen, who worried about smoke's effects on her own family, announced to Parliament that securing London's coal supply was one of her government's highest priorities. The average Londoner therefore, according to one hostile observer, concluded that though coal smoke was to blame for 'polluting' the city's air, yet a prohibition of the coal trade was clearly 'not to be expected'.[17] Londoners, in response, learned to live with an environment they did not feel they could change. These accommodations took the form of suburban villas and country retreats for the rich, brief excursions out of town for the middling, and very few attractive choices indeed for the poor and the vulnerable. Hymns to the countryside entrenched and popularized a vision of the city as an unavoidably dirty place, something that the virtuous would shun rather than reform.

The polluted city was thus painted as the natural home of the human passions, of lust after money and power and novelty. London's smoky air, in this sense, was seen as a natural outgrowth of human nature. By the end of the early modern period the smoke of London seemed an inevitable result of the

---

[16] Mandeville is discussed above in Chapter 8.
[17] 'Orvietan', f. 8v. Discussed above in Chapter 1.

city's economic and political greatness, its status as the seat of liberty, trade, and modernity. It offered the world a model in which development, power, and environmental degradation would progress together, at least until some future technological innovation finally allows a style of growth that would not entail such tremendous costs.

Bibliography

MANUSCRIPTS

Bedfordshire and Luton Archives and Record Service

L   Lucas MSS
X 800/7   Antonie Papers

Bodleian Library, University of Oxford

Bankes MSS
Carte MSS
Clarendon MSS
Rawlinson MSS

British Library, London

Additional MSS
Cotton MSS
Egerton MSS
Lansdowne MSS
Sloane MSS
Stowe MSS

Cambridge University Library Manuscripts

Ch(H) Political   Cholmondeley (Houghton), Walpole Papers

Cambridgeshire Record Office

P23   St Andrew's Parish Records
P26   St Botolph's Parish Records
P28   St Edward King and Martyr Parish Records
P30   Great St Mary's Parish Records
P52   Croxton Parish Records

239

Cumbrian Record Office

D LONS/W1   Lowthers of Whitehaven Papers

Folger Library

MS L.a.   Bagot Papers

Gloucestershire Record Office

GBR/B3   Gloucester Borough Minutes
P170   Hawkesbury Parish

Guildhall Library, London

MS 3018   St Dunstan's in the West, Wardmote Presentments
MS 3047   Tylers' and Bricklayers' Company Search Books
MS 5442   Brewers Company Warden's Accounts
MS 5445   Brewers Company Court Minute Books
MS 5491   Brewers Company Charity Account Books
MS 5570   Fishmongers Company Court Books
MS 12,830   Christ's Hospital, Carmen's Records
MS 17,872   Ironmongers Company Charities and Estates
MS 25,240   St Paul's Cathedral Dean and Chapter Estates
MS 34,010   Merchant Taylors Company Court of Assistants

Henry E. Huntington Library, San Marino CA

EL   Ellesmere MSS
ST   Stowe MSS

Hertfordshire Archives and Local Studies

AH   Ashridge Collection, Egerton Family Estates and Letters

Kent History and Library Centre

U269/1   Cranfield Papers

Lambeth Palace Library

MS 748   Coal Accounts

London Metropolitan Archives

CLA/47/LJ/1   City of London, Sessions of the Peace
CLA/43/1   City of London, Manor Records

COL/CHD/PR/3/11   City of London, Chamberlain's Department, Poor Relief
COL/AD/1   Corporation of London, Letterbooks
COL/AD/4   Corporation of London, Wardmote Presentments
COL/CA/1/1   Corporation of London, Court of Aldermen Repertories
COL/CC/1/1   Corporation of London, Court of Common Council Journals
COL/RMD/PA/1   Remembrancer's Department Papers
COL/SJ/27   Corporation of London, Viewers Reports
E/BER/CG/L22   Bedford Estate, Covent Garden, Leases
MJ/SBR   Middlesex Sessions of the Peace
P92/SAV   St Saviour's, Southwark Parish Records
LMA WJ/SP   Westminster Quarter Sessions

The National Archives, London

ADM 3   Admiralty: Minutes
ADM 106   Navy Board: Records
C 8   Court of Chancery: Six Clerks Office: Pleadings before 1714
C 11   Court of Chancery: Six Clerks Office: Pleadings 1714 to 1758
C 66   Chancery and Supreme Court of Judicature: Patent Rolls
C 114   Chancery Master's Exhibits
C 115   Chancery Master's Exhibits, Duchess of Norfolk's Deeds (Scudamore MSS)
E 134   Exchequer: King's Remembrancer: Depositions Taken by Commission
E 214   Exchequer: King's Remembrancer: Modern Deeds
KB 27   Court of King's Bench: Plea and Crown Sides: Coram Rege Rolls
LC 5   Lord Chamberlain's Department: Miscellaneous Records
LS 13   Lord Steward's Department: Miscellaneous Books
PC 2   Privy Council Registers
PRO 30   Domestic Records of the Public Record Office, Gifts, Notes, and Transcripts: Original Records Acquired as Gifts or on Deposit
PRO 31   Domestic Records of the Public Record Office, Gifts, Notes, and Transcripts: Transcripts of Records and Scholars' Notes
PROB 11   Prerogative Court of Canterbury and related Probate Jurisdictions: Will Registers
SP 12   State Papers Domestic, Elizabeth I
SP 14   State Papers Domestic, James I
SP 16   State Papers Domestic, Charles I
SP 29   State Papers Domestic, Charles II
SP 42   State Papers Naval
STAC 8   Court of Star Chamber: Proceedings, James I

Norfolk Record Office

PD629   Holme-next-the-Sea Parish Records

Northamptonshire Record Office

E(B)   Ellesmere Brackley Collection

Parliamentary Archives, London

HL/PO/JO/10   House of Lords, Main Papers

Tyne and Wear Archives

GU.TH/21   Trinity House, Newcastle Cash Books

City of Westminster Archive Centre

Acc. 1815   Craven Estate Map
1049   Grosvenor Papers
E   St Margaret Parish Records

PRINTED PRIMARY SOURCES

Periodicals

The Annual Register
The Connoisseur
Exact Diurnall
The Gentleman's Magazine
Le Journal des Sçavans
Lloyd's Evening Post
London Advertiser and Literary Gazette
London Evening Post
The London Gazette
The London Magazine
Mercurius Aulicus
The Tatler
Weekly Account
Weekly Journal or British Gazetteer

Books and pamphlets

An Account of a Dangerous Combination and Monopoly Upon the Collier-Trade (1698).
An Account of the Constitution and Present State of Great Britain (1759).
Acts of the Privy Council of England: New Series, 46 vols. (1890–1964).
Additional Collections Towards the History and Antiquities of the Town and County of Leicester (1790).
An Answer to the Coal-Traders and Consumptioners Case (1689).
Arbuthnot, John. *An Essay Concerning the Effects of Air on Human Bodies* (1733).
Augustine. *Of the Citie of God with the Learned Comments of Io. Lodouicus Viues* (1620).
Baker, J. H. and S. F. C. Milsom, eds. *Sources of English Legal History: Private Law to 1750* (1986).

Baker, Richard. *A Chronicle of the Kings of England, from the Time of the Romans Goverment [sic] unto the Raigne of our Soveraigne Lord, King Charles* (1643).

Baker, Thomas. *Hampstead Heath. A Comedy. As It was Acted at the Theatre Royal in Drury Lane* (1706).

'The Ballad of Gresham College' Stimson, Dorothy, ed. *Isis* 18 (July 1933), 103–17.

Betts, John. *De Ortu et Natura Sanguinis* (1669).

Bibliotheca Digbeiana, sive, Catalogus Librorum in Variis Linguis Editorum quos post Kenelmum Digbeium Eruditiss (1680).

Bickerstaffe, Isaac. *The Plays of Isaac Bickerstaffe*, Peter A. Tasch, ed. (New York, 1981).

Bidwell, William B. and Maija Jansson, eds. *Proceedings in Parliament 1626*, 4 vols. (New Haven, 1991).

Birch, Thomas. *The Heads of Illustrious Persons of Great Britain* (1743).

Birch, Thomas, ed., *A Collection of the Yearly Bills of Mortality, from 1657 to 1758 Inclusive* (1759).

Birch, Thomas and Robert Folkestone Williams, eds. *The Court and Times of Charles the First*, 2 vols. (1848).

Blackstone, William. *Commentaries on the Laws of England* (Oxford, 1765–9).

Boate, Arnold. 'To the Second Letter of the Animadversor' in Samuel Hartlib, ed. *Samuel Hartlib, His Legacy of Husbandry* (1662).

Bohun, Ralph. *Discourse of Winds* (1671).

Bolton, Solomon. *The Present State of Great Britain, and Ireland ... Begun by Mr. Miege; and Now Greatly Improved, Revised and Completed to the Present Time, By Mr. Bolton* (10th ed., 1745).

Bond, Donald F., ed. *The Spectator* (Oxford, 1965).

Boothman, Lyn and Sir Richard Hyde Park, eds. *Savage Fortune: An Aristocratic Family in the Early Seventeenth Century* (Woodbridge, 2006).

Boswell, James. *Boswell's Life of Johnson*, George Birkbeck Hill, ed. 6 vols. (Oxford, 1887).

⸻ *The Correspondence of James Boswell with James Bruce and Andrew Gibb, Overseers of the Auchinleck Estate*, Nellie Pottle Hankins and John Strawhorn, eds. (New Haven, 1998).

Botelho, Lynn, ed. *Churchwardens' Accounts of Cratfield 1640–1660* (Woodbridge, 1999).

Bowman, William. *An Impartial View of the Coal-Trade* (1743).

Boyer, Abel. *The History of King William the Third*, 3 vols. (1702–3).

Boyle, Robert. '*Some Considerations touching the Vsefulnesse of Experimental Naturall Philosophy*' in Hunter and Davis, eds. *Works* (1663).

⸻ 'An Experimental Discourse of Some Little Observed Causes of the Insalubrity and Salubrity of the Air and its Effects' in Hunter and Davis, eds. *Works* (1685).

⸻ 'The General History of Air' in Hunter and Davis, eds. *Works*.

⸻ *The Works of Robert Boyle*, 14 vols. Michael Hunter and Edward B. Davis, eds. (2000).

Breval, John. *The Play is the Plot* (1718).

A Briefe Declaration for what Manner of Speciall Nusance Concerning Private Dwelling Houses . . . (1639).

Brome, Alexander. *Songs and Other Poems* (1664).

Bromley, J. S. ed. *The Manning of the Royal Navy: Selected Public Pamphlets 1693–1873* (1974).

Brooke, Humphrey. *YTIEINH, or a Conservatory of Health* (1650).

Brookes, Richard. *The General Practice of Physic* (1754).

Bullein, William. *The Government of Health* (1595).

Burrow, James. *Reports of Cases Argued and Adjudged in the Court of King's Bench, During the Time of Lord Mansfield's Presiding in that Court* (4th ed., 1790).

Burton, John. *A Treatise on the Non-Naturals. In which the Great Influence They Have on Human Bodies Is Set forth, and Mechanically Accounted for* (York, 1738).

Byron, George Gordon. *Byron. Poetical Works*, Frederick Page, ed. (Oxford, 1970).

Calendar of State Papers Domestic – Anne, 4 vols. (1916–2006).

Calendar of State Papers Domestic – Charles I, 23 vols. (1858–97).

Calendar of State Papers Domestic – Charles II, 28 vols. (1860–1939).

Calendar of State Papers Domestic – Edward, Mary and Elizabeth, 1547–80 (1856).

Calendar of State Papers Domestic – Interregnum, 13 vols. (1875–86).

Calendar of State Papers Domestic – James I, 4 vols. (1857–9).

Calendar of State Papers Domestic – William and Mary, 11 vols. (1895–1937).

Calendar of State Papers Foreign Series, of the Reign of Elizabeth, 23 vols. (1863–1950).

Calendar of State Papers Relating to English Affairs in the Archives of Venice, 38 vols. (1864–1947).

Camden, William. *Britain, or A Chorographicall Description of the Most Flourishing Kingdomes, England, Scotland, and Ireland, and the Ilands Adjoyning, Out of the Depth of Antiquitie* (1637 ed.).

Carlile, James. *The Fortune Hunters, or, Two Fools Well Met* (1689).

Carter, Elizabeth. *Elizabeth Carter, 1717–1806: An Edition of Some Unpublished Letters*, Gwen Hampshire, ed. (Cranbury, NJ, 2005).

The Case of the Glass-Makers in and about the City of London [n.d., c. 1711].

The Case of the Glass-Makers, Sugar-Bakers, and Other Consumers of Coals (1740).

The Case of the Owners and Masters of Ships Imployed in the Coal-Trade (1729).

The Case of the Poor Skippers and Keel-Men of New-Castle, Truly Stated (n.d., c. 1711).

A Catalogue of the Library, Antiquities, &c. of the Late Learned Dr. Woodward (1728).

Centlivre, Susannah. *The Platonick Lady* (1707).

Chamberlayne, Edward. *Angliae Notitia: Or the Present State of England* (1702 ed.).

Chapman, George, Ben Jonson, and John Marston. *Eastward Ho*, in James Knowles, ed. *The Roaring Girl and Other City Comedies* (Oxford, 2001).

Cheyne, George. *The English Malady: Or a Treatise of Nervous Diseases of All Kinds* (1733).

——— *An Essay of Health and Long Life* (10th ed., 1745).

Clavering, James. *The Correspondence of Sir James Clavering*, H. T. Dickinson, ed. (Gateshead, 1963).

Cogan, Thomas. *The Haven of Health* (1634 ed.).

Colman, George. *The Connoisseur. By Mr. Town, Critic and Censor-General*, 4 vols. (1757–60).

(Cooke, William). 'James Smith of Tewkesbury' *The Art of Living in London* (1785 ed.).

Cooper, Anthony Ashley, Earl of Shaftesbury, *Letters from the Right Honourable the Late Earl of Shaftesbury, to Robert Molesworth, Esq.* (1721).

Cowley, Abraham. *The Works of Mr. Abraham Cowley* (1668).

Cox, Thomas. *Magna Britannia Antiqua & Nova*, 6 vols. (1738).

Davenant, Charles. *Discourses upon the Publick Revenues, and Trade of England* (1698).

—— *Essay upon the Probable Methods of Making a People Gainers in the Ballance of Trade* (1699).

Davenant, William. *The Shorter Poems, and the Songs from the Plays and Masques*, A. M. Gibbs, ed. (Oxford, 1972).

—— *The Works of Sir William Davenant*, 2 vols. (1673, reissued; New York, 1968).

Defoe, Daniel. *Tour Thro' the Whole Island of Great Britain*, 2 vols. (1968).

De la Bédoyère, Guy, ed. *Particular Friends: The Correspondence of Samuel Pepys and John Evelyn* (Woodbridge, 1997).

De Saussure, Cesar. *A Foreign View of England in 1725–1729; The Letters of Monsieur Cesar de Saussure to his Family*, Madame van Muyden, ed. (1995).

Dekker, Thomas. *Nevves from Graues-End Sent to Nobody* (1604).

—— *A Strange Horse-Race at the End of Which, Comes in the Catch-Poles Masque* (1613).

Delaune, Thomas. *The Present State of London* (1681).

Dendy, F.W., ed. *Extracts from the Records of the Company of Hostmen of Newcastle-upon-Tyne* (Durham, 1901).

Denham, John. *Cooper's Hill. A Poeme* (1642).

Digby, Kenelm. *A Late Discourse Made in a Solemne Assembly of Nobles and Learned Men at Montpellier in France* (1658).

Dilke, Thomas. *The Lover's Luck* (1696).

Dodsley, Robert. *The footman: An Opera* (1732).

Drayton, Michael. *The Second Part, or a Continuance of Poly-Olbion* (1622).

Dryden, John. *The Poems of John Dryden*, 5 vols., Paul Hammond, ed. (1995).

Duck, Stephen. *Poems on Several Occasions by the Reverend Mr. Stephen Duck* (3rd ed., 1753).

Dudley, Dud. *Metallum Martis OR, IRON Made with Pit-coale, Sea-coale, &c.* (1665).

Dugdale, William. *The History of St Pauls Cathedral in London, from its Foundation untill these Times* (1658).

Ellis, J. M., ed. *The Letters of Henry Liddell to William Cotesworth* (Leamington Spa, 1987).

England's Remarques (1682)

The Englishmans Docter. Or, The Schoole of Salerne (1607).

Eward, Suzanne, ed. *Gloucester Cathedral Chapter Act Book 1616–1687* (Bristol, 2007).

Evelyn, John. *Parallel of the Antient Architecture with the Modern* (1664).

—— *Sylva* (4th ed., 1706).

——*Fumifugium Or, the Inconvenience of the Aer and Smoake of London Dissipated* (1772)

—— *The Diary of John Evelyn*, 6 vols., E. S. De Beer, ed. (Oxford, 1955).

—— *The Writings of John Evelyn*, Guy De la Bédoyère, ed. (Woodbridge, 1995).

Farley, Henry. *The Complaint of Paules to all* (1616)

—— *St. Paules-Church her bill for the Parliament* (1621).

—— *Portland-stone in Paules-Church-yard Their Birth, Their Mirth, Their Thankefulnesse, Their Aduertisement* (1622).

Farquhar, George. *The Works of George Farquhar*, 2 vols., Shirley Strum Kenny, ed. (Oxford, 1988).

A Farther Case Relating to the Poor Keel-men of Newcastle (n.d., c. 1711).

Firth, C. H. and B. S. Rait, eds. *Acts and Ordinances of the Interregnum, 1642–1660*, 3 vols. (1911).

Foote, Samuel. *The Plays of Samuel Foote*, Paula R. Backscheider and Douglas Howard, eds. (New York, 1983).

The Foreigner's Guide: Or, a Necessary and Instructive Companion Both for the Foreigner and Native (1729).

Foster, Elizabeth Read. *Proceedings in Parliament 1610*, 2 vols. (New Haven, 1966)

Fothergill, John. 'Further Remarks on the Treatment of Consumptions, &c. by John Fothergill, M.D. F.R.S' in *Medical Observations and Inquiries. By a Society of Physicians in London*, 6 vols. (1757–84).

Franklin, Benjamin. *The Works of Benjamin Franklin; Containing Several Political and Historical Tracts*, 10 vols., Jared Sparks, ed. (Boston, 1844).

Fransham, John. *The World in Miniature: Or, the Entertaining Traveller* (1740).

Frauds and Abuses of the Coal-Dealers Detected and Exposed: In a Letter to an Alderman of London (3rd ed., 1747).

Fuller, Thomas. *Exanthematologia: Or, An Attempt to Give a Rational Account of Eruptive Fevers* (1730).

Garrick, David. *The Country Girl. A Comedy. (Altered from Wycherley)* (1766).

The Gentleman's Assistant, Tradesman's Lawyer, and Country-man's Friend (1709).

Glapthorne, Henry. *The Lady Mother* (Oxford, 1958).

Graunt, John. *Natural and Political Observations Mentioned in a Following Index, and Made upon the Bills of Mortality* (1662).

Grey, Anchitell. *Debates of the House of Commons, from the Year 1667 to the Year 1694*, 10 vols. (1763).

Grierson, H. J. C. and G. Bullough, eds. *The Oxford Book of Seventeenth Century Verse* (Oxford, 1938).

Grosley, Pierre-Jean. *A Tour to London: Or, New Observations on England, and its Inhabitants*, Thomas Nugent, trans. (1772).

Gwynn, John. *London and Westminster Improved, Illustrated by Plans* (1765).

Hampshire, Gwen, ed. *Elizabeth Carter, 1717–1806: An Edition of Some Unpublished Letters* (Cranbury, NJ, 2005).

Hanway, Jonas. *Serious Considerations on The Salutary Design of an Act of Parliament for a Regular, Uniform Register of the Parish Poor* (1762).

—— *Letters to the Guardians of the Infant Poor* (1767).

Hardy, W. J., ed. *Middlesex County Records. Calendar of Sessions Books 1689–1709* (1905).

Harrison, G. B., ed. *Advice to his Son: By Henry Percy Ninth Earl of Northumberland (1609)* (1930).

Harrison, William. *Harrison's Description of England in Shakspere's Youth*, Frederick J. Furnivall, ed. (1877).

Harvey, Gideon. *Morbus Anglicus: Or, The Anatomy of Consumptions* (1666).

—— *The Art of Curing Diseases by Expectation* (1689).

Harvey, William. *The Works of William Harvey, M.D. Physician to the King, Professor of Anatomy and Surgery to the College of Physicians*, Robert Willis, ed. and trans. (1847).

Hell upon Earth: Or The Town in an Uproar (1729).

Her Majesties Most Gracious Speech to Both Houses of Parliament, on Tuesday the Ninth Day of November, 1703 (1703).

Heylyn, Peter. *Eroologia Anglorum. Or, An Help to English History* (1641).

—— *Cyprianus anglicus, or, The History of the Life and Death of the Most Reverend and Renowned Prelate William, by Divine Providence Lord Archbishop of Canterbury* (1668).

Hill, John. *The Old Man's Guide to Health and Longer Life: With Rules for Diet, Exercise, and Physick* (Dublin, 1760).

His Majesties Declaration to his City of London, Upon Occasion of the Calamity by the Lamentable Fire (1666).

Hodges, Sir William. *Great Britain's Groans: Or, An Account of the Oppression, Ruin, and Destruction of the Loyal Seamen of England* (1695).

Hooke, Nathaniel. *The Secret History of Colonel Hooke's Negotiations in Scotland, in Favour of the Pretender; in 1707* (1760).

Hooke, Robert. *The Posthumous Works of Robert Hooke: With an Introduction by Richard S. Westfall*, The Sources of Sciences, no. 73 (1705; reprint, New York, 1969).

Hopwood, Charles Henry, ed. *Middle Temple Records: Minutes of Parliament Vol. III 1650–1703* (1905).

Houghton, John. *Collection for Improvement of Agriculture and Trade* n. 1 (1692) – n. 305 (1698).

——— *Husbandry and Trade Improv'd: Being a Collection of many Valuable Materials Relating to Corn, Cattle, Coals, Hops, Wool, &c.* (1727).

Howell, James. *Londinopolis* (1657).

——— *Epistolæ Ho-Elianæ. Familiar Letters, Domestick and Foreign* (7th ed., 1705).

Howes, Edmond. *The annales, or a generall chronicle of England, begun first by maister Iohn Stow, and after him continued ... vnto the ende of this present yeere 1614* (1615).

Hull, Charles Henry. *The Economic Writings of Sir William Petty, Together with the Observations Upon the Bills of Mortality, More Probably by Captain John Graunt* (Cambridge, 1899).

Hutton, Richard. *The Reports of that Reverend and Learned Judge, Sir Richard Hutton Knight; Sometimes on of the Judges of the Common Pleas* (1656).

Huygens, Lodewijck. *The English Journal 1651–1652*, A. G. H. Bachrach and R. G. Collmer, eds. and trans. (Leiden, 1982).

Inderwick, F. A., ed. *A Calendar of the Inner Temple Records Vol. II, James I. (1603) – Restoration (1660)* (1898).

Isham, Sir Gyles, ed. *The Correspondence of Bishop Brian Duppa and Sir Justinian Isham, 1650–1660* (Northampton, 1955).

Jacob, Giles. *A New Law-Dictionary: Containing the Interpretation and Definition of Words and Terms Used in the Law* (1756).

Jansson, Maija, ed. *Proceedings in the Opening Session of the Long Parliament House of Commons*, 7 vols. (Rochester, 2000).

Jeaffreson, John Cordy, ed. *Middlesex County Records Volume II. 3 Edward VI to 22 James I* (1887).

Jenner, Charles. *Town Eclogues* (1772).

Jenyns, Soame. *Poems, By * * * * ** (1752).

Jevons, W. Stanley. *The Coal Question: An Inquiry Concerning the Progress of the Nation, and the Probably Exhaustion of our Coal-mines*, A. W. Flux, ed. (3rd ed., 1906).

Johnson, Joshua. *Joshua Johnson's Letterbook 1771–1774: Letters from a Merchant in London to his Partners in Maryland*, Jacob M. Price, ed. (1979).

Johnson, Robert C., Mary Frear Keeler, Maija Jansson Cole, and William B. Bidwell, eds. *Commons Debates 1628*, 4 vols. (New Haven, 1977–1983).

Johnson, Samuel. *Works of the English Poets. Volume LVII* (1790).

————— *The Yale Edition of the Works of Samuel Johnson*, 23 vols. (New Haven, 1958–).

Jones, Inigo and William Davenant. *Britannia Triumphans a masque, presented at White Hall, by the Kings Majestie and his lords, on the Sunday after Twelfth-night, 1637* (1638).

Journals of the House of Commons (1802–).

Journals of the House of Lords (1767–).

Kalm, Pehr. *Kalm's Account of his Visit to England on his Way to America in 1748*, Joseph Lucas, trans. (1892).

Kayll, Robert. *Trade's Increase* (1615).

Keeler, Mary Frear, Maija Jansson Cole, and William B. Bidwell, eds. *Proceedings in Parliament 1628, Volume 6: Appendixes and Indexes* (New Haven, 1983).

King, John. *A Sermon at Paules Crosse, on behalfe of Paules Church, March 26 1620* (1620).

Kinnaston, Francis. *Leoline and Sydanis* (1642).

Larkin, John F., ed. *Stuart Royal Proclamations; Vol. II Royal Proclamations of King Charles I 1625–1646* (Oxford, 1983).

Laud, William. *The Works of the Most Reverend Father in God, William Laud, D.D.*, 7 vols. (Oxford, 1847–60).

Le Clerc, Jean. *The Life and Character of Mr. John Locke* (1706).

Le Hardy, William, ed. *County of Middlesex. Calendar to the Sessions Records*, 4 vols. (1935–41).

A Letter to Sir William Strickland, Bart. Relating to the COAL TRADE (1730).

Lithgow, William. *The Present surveigh of London and Englands State* (1643).

Livock, D. M. ed. *City Chamberlains' Accounts in the Sixteenth and Seventeenth Centuries* (Bristol, 1966).

Locke, John. *The Correspondence of John Locke*, 8 vols., E. S. De Beer, ed. (Oxford, 1976–89).

de Longeville Harcouet, M. *Long Livers: A Curious History of Such Persons of both Sexes Who Have liv'd Several Ages, and Grown Young Again*, Robert Samber, trans. (1722).

Lowndes, Thomas. *A State of the Coal-Trade to Foreign Parts, by Way of Memorial to a Supposed very Great Assembly* (1745).

McIlwain, Charles Howard, ed. *Political Works of James I* (Cambridge, MA, 1918).

Macky, John. *A Journey through England, In Familiar Letters From a Gentleman Here to his Friend Abroad* (1714).

Malynes, Gerald. *Consuetudo, vel lex mercatoria, or the Ancient Law-merchant* (1622).

Mandeville, Bernard. *The Fable of the Bees: And Other Writings*, E. J. Hundert, ed. (Indianapolis, 1977).

Mathew, Francis. *A Mediterranean Passage by Water, from London to Bristol* (1670).

Maynard, Sir John. *The Copy of a Letter Addressed to the Father Rector at Brussels* (1643).

Memoirs of the Life and Character of Mr. John Locke (1742).

Merriman, R. D., ed. *Queen Anne's Navy: Documents Concerning the Administration of the Navy of Queen Anne, 1702–1714* (1961).

Middleton, Thomas. *The Triumphs of Truth* (1613).

Miege, Guy. *The New State Of England Under Their Majesties K. William and Q. Mary* (1693).

————— *The Present State of Great Britain and Ireland* (1723 ed.).

Milton, John. *The Complete English Poems of John Milton*, John D. Jump, ed. (New York, 1964).

The Mischief of the Five Shillings Tax upon COAL (1698)

M.Misson's *Memoirs and Observations in his Travels over England* (1719).

Moffett, Thomas. *Health's Improvement or, Rules Comprizing and Discovering the Nature, Method and Manner of Preparing all Sorts of Foods used in this Nation* (1655 ed.).

Montagu, Mary Wortley. *Lady Mary Wortley Montagu: Essays and Poems and Simplicity, A Comedy*, Robert Halsband and Isobel Grundy, eds. (Oxford, 1977).

(Moore, Edward), ed. 'Adam Fitz-Adam'. *The World* (Dublin, 1755–7).

Morris, Corbyn. *Observations on the Past Growth and Present State of the City of London* (1751).

Morton, Richard. *Phthisiologia, or, A Treatise of Consumptions Wherein the Difference, Nature, Causes, Signs, and Cure of all Sorts of Consumptions are Explained* (1694).

Moryson, Fynes. *Itinerary* (1617).

Motteux, Peter Anthony. *Love's a Jest. A Comedy: Acted at the New Theatre in Little-Lincoln's Inn-Fields* (1696).

Mottley, John. *A Survey of the Cities of London and Westminster, and the Borough of Southwark*, 2 vols. (1754–5).

Mountagu, Edward. *The Journal of Edward Mountagu First Earl of Sandwich Admiral and General at Sea 1659–1665*, R. C. Anderson, ed. (1929).

Muralt, Louis Béat de. *The Customs and Character of the English and French Nations* (1728).

Nixon, George. *An Enquiry into the Reasons of the Advance of the Price of Coals, Within Seven Years Past* (1739).

Noorthouck, John. *A New History of London, Including Westminster and Southwark* (1773).

Notestein, Wallace, Frances Helen Relf, and Hartley Simpson, eds. *Commons Debates 1621*, 7 vols. (New Haven, 1935).

Nourse, Timothy. *Campania Fœlix; or A Discourse of the Benefits and Improvements of Husbandry* (1700).

Oldham, James. *The Mansfield Manuscripts and the Growth of the English Law in the Eighteenth Century*, 2 vols. (Chapel Hill, 1992).

An Order concerning the Price of Coales and the Disposing Thereof, Within the City of London, and the Suburbs, &c. Die Jovis 8. Junii 1643.

An Ordinance ... That no Wharfinger, Woodmonger, or Other Seller of New-Castle Coales ... Shall after the Making Hereof Sell any New-Castle Coales, above the Rate of 23s. The Chaldron (1643).

Ordonaux, John, trans. and ed. *Code of Health of the School of Salernum* (Philadelphia, 1871).

Palmer, Geoffrey. *Les Reports de Sir Gefrey Palmer, Chevalier & Baronet; Attorney General a Son Tres Excellent Majesty le Roy Charles Le Second* (1678).

The Parliamentary Debates from the Year 1803 to the Present Time [Hansard], 41 vols. (1819).

The Parliamentary Debates. New Series, 25 vols. (1820–30).

The Parliamentary History of England, from Its Earliest Period to the Year 1803, 36 vols. (1806–20).

Pechey, John. *The Store-House of Physical Practice being a General Treatise of the Causes and Signs of all Diseases Afflicting Human Bodies* (1695).

—— ed. and trans., *The Whole Works of that Excellent Physician, Dr Thomas Sydenham* (1696).

Penn, William. *A Collection of the Works of William Penn*, 2 vols. (1726).

Pennington, Montagu, ed. *Letters from Mrs. Elizabeth Carter, to Mrs. Montagu, Between the Years 1755 and 1800. Chiefly Upon Literary and Moral Subjects*, 3 vols. (1817).

Pepys, Samuel. *The Diary of Samuel Pepys*, 10 vols., Robert Latham and William Matthews, eds. (Berkeley, 1974).

—— *Samuel Pepys and the Second Dutch War: Pepys's Navy White Book and Brooke House Papers*, Robert Latham, ed., transcribed by William Matthew and Charles Knighton. (Aldershot, 1995).

Petty, William. *A Treatise of Taxes and Contributions* (1662).

Philalethes. *A Free and Impartial Enquiry into the Reason of the Present Extravagent Price of Coal* (1729).

Pitcairn, Archibald. *The Whole Works of Dr. Archibald Pitcairn* (1727).

Pix, Mary. *The Innocent Mistress a Comedy* (1697).

A Plan for the Better Regulating the Coal Trade to London, by Preventing the Fluctuation of the Price (n.d., c. 1750).

Plat, Hugh. *A Discoverie of Certain English Wants* (1595).

—— *A New, Cheape and Delicate Fire of Cole-Balles* (1603).

Pöllnitz, Karl Ludwig, Freiherr von, *The Memoirs of Charles-Lewis, Baron de Pollnitz* (1739).

Popple, William. *The Double Deceit, or A Cure for Jealousy* (1736).

Powell, J. R. and E. K. Timings, eds. *The Rupert and Monck Letter Book 1666* (1969).

The Present State of England. Part III and Part IV Containing; I. An Account of the Riches, Strength, Magnificence, Natural Production, Manufactures of this Island ... II. The Trade and Commerce within It Self (1683).

Pulteney, William (Earl of Bath). *Some Considerations on the National Debts, the Sinking Fund, and the State of Publick Credit: In a Letter to a Friend in the Country* (1729).

Raithby, John. *The Statutes Relating to the Admiralty, Navy, Shipping, and Navigation of the United Kingdom* (1823).

R.C. [Rooke Church], *An olde thrift nevvly reuiued* (1612).

Reasons Humbly Offer'd for Continuing the Clause against Mixing at the Staiths the Coals of Different Collieries (1711).

Reasons Humbly Offer'd to the Honourable the House of Commons, against the Bill for Laying a Duty of 5s per Chaldron upon Coals (n.d. c. 1695).

Reasons Humbly Offered against the Bill for Laying Certain Duties on Glass Wares (n.d., c. 1700).

Reasons Humbly Offered; to Shew, that a Duty upon In-Land Coals, will be no Advantage to His Majesty (n.d., c. 1725).

Reasons, Humbly Offered to the Honourable House of Commons, by the Dyers, against Laying a Further Duty upon Coals (n.d., c. 1696?).

The Records of the Honorable Society of Lincoln's Inn. The Black Books Vol. III From A.D. 1660 to A.D. 1775 W. Pailey Baildon, ed. (1899).

Revolution Politicks: Being a Compleat Collection of all the Reports, Lyes, and Stories Which were the Fore-Runners of the Great Revolution in 1688 (1733).

Robbins, Caroline, ed. *The Diary of John Millward* (Cambridge, 1938).

Royal Commission on Historical Manuscripts. *The Manuscripts of His Grace the Duke of Rutland ... Preserved at Belvoir Castle*, 4 vols. (1888–905)

—— *The Manuscripts of S.H. le Fleming, Esq. of Rydal Hall* (1890)

—— *The Manuscripts of the Corporations of Southampton and King's Lynn* (1887).

—— *The Manuscripts of the Duke of Beaufort, K.G., The Earl of Donoughmore, and Others* (1891).

—— *The Manuscripts of the Earl Cowper, Preserved at Melbourne Hall*, 3 vols. (1888–9).

—— *The Manuscripts of the Earl of Buckinghamshire, the Earl of Lindsey, etc.* (1895).

—— *The Manuscripts of the Marquess Townshend* (1887).

—— *Calendar of the Manuscripts of the Earl of Bath, Preserved at Longleat, Wiltshire*, 5 vols. (1904–80).

—— *Calendar of the Stuart Papers Belonging to His Majesty the King, Preserved at Windsor Castle*, 7 vols. (1902–23).

Rushworth, John, ed. *Historical Collections. The Second Volume of the Second Part* (1721).

Rutt, John, ed. *Diary of Thomas Burton* (1828).

Salkeld, William, ed. *Reports of Cases Adjudg'd in the Court of King's Bench; with Some Special Cases in the Court of Chancery, Common Pleas and Exchequer, from the First Year of K. William and Q. Mary, to the Tenth Year of Queen Anne*, 2 vols. (1717).

Salmon, Thomas. *Modern History: Or, the Present State of all Nations*, 5 vols. (Dublin, 1739).

Sancroft, William, *Familiar Letters of Dr. William Sancroft, Late Lord Archbishop of Canterbury* (1757).

Sea-Coale, Char-Coale, and Small-Coale: Or a Discourse betweene A New-Castle Collier, a Small-Coale-Man, and a Collier of Croydon (1643/4).

The Second Part of the Present State of England Together with Divers Reflections upon the Antient State Thereof (1671).

A Series of Letters Between Mrs. Elizabeth Carter and Miss Catherine Talbot from the Year 1740 to 1770. To Which Are Added, Letters from Mrs. Elizabeth Carter to Mrs. Vesey, Between the Years 1763 and 1767, 4 vols. (1809).

Shelley, Percy Bysshe. *Shelley's Poetry and Prose*, Donald H. Reiman and Sharon B. Powers, eds. (New York, 1977).

Short, Thomas. *A Comparative History of the Increase and Decrease of Mankind in England, and Several Countries Abroad* (1767).

Shower, Bartholomew. *The Second Part of the Reports of Cases and Special Arguments, Argued and Adjudged in the Court of King's Bench* (1720).

Sinclair, George. *Natural Philosophy Improven by New Experiments* (Edinburgh, 1683).

Smith, Adam. *An Inquiry into the Nature and Causes of the Wealth of Nations*, R. H. Campbell and A. S. Skinner, eds. (Oxford, 1976).

Some Considerations Humbly offered to the Honourable House of Commons Against Passing the Bill for Laying a Further Duty of Coals (n.d., c. 1695)

Some Memorials of the Controversie with the Wood-Mongers, or, Traders in Fuel (1680).

Spenser, Edmund. *Prothalamion or a spousall verse* (1596).

Sprat, Thomas. *History of the Royal Society by Thomas Sprat*, Jackson I. Cope and Harold Whitmore Jones, eds. (St. Louis, 1959).

Standish, Arthur. *The Commons Complaint* (1611).

The Statutes at Large, from the Second to the 9th Year of George. [VOL XVI] (Cambridge, 1765).

The Statutes at Large, from the 15th to the 20th Year of King George II. [Vol. XVIII] (1765)

The Statutes of the Realm, 11 vols. (1810–28)

Stedman, Thomas, ed. *Letters to and From the Rev. Philip Doddridge, D.D.* (1790).

Stocks, Helen, ed. *Records of the Borough of Leicester. Being a series of Extracts from the Archives of the Corporation of Leicester, 1603–1688* (Cambridge, 1923).

Stow, John. *A Survey of London by John Stow,* 2 vols., Charles Lethbridge Kingsford, ed. (Oxford, 1908).

Stukeley, William. *The Family Memoirs of the Rev. William Stukeley, M.D.* (Durham, 1887).

Sturtevant, Simon. *Metallica* (1612).

Summers, Montague, ed. *Roscius Anglicanus by John Downes* (n.d.).

Tate, Nahum. *Cuckolds-haven* (1685).

Taylor, John. *The Old, Old, Very Old Man* (1635).

Thirsk, Joan and J. P. Cooer, eds. *Seventeenth Century Economic Documents* (Oxford, 1972).

Tremaine, Sir John, John Rice, and Thomas Vickers. *Pleas of the Crown in Matters Criminal and Civil* (Dublin, 1793).

Tryon, Thomas. *Monthly Observations for the Preserving of Health with a Long and Comfortable Life* (1688).

————*Miscellania: Or, A Collection of Necessary, Useful, and Profitable Tracts* (1696).

————*Tryon's Letters upon Several Occasions* (1700).

Tucker, Josiah. *A Brief Essay on the Advantages and Disadvantages, Which Respectively Attend France and Great Britain, with Regard to Trade.* (1749).

————*The Case of the Importation of Bar-Iron from our own Colonies of North America* (1756).

————*Instructions for Travellers* (Dublin, 1758).

Shadwell, Thomas, *The Complete Works of Thomas Shadwell,* Montague Summers, ed. (1968, ed.).

Uffenbach, Zacharias Conrad von. *London in 1710 From the Travels of Zacharias Conrad von Uffenbach,* eds. and trans. W. H. Quarrell and Margaret Mare (1934).

(Vanbrugh, John) Colley Cibber. *The Provok'd Husband, or, A Journey to London* (1728).

Vanbrugh, John. *The Provok'd Wife a Comedy: As It is Acted at the New Theatre in Little Lincolns-Inn-Fields* (1698).

Venner, Tobias. *Via Recta ad Vitam Longam* (1650).

Walpole, Robert. *Some Considerations Concerning the Publick Funds: The Publick Revenues, and The Annual Supplies, Granted by Parliament* (1733).

Ward, Ned. *British Wonders: Or, a Poetical Description of the Several Prodigies and Most Remarkable Accidents that have happen'd in Britain since the Death of Queen Anne* (1717).

Warwick, Sir Philip. *Memoires of the Reign of King Charles I* (1701).

Watts, Isaac. *Horæ lyricæ. Poems, Chiefly of the Lyric Kind* (1715 ed.)

Well-willer to the prosperity of this famous Common-wealth. *The Two Grand Ingrossers of Coles: viz. the Wood-Monger, and the Chandler* (1653).

Willis, Thomas. *Dr. Willis's Practice of Physick, Being the Whole Works of that Renowned and Famous Physician* (1684).

Wiseman, Richard. *Severall Chirurgicall Treatises* (1676).

The Woman turn'd Bully (1675).

Wordsworth, William. *The Works of William Wordsworth* (Ware, Herts., 1994).
Wren, Christopher. *Life and Works of Sir Christopher Wren. From the Parentalia or Memoirs by His Son Christopher* (1903).
Wroth, William and Arthur Edgar Wroth. *The London Pleasure Gardens of the Eighteenth Century* (1896).
The York-Buildings Dragons (1726).
Young, Arthur. *The Farmer's Letters to the People of England* (1768).
——*Travels During the Years 1787, 1788, and 1789* (Dublin, 1793).
——*The Autobiography of Arthur Young, With Selections from His Correspondence*, Matilda Betham-Edwards, ed. (Cambridge, 2012).
Young, Edward. *Love of Fame, The Universal Passion. In Seven Characteristical Satires* (4th ed., 1741).

SECONDARY SOURCES

Allen, Robert C. *The British Industrial Revolution in Global Perspective* (Cambridge, 2009).
Andriesse, C. D. *Huygens: The Man Behind the Principle* (Cambridge, 2011).
Appadurai, Arjun, ed. *The Social Life of Things: Commodities in Cultural Perspective* (Cambridge, 1986).
Appuhn, Karl. *A Forest on the Sea: Environmental Expertise in Renaissance Venice* (Baltimore, 2009).
Archer, Ian. *The Pursuit of Stability: Social Relations in Elizabethan London* (Cambridge, 1991).
Ashton, T. S. and Joseph Sykes. *The Coal Industry of the Eighteenth Century* (Manchester, 1929).
Ashworth, William J. *Customs and Excise: Trade, Production and Consumption in England 1640–1845* (Oxford, 2003).
Baker, T. F. T. *A History of Middlesex, Volume VII, Acton, Chiswick, Ealing, and Willesden Parishes* (Oxford, 1982).
——*A History of Middlesex Vol. IX, Hampstead and Paddington Parishes* (Oxford, 1989).
Barber, Alex. '"It is Not Easy What to Say of our Condition, Much Less to Write It": The Continued Importance of Scribal News in the Early 18th Century' *Parliamentary History* 32:2 (2013), 293–316.
Barnes, T. G. 'The Prerogative and Environmental Control of London Building in the Early Seventeenth Century' *California Law Review* LVIII (1970), 1332–63.
Barrell, Rex A. *Anthony Ashley Cooper Earl of Shaftesbury (1671–1713) and 'Le Refuge Français' – Correspondence* (Lewiston, MA, 1989).
Barton, Gregory A. *Empire Forestry and the Origins of Environmentalism* (Cambridge, 2002).
Batchelor, Robert. *London: The Selden Map and the Making of a Global City, 1549–1689* (Chicago, 2014).
Baugh, Daniel A. *British Naval Administration in the Age of Walpole* (Princeton, 1965).
——'The Eighteenth Century Navy as a National Institution 1690–1815' in J. R. Hill and Bryan Ranft, eds. *The Oxford Illustrated History of the Royal Navy* (Oxford, 1995), 120–60.
van Bavel, Bas and Oscar Gelderblom, 'The Economic Origins of Cleanliness in the Dutch Golden Age' *Past and Present* 205 (2009), 41–69.

Bending, Stephen. *Green Retreats: Women, Gardens and Eighteenth-Century Culture* (Cambridge, 2013).

Benedict, Philip. *Cities and Social Change in Early Modern France* (1989).

Bennett, Judith M. *Ale, Beer, and Brewsters in England: Women's Work in a Changing World, 1300-1600* (New York, 1996).

Bernhardt, Christoph and Geneviève Massard-Guilbaud, eds. *Le Démon Moderne: La Pollution dans les Sociétés Urbaines et Industrielle d'Europe/ The Modern Demon: Pollution in Urban and Industrial European Societies* (Clermont-Ferrand, 2002).

Bernofsky, Karen. 'Respiratory Health in the Past: A Bioarchaeological Study of Chronic Maxillary Sinusitis and Rib Periostitis from the Iron Age to the Post Medieval Period in Southern England' (Durham University: PhD Thesis, 2010).

Blackbourn, David. *The Conquest of Nature: Water, Landscape, and the Making of Modern Germany* (New York, 2006).

Bohstedt, John. *The Politics of Provisions: Food Riots, Moral Economy, and Market Transition in England, c. 1550–1850* (Farnham, 2010).

Borlik, Todd. *Ecocriticism and Early Modern English Literature: Green Pastures* (New York, 2011).

Borsay, Peter. *The English Urban Renaissance: Culture and Society in the Provincial Town, 1660–1760* (Oxford, 1989).

Boulton, Jeremy. 'The Poor among the Rich: Paupers and the Parish in the West End, 1600–1724' in Paul Griffiths and Mark S. R. Jenner, eds. *Londinopolis: Essays in the Cultural and Social History of Early Modern London* (Manchester, 2000), 197–226.

Boyar, Ebru. 'The Ottoman City: 1500–1800' in Peter Clark, ed. *The Oxford Handbook of Cities in World History* (Oxford, 2013), 275–91.

Brenner, J. F. 'Nuisance Law and the Industrial Revolution' *Journal of Legal Studies* 3 (1974), 403–33.

Brewer, John. *The Sinews of Power: War, Money, and the English State, 1688–1783* (Cambridge, 1988).

——— *The Pleasures of the Imagination: English Culture in the Eighteenth Century* (New York, 1997).

Brimblecombe, Peter. 'London Air Pollution, 1500–1900' *Atmospheric Environment* 11 (1977), 1157–62.

———*The Big Smoke: A History of Air Pollution in London Since Medieval Times* (1987).

Brimblecombe, Peter and Carlotta M. Grossi. 'Millennium-Long Damage to Building Materials in London' *Science of The Total Environment* 407 (February 2009), 1354–61.

Broad, John. 'Parish Economies of Welfare, 1650–1834' *The Historical Journal* 42 (1999), 985–1006.

Brooks, C. W. *Pettyfoggers and Vipers of the Commonwealth: The 'Lower Branch' of the Legal Profession in Early Modern England* (Cambridge, 1986).

Brown, Peter. *Through the Eye of a Needle: Wealth, the Fall of Rome, and the Making of Christianity in the West, 350–550* (Princeton, 2012).

Bruce, Scott G., ed. *Ecologies and Economies in Medieval and Early Modern Europe: Studies in Environmental History for Richard C. Hoffmann* (Leiden, 2010).

Bruijn, J. R. 'Dutch Privateering during the Second and Third Anglo-Dutch Wars' *Acta Historiae Neerlandicae: Studies on the History of the Netherlands* 11 (1978), 79–93.

Burke III, Edmund. 'The Big Story: Human History, Energy Regimes, the Environment' in Edmund Burke III and Kenneth Pomeranz, eds. *The Environment and World History* (Berkeley, 2009), 33–53.

Bushaway, Bob. *By Rite: Custom, Ceremony and Community in England 1700–1880* (1982).

Capp, Bernard. *Cromwell's Navy: The Fleet and the English Revolution, 1648–1660* (Oxford, 1989).

Carpenter, Audrey T. *John Theophilus Desaguliers: A Natural Philosopher, Engineer and Freemason in Newtonian England* (2011).

Cavallo, Sandra and Tessa Storey. *Healthy Living in Late Renaissance Italy* (Oxford, 2013).

Cavert, William. 'The Environmental Policy of Charles I: Coal Smoke and the English Monarchy, 1624-40' *Journal of British Studies* 53:2 (2014), 310–33.

——'The Politics of Fuel Prices' (forthcoming).

——'Industrial Coal Consumption in Early Modern London', *Urban History* (forthcoming).

——'Villains of the Fuel Trade' (forthcoming).

Clark, Gregory and David Jacks. 'Coal and the Industrial Revolution, 1700–1869' *European Review of Economic History* 11 (April 2007), 39–72.

Clark, Peter. *The English Alehouse: A Social History 1200–1830* (1983).

——'"The Ramoth-Gilead of the Good" Urban Change and Political Radicalism at Gloucester 1540–1640' in P. Clark, A. Smith, and N. Tyacke, eds. *The English Commonwealth 1547–1640: Essays in Politics and Society Presented to Joel Hurstfield* (Leicester, 1979), 167–87.

——'The Civic Leaders of Gloucester 1580–1800' in Peter Clark, ed. *The Transformation of English Provincial Towns, 1600–1800* (1984), 311–45.

Claydon, Tony. *William III and the Godly Revolution* (Cambridge, 1996).

Coates, Ben. *The Impact of the English Civil War on the Economy of London, 1642–50* (Aldershot, 2004).

Cockayne, Emily. *Hubbub: Filth, Noise, and Stench in England, 1600–1770* (New Haven, 2007).

Coffman, D'Maris. *Excise Taxation and the Origins of Public Debt* (Basingstoke, 2013).

Coffman, D'Maris, Adrian Leonard, and Larry Neal. *Questioning Credible Commitment: Perspectives on the Rise of Financial Capitalism* (Cambridge, 2013).

Cogswell, Thomas. *The Blessed Revolution: English Politics and the Coming of War, 1621–1624* (Cambridge, 1989).

Coleman, D. C. 'The Economy of Kent under the Later Stuarts' (University of London: PhD Dissertation, 1951).

——'Naval Dockyards under the Later Stuarts' *Economic History Review* 2nd ser. 6:2 (1953), 134–55.

——*Industry in Tudor and Stuart England* (1975).

——'The Coal Industry: A Rejoinder' *Economic History Review* 30 (1977), 343–5.

——*The Economy of England 1450–1750* (Oxford, 1977).

Coquillette, Daniel R. 'Mosses from an Olde Manse: Another Look at Some Historic Property Cases about the Environment' *Cornell Law Review* 74 (1979), 765–72.

Corbin, Alain. *The Foul and the Fragrant* (Cambridge, MA, 1994).

Cowan, Brian. *The Social Life of Coffee: The Emergence of the British Coffeehouse* (New Haven, 2005).

———'Geoffrey Holmes and the Public Sphere: Augustan Historiography from the Post-Namierite to the Post-Habermasian' *Parliamentary History* 28:1 (February 2009), 166–78.

Cronon, William. *Changes in the Land: Indians, Colonists, and the Ecology of New England* (New York, 1983).

———'The Trouble with Wilderness: Or, Getting Back to the Wrong Nature' in William Cronon, ed. *Uncommon Ground: Rethinking the Human Place in Nature* (New York, 1996), 69–90.

Crosby, Alfred. *The Columbian Exchange: Biological and Cultural Consequences of 1492* (Westport, CT, 1972).

Cruickshank, Dan and Neil Burton. *Life in the Georgian City* (1990).

Cruickshanks, Eveline, Stuart Handley, and D. W. Hayton, eds. *The House of Commons, 1690–1715*, 5 vols. (Cambridge, 2002).

Dabhoiwala, Faramerz. *The Origins of Sex: A History of the First Sexual Revolution* (Oxford, 2012).

Dale, Hylton B. *The Fellowship of the Woodmongers: Six Centuries of the London Coal Trade* (n.pub. info, *c.* 1923).

Darley, Gillian. *John Evelyn: Living for Ingenuity* (New Haven, 2006).

Davidson, Andrew, 'John Jones', in Andrew Thrush and John P. Ferris, eds. *The House of Commons 1604–1629* (Cambridge, 2010).

Davis, James. *Medieval Market Morality: Life, Law, and Ethics in the English Marketplace* (Cambridge, 2012).

De Krey, Gary. *London and the Restoration, 1659–1683* (Cambridge, 2005).

———*A Fractured Society: The Politics of London in the First Age of Party, 1688–1715* (Oxford, 1985).

Denton, Peter. '"Puffs of Smoke, Puffs of Praise": Reconsidering John Evelyn's *Fumifugium* (1661)' *Canadian Journal of History* 35 (2000) 441–51.

Dickson, P. G. M. *The Financial Revolution in England: A Study in the Development of Public Credit, 1688–1756* (1967).

Dobbs, Betty Jo. 'Studies in the Natural Philosophy of Sir Kenelm Digby' *Ambix* 18 (1971), 1–25.

Douglas, Mary. *Purity and Danger: An Analysis of Concepts of Pollution and Taboo* (New York, 2006).

Downes, Kerry. 'Wren and Whitehall in 1664' *The Burlington Magazine* 113:815 (February 1971), 89–93.

Dupuis, Melanie, ed. *Smoke and Mirrors: The Politics and Culture of Air Pollution* (New York, 2004).

Earle, Peter. *The Making of the English Middle Class: Business, Society and Family Life in London, 1660–1730* (Berkeley, 1989).

Edmond, Mary. *Rare Sir William Davenant: Poet Laureate, Playwright, Civil War General, Restoration Theatre Manager* (Manchester, 1987).

Elvin, Mark. *Retreat of the Elephants: An Environmental History of China* (New Haven, 2004).

Fewster, J. M. *The Keelmen of Tyneside: Labour Organization and Conflict in the North-east, 1600–1830* (Woodbridge, 2011).

Fincham, Kenneth and Peter Lake. 'The Ecclesiastical Policies of James I and Charles I' in Kenneth Fincham, ed. *The Early Stuart Church 1603–1642* (Basingstoke, 1993), 23–50.

Findlen, Paula, ed. *Early Modern Things: Objects and Their Histories* (New York, 2012).

Flinn, Michael W. *The History of the British Coal Industry. Volume 2 1700-1830: The Industrial Revolution* (Oxford, 1984).

Forbes, Thomas R. 'Weaver and Cordwainer: Occupations in the Parish of St. Giles without Cripplegate, London, in 1654–1693 and 1729–1743' *Guildhall Studies in London History* 4 (1980), 119–32.

Fournier, Patrick. 'De la souillure à la pollution, un essai d'interpretation des origines de l'idee de pollution' in Bernhardt and Massard-Guilbaud, eds. *Le Démon Moderne*, 33–56.

Freese, Barbara. *Coal: A Human History* (Cambridge, MA, 2002).

Fressoz, Jean-Baptiste. *L'Apocalypse Joyeuse: Une Histoire du Risque Technologique* (Paris, 2012).

Galloway, James, Derek Keene, and Margaret Murphy. 'Fuelling the City: Production and Distribution of Firewood in London's Region, 1290–1400' *Economic History Review* 49 (1996), 447–72.

Geltner, Guy. 'Healthscaping a Medieval City: Lucca's Curia viarum and the Future of Public Health History' *Urban History* 40:3 (2013), 395–415.

Gibson, John. *Playing the Scottish Card* (Edinburgh, 1988).

Glacken, Clarence J. *Traces on the Rhodian Shore: Nature and Culture in Western Thought from Ancient Times to the End of the Eighteenth Century* (Berkeley, 1967).

Glass, D. V. 'Two Papers on Gregory King' in D. V. Glass and D. E. C. Eversley, eds. *Population in History: Essays in Historical Demography. Volume I: General and Great Britain* (1965), 159–220.

Golinski, Jan. *British Weather and the Climate of Enlightenment* (Chicago, 2007).

Gowing, Laura. *Common Bodies: Women, Touch and Power in Seventeenth-Century England* (New Haven, 2003).

Greengrass, Mark, Michael Leslie, and Timothy Raylor, eds. *Samuel Hartlib and the Universal Reformation: Studies in Intellectual Communication* (Cambridge, 1994).

Gregg, Edward. *Queen Anne* (New Haven, 2001).

Greig, Hannah. *The Beau Monde: Fashionable Society in Georgian London* (Oxford, 2013).

Griffiths, Paul. *Lost Londons: Crime, Change, and Control in the Capital City 1550–1640* (Cambridge, 2008).

Grove, Richard H. *Green Imperialism: Colonial Expansion, Tropical Island Edens, and the Origins of Environmentalism, 1600–1860* (Cambridge, 1995).

Gugliotta, Angela. '"Hell with the Lid Taken Off:" A Cultural History of Pollution – Pittsburgh' (University of Notre Dame: PhD Dissertation, 2004).

Guillery, Peter. *The Small House in Eighteenth century London* (2004).

Habermas, Jürgen. *The Structural Transformation of the Public Sphere: An Inquiry into a Category of Bourgeois Society*, trans. Thomas Burger (Cambridge, MA, 1989).

Hamlin, Christopher. 'Public Sphere to Public Health: The Transformation of 'Nuisance" in Steve Sturdy, ed. *Medicine, Health, and the Public Sphere in Britain 1600–2000* (London, 2002), 189–204.

Hammersley, G. 'Crown Woods and Their Exploitation in the Sixteenth and Seventeenth Centuries' *Bulletin of the Institute for Historical Research* 30 (1957), 136–61.

———'The Charcoal Iron Industry and Its Fuel, 1540–1750' *Economic History Review* 26 (1973), 593–613.

Hancock, David. *Citizens of the World: London Merchants and the Integration of the British Atlantic Community, 1735–1785* (Cambridge, 1995).

Hanson, Craig Ashley. *The English Virtuoso: Art, Medicine, and Antiquarianism in the Age of Empiricism* (Chicago, 2009).

Harding, Vanessa. 'The Population of London, 1550–1700: A Review of the Published Evidence' *London Journal* 15 (1990), 111–28.

Harkness, Deborah. *The Jewel House* (New Haven, 2007).

Harris, Tim. *The Politics of the London Crowd in the Reign of Charles II* (Cambridge, 1987).

Hatcher, John. *The History of the British Coal Industry. Volume 1, Before 1700; Towards the Age of Coal* (Oxford, 1993).

Hattendorf, John B. *England in the War of the Spanish Succession: A Study in the English View and Conduct of Grand Strategy, 1702–1712* (New York, 1987).

Hausman, William John. 'Public Policy and the Supply of Coal to London, 1700–1770' (University of Illinois at Urbana-Champaign: PhD Thesis, 1976).

Healy, Simon. 'The Tyneside Lobby on the Thames: Politics and Economic Issues, c. 1580–1630' in Diana Newton and A. J. Pollard, eds. *Newcastle and Gateshead before 1700* (Chichester, 2009), 219–40.

Hill, Marquita K. *Understanding Environmental Pollution* (Cambridge, 2010).

Hiltner, Ken. *What Else is Pastoral? Renaissance Literature and the Environment* (Ithaca, 2011)

Hindle, Steve. *On the Parish? The Micro-Politics of Poor Relief in Rural England c. 1550–1750* (Oxford, 2004).

―――― *A History of the County of Essex*, 11 vols. (1903–2012).

―――― *A History of the County of Middlesex*, 13 vols. (1911–2009).

―――― *A History of the County of Staffordshire*, 6 vols. (1906–2014).

Hoffmann, Richard C. *An Environmental History of Medieval Europe* (Cambridge, 2014).

Hoppit, Julian. 'The Nation, the State, and the First Industrial Revolution' *Journal of British Studies*, 50:2 (April 2011), 307–31.

Howard, Jean E. *Theater of a City: The Spaces of London Comedy, 1598–1642* (Philadelphia, 2007).

Hoyle, Richard W., ed. *Custom, Improvement and the Landscape in Early Modern Britain* (Aldershot, 2011).

Hubbard, Eleanor. *City Women: Money, Sex, and the Social Order in Early Modern London* (Oxford, 2012).

Hughes, Edward. *North Country Life in the Eighteenth Century: The North East, 1700–1750* (Oxford, 1952).

Hughes, J. Donald. *An Environmental History of the World: Humankind's Changing Role in the Community of Life* (Abingdon, 2009).

Hundert, E. J. *The Enlightenment's Fable: Bernard Mandeville and the Discovery of Society* (Cambridge, 1994).

Hunt, Margaret. *The Middling Sort: Commerce, Gender, and the Family in England, 1680–1780* (Berkeley, 1996).

Hunter, Michael. 'John Evelyn in the 1650s: A Virtuoso in Quest of a Role' in Therese O'Malley and Joachim Wolscke-Bulmahn, eds. *John Evelyn's 'Elysium Britannicum' and European Gardening* (Washington, 1998), 79–106.

―――― *Boyle: Between God and Science* (New Haven, 2010).

Jackson, Lee. *Dirty Old London: The Victorian Fight Against Filth* (New Haven, 2014).

Jenner, Mark S. R. 'Early Modern Conceptions of Cleanliness and Dirt as Reflected in the Environmental Regulation of London, c. 1530–1700' (Oxford: D. Phil. Thesis, 1992).

———'"Another *epocha*"? Hartlib, John Lanyon and the Improvement of London in the 1650s' in Mark Greengrass, Michael Leslie, and Timothy Raylor, eds. *Samuel Hartlib and the Universal Reformation: Studies in Intellectual Communication* (Cambridge, 1994), 343–56.

——— 'The Politics of London Air: John Evelyn's *Fumifugium* and the Restoration' *The Historical Journal*, 38 (1995), 535–51.

——— 'Death, Decomposition and Dechristianisation? Public Health and Church Burial in Eighteenth-Century England' *English Historical Review* 120 (2005), 615–32.

——— 'Follow Your Nose? Smell, Smelling, and their Histories' *American Historical Review* 116 (2011), 335–51.

——— 'Polite and Excremental Labour: Selling Sanitary Services in London, 1650–1830', paper presented at the Cambridge Early Medicine Seminar, November 2013.

——— 'Print, Publics, and the Rebuilding of London: The Presumptuous Proposal of Valentine Knight' paper presented at the Institute for Historical Research, British History in the Seventeenth Century Seminar, January 2014.

Jones, D. W. *War and Economy in the Age of William III and Marlborough* (Oxford, 1988).

Jonsson, Fredrik Albritton. *Enlightenment's Frontier: The Scottish Highlands and the Origins of Environmentalism* (New Haven, 2013).

Jørgensen, Dolly. '"All Good Rule of the Citee": Sanitation and Civic Government in England, 1400–1600' *Journal of Urban History* 36:3 (2010), 300–15.

Joy, Neill R. 'Politics and Culture: The Dr. Franklin – Dr. Johnson Connection, with an Analogue' *Prospects* 23 (1998), 59–105.

Kander, Astrid, Paolo Malanima, and Paul Warde. *Power to the People: Energy in Europe over the Last Five Centuries* (Princeton, 2014).

Kasuga, Ayuka. 'Views of Smoke in England, 1800–1830' (University of Nottingham: PhD Thesis, 2013).

——— 'The Introduction of the Steam Press: A Court Case on Smoke and Noise Nuisances in a London Mansion, 1824' *Urban History* 42:3 (August 2015), 405–23.

Kent, Joan and Steve King. 'Changing Patterns of Poor Relief in some Rural English Parishes Circa 1650–1750' *Rural History* 14:2 (October 2003), 119–56.

Keynes, Geoffrey. *John Evelyn: A Study in Bibliophily and a Bibliography of his Writings* (Cambridge, 1937).

Knights, Mark. *Politics and Opinion in Crisis, 1678–81* (Cambridge, 1994).

Kohl, Benjamin and Ronald G. Witt. *The Earthly Republic: Italian Humanists on Government and Society* (Manchester, 1978).

Kyle, Chris. 'Prince Charles in the Parliaments of 1621 and 1624' *Historical Journal* 41:3 (September 1998), 603–24.

——— *Theater of State: Parliament and Political Culture in Early Stuart Britain* (Stanford, 2012).

Lake, Peter. 'The Laudian Style: Order, Uniformity, and the Pursuit of the Beauty of Holiness in the 1630s' in Kenneth Fincham, ed. *The Early Stuart Church, 1603–1642* (Basingstoke, 1993), 161–85.

——— and Steven Pincus, eds. *The Politics of the Public Sphere in Early Modern England* (Manchester, 2007).

Larwood, Jacob and John Camden Hotten. *The History of Sign-Boards, From the Earliest Times to the Present Day* (1866).

Le Roux, Thomas. *Le Laboratoire des Pollutions Industrielles. Paris 1770–1830* (Paris, 2011).

Levine, David and Keith Wrightson. *The Making of an Industrial Society: Whickham, 1560–1765* (Oxford, 1991).

Loengard, Janet 'The Assize of Nuisance: Origins of an Action at Common Law' *The Cambridge Law Journal* 47 (1978), 144–66.

Luu, Lien Bich. *Immigrants and the Industries of London, 1500–1700* (Aldershot, 2005).

McCann, James C. *Maize and Grace: Africa's Encounter with a New World Crop, 1500–2000* (Cambridge, MA, 2005).

McClain, James. 'Japan's Pre-Modern Urbanism' in Peter Clark, ed. *The Oxford Handbook of Cities in World History* (Oxford, 2013), 328–45.

McCloskey, Deirdre. *Bourgeois Dignity: Why Economics Can't Explain the Modern World* (Chicago, 2010).

McColley, Diane Kelsey. *Poetry and Ecology in the Age of Milton and Marvell* (Aldershot, 2007).

McCormick, Ted. *William Petty and the Ambitions of Political Arithmetic* (Oxford, 2009).

McHenry, Robert W. 'Dryden and the 'Metropolis of Great Britain"' in W. Gerald Marshall, ed. *The Restoration Mind* (1997), 177–92.

McIntosh, Marjorie. *Poor Relief in England, 1350–1600* (Cambridge, 2012).

McKellar, Elizabeth. *Landscapes of London: The City, The Country and the Suburbs, 1660–1840* (New Haven, 2013).

McLaren, J. P. S. 'Nuisance Law and the Industrial Revolution – Some Lessons from Social History' *Oxford Journal of Legal Studies* 3, no. 2 (1973), 155–221.

McNeill, J. R. *Something New Under the Sun: An Environmental History of the Twentieth-Century World* (New York, 2001).

——— *Mosquito Empires: Ecology and War in the Greater Caribbean, 1620–1914* (Cambridge, 2010).

Mallainathan, Sendhil and Eldar Shafir. *Scarcity: Why Having Too Little Means So Much* (New York, 2013).

Manley, Lawrence, ed. *The Cambridge Companion to the Literature of London* (Cambridge, 2011).

Manning, Roger B. *Village Revolts: Social Protest and Popular Disturbances in England, 1509–1640* (Oxford, 1988).

Marin, Brigitte. 'Town and Country in the Kingdom of Naples' in S. R. Epstein, ed. *Town and Country in Europe, 1300–1800* (Cambridge, 2004), 316–31.

Melville, Elinor G. K. *A Plague of Sheep: Environmental Consequences of the Conquest of Mexico* (Cambridge, 1994).

Merritt, Julia. 'Puritans, Laudians, and the Phenomenon of Church-Building in Jacobean London' *Historical Journal* 41 (1998), 935–60.

——— *The Social World of Early Modern Westminster: Abbey, Court and Community, 1525–1640* (Manchester, 2005).

——— *Westminster 1640–60: A Royal City in a Time of Revolution* (Manchester, 2013).

Mikhail, Alain. *Nature and Empire in Ottoman Egypt: An Environmental History* (Cambridge, 2011).

Mokyr, Joel. *The Enlightened Economy: An Economic History of Britain 1700–1850* (New Haven, 2009).

Morag-Levine, Noga. *Chasing the Wind: Regulating Air Pollution in the Common Law State* (Princeton, 2003).

Moran, Bruce T. *Distilling Knowledge: Alchemy, Chemistry, and the Scientific Revolution* (Cambridge, MA, 2006).

Mosley, Stephen. *The Chimney of the World: A History of Smoke Pollution in Victorian and Edwardian Manchester* (Cambridge, 2001).

Mukherjee, Ayesha. *Penury Into Plenty: Dearth and the Making of Knowledge in Early Modern England* (2014).

Muldrew, Craig. *The Economy of Obligation: The Culture of Credit and Social Relations in Early Modern England* (New York, 1998).

—— *Food, Energy, and the Creation of Industriousness: Work and Material Culture in Agrarian England, 1550–1780* (Cambridge, 2011).

Munro, Ian. *The Figure of the Crowd in Early Modern London: The City and Its Double* (Basingstoke, 2005).

Naquin, Susan. *Peking Temples and City Life, 1400–1900* (Berkeley, 2000).

Nash, Roderick. *Wilderness and the American Mind* (New Haven, 1967).

Neeson, Janet. *Commoners: Commons Right, Enclosure and Social Change in England 1700–1820* (Cambridge, 1993).

Nef, J. U. *The Rise of the British Coal Industry*, 2 vols. (1932).

Newman, Karen. *Cultural Capitals: Early Modern London and Paris* (Princeton, 2007).

Newman, William R. *Atoms and Alchemy: Chymistry and the Experimental Origins of the Scientific Revolution* (Chicago, 2006).

Nixon, Rob. *Slow Violence and the Environmentalism of the Poor* (Cambridge, MA, 2011).

North, Douglass C. and Barry R. Weingast. 'Constitutions and Commitment: The Evolution of Institutions Governing Public Choice in Seventeenth-Century England' *Journal of Economic History* 49 (1989), 803–32.

Ogborn, Miles. *Spaces of Modernity: London's Geographies, 1680–1780* (1998).

—— *Global Lives: Britain and the World 1550–1800* (Cambridge, 2008).

Ogilvie, Sheilagh and Merkus German. *European Proto-Industrialization: An Introductory Handbook* (Cambridge, 1996).

O'Neill, Lindsay. *The Opened Letter: Networking in the Early Modern British World* (Philadelphia, 2014).

Ormrod, David. *The Rise of Commercial Empires: England and the Netherlands in the Age of Mercantilism 1650–1770* (Cambridge, 2003).

Owen, John Hely. *War at Sea under Queen Anne 1702–1708* (Cambridge, [orig. 1938], 2010).

Parker, Geoffrey. *Global Crisis: War, Climate Change, and Catastrophe in the Seventeenth Century* (New Haven, 2012).

Parthasarathi, Prasannan. *Why Europe Grew Rich and Asia Did Not: Global Economic Divergence, 1600–1850* (Cambridge, 2011).

Peacey, Jason. *Print and Politics in the English Revolution* (Cambridge, 2013).

Peck, Linda Levy. *Consuming Splendor: Society and Culture in Seventeenth Century England* (Cambridge, 2005).

Perdue, Peter. *China Marches West: The Qing Conquest of Central Eurasia* (Cambridge, MA, 2005).

Perez, Louis G. *Daily Life in Early Modern Japan* (Westport, CT, 2002).

Pettit, Philip A. J. *The Royal Forests of Northamptonshire: A Study in their Economy 1558–1714* (Gateshead, 1968).

Pincus, Steven. '"Coffee Politicians Does Create": Coffeehouses and Restoration Political Culture' *Journal of Modern History* 67:4 (December 1995), 807–34.

—— *Protestantism and Patriotism: Ideologies and the Making of English Foreign Policy, 1650–1668* (Cambridge, 1996).

———— 'John Evelyn: Revolutionary' in Frances Harris and Michael Hunter, eds. *John Evelyn and His Milieu* (2003), 185–220.

————*1688: The First Modern Revolution* (New Haven, 2009).

Pomeranz, Kenneth. *The Great Divergence: China, Europe, and the Making of the Modern World Economy* (Princeton, 2000).

Ponting, Clive. *A New Green History of the World: The Environment and the Collapse of Great Civilisations* (2011).

Porter, Roy. *English Society in the Eighteenth Century* (1991).

———— *Enlightenment: Britain and the Creation of the Modern World* (2000).

———— *London: A Social History* (Cambridge, MA: Harvard University Press, 2001).

Porter, Stephen. *The Great Fire of London* (Stroud, 1996).

Principe, Lawrence. *The Secrets of Alchemy* (Chicago, 2013).

———— 'Sir Kenelm Digby and His Alchemical Circle in 1650s Paris: Newly Discovered Manuscripts' *Ambix* 60:1 (February 2013), 3–24.

Provine, D. M. 'Balancing Pollution and Property Rights: A Comparison of the Development of English and American Nuisance Law' *Anglo-American Law Review* 7 (1978), 31–56.

Rackham, Oliver. *History of the Countryside* (Dent, 1986).

Radkau, Joachim. *Nature and Power: A Global History of the Environment*, trans. Thomas Dunlap (Cambridge, 2008).

Rawcliffe, Carole. *Urban Bodies: Communal Health in Late Medieval English Towns and Cities* (Woodbridge, 2013).

Raychaudhuri, Tapan and Irfan Habib, eds. *The Cambridge Economic History of India. Vol. 1: c. 1200-c. 1750* (Cambridge, 1982).

Reinke-Williams, Tim. *Women, Work and Sociability in Early Modern London* (Basingstoke, 2014).

Richards, John F. *The Unending Frontier: An Environmental History of the Early Modern World* (Berkeley, 2006).

Righter, Robert W. *The Battle over Hetch Hetchy: America's Most Controversial Dam and the Birth of Modern Environmentalism* (Oxford, 2005).

Ritvo, Harriet. *The Dawn of Green: Manchester, Thirlmere, and Modern Environmentalism* (Chicago, 2009).

Roberts, Charlotte and Keith Manchester. *The Archaeology of Disease* (Ithaca, NY, 2007).

Robertson, J. 'Stuart London and the Idea of Royal Capital City' *Renaissance Studies* 15:1 (March 2001), 37–58.

Robertson, John. *The Case for Enlightenment: Scotland and Naples 1680–1760* (Cambridge, 2005).

Roche, Daniel. *A History of Everyday Things: The Birth of Consumption in France* (Cambridge, 2000).

Rodger, N. A. M. *The Command of the Ocean: A Naval History of Britain 1649–1815* (New York, 2006).

Rogers, Pat, ed. *The Samuel Johnson Encyclopedia* (Westport, CT, 1996).

Rome, Adam W. 'Coming to Terms with Pollution: The Language of Environmental Reform, 1865–1915' *Environmental History* 1 (1996), 6–28.

Røstvig, Maren-Sofie. *The Happy Man: Studies in the Metamorphoses of a Classical Ideal. Volume II 1700–1760* (Oslo, 1971).

Rowe, William T. 'China: 1300–1900' in Peter Clark, ed. *The Oxford Handbook of Cities in World History* (Oxford, 2013), 310–27.

Rusnock, Andrea. *Vital Accounts: Quantifying Health and Population in Eighteenth-Century England and France* (Cambridge, 2002).
—— 'Hippocrates, Bacon, and Medical Meteorology at the Royal Society, 1700–1750' in David Cantor, ed. *Reinventing Hippocrates* (Aldershot, 2002), 136–153.
Schaffer, Simon. 'Measuring Virtue: Eudiometry, Enlightenment, and Pneumatic Medicine' in Roger French and Andrew Cunningham, eds. *The Medical Enlightenment of the Eighteenth Century* (Cambridge, 1990), 281–318.
Schama, Simon. *The Embarrassment of Riches: An Interpretation of Dutch Culture in the Golden Age* (Berkeley, 1988).
Selwood, Jacob. *Diversity and Difference in Early Modern London* (Farnham, 2010).
Sharpe, Kevin. 'The Image of Virtue: The Court and Household of Charles I, 1625–1642' in *Politics and Ideas in Early Stuart England: Essays and Studies* (1989), 147–73.
—— *Image Wars: Promoting Kings and Commonwealths in England, 1603–1660* (New Haven, 2010).
—— *Rebranding Rule: The Restoration and Revolution Monarchy, 1660–1714* (New Haven, 2013).
Sharpe, William. 'London and Nineteenth-Century Poetry' in Lawrence Manley, ed. *The Cambridge Companion to the Literature of London* (Cambridge, 2011), 119–41.
Skelton, Leona, *Sanitation in Urban Britain, 1560–1700* (2015).
Slack, Paul. *The Impact of Plague in Tudor and Stuart England* (Oxford, 1990).
—— 'Great and Good Towns 1540–1700' in Peter Clark, ed. *The Cambridge Urban History of Britain. Volume II 1540–1840* (Cambridge, 2000), 347–76.
—— *The Invention of Improvement: Information and Material Progress in Seventeenth-Century England* (Oxford, 2015).
Smil, Vaclav. *General Energetics: Energy in the Biosphere and Civilization* (New York, 1991).
Smuts, Malcolm. 'The Court and its Neighborhood: Royal Policy and Urban Growth in the Early Stuart West End' *Journal of British Studies* 30:2 (1991), 117–49.
Snell, K. D. M. *Annals of the Labouring Poor: Social Change and Agrarian England, 1660–1900* (Cambridge, 1985).
Spencer, J. R. 'Public Nuisance – A Critical Examination' *Cambridge Law Journal* 48 (1989), 55–84.
Stewart, Larry. *The Rise of Public Science: Rhetoric, Technology, and Natural Philosophy in Newtonian Britain, 1660–1750* (Cambridge, 1992).
Stobbart, Jon. *Sugar and Spice: Grocers and Groceries in Provincial England 1650–1830* (Oxford, 2012).
Stolberg, Michael. *Ein Recht auf saubere Luft? Umweltkonflikte am Beginn des Industriezeitalters* (Erlangen, 1994).
Stone, Lawrence. 'The Residential Development of the West End of London in the Seventeenth Century' in Barbara Malament, ed. *After the Reformation: Essays in Honor of J.H. Hexter* (Manchester, 1980), 167–212.
Stradling, David. *Smokestacks and Progressives: Environmentalists, Engineers, and Air Quality in America, 1881–1951* (Baltimore, 2002).
Studnicki-Gizbert, Daviken *Survey of London*, 49 vols. (1900–).
Studnicki-Gizbert, Daviken and David Schecter. 'The Environmental Dynamics of a Colonial Fuel-Rush: Silver Mining and Deforestation in New Spain, 1522–1810' *Environmental History* 15:1 (2010), 94–119.

Sweet, Rosemary. *The Writing of Urban Histories in Eighteenth-Century England* (Oxford, 1997).

Szechi, Daniel. *1715: The Great Jacobite Rebellion* (New Haven, 2006).

Tarr, Joel. *The Search for the Ultimate Sink: Urban Pollution in Historical Perspective* (Akron, OH, 1996).

Te Brake, William H. 'Air Pollution and Fuel Crises in Preindustrial London, 1250–1650' *Technology and Culture* 16 (1975), 337–59.

Temin, Peter and Hans-Joachim Voth. *Prometheus Shackled: Goldsmith Banks and England's Financial Revolution after 1700* (Oxford, 2012).

Thick, Malcolm. *The Neat House Gardens: Early Market Gardening Around London* (Totnes, 1998).

Thomas, Keith. *Man and the Natural World: Changing Attitudes in England 1500–1800* (New York, 1983).

———— 'Cleanliness and Godliness in Early Modern England' in Anthony Fletcher and Peter Roberts, eds. *Religion, Culture, and Society in Early Modern Britain: Essays in Honour of Patrick Collinson* (Cambridge, 1994), 56–83.

Thorsheim, Peter. *Inventing Pollution: Coal, Smoke, and Culture in Britain since 1800* (Athens, OH, 2006).

Thrupp, Sylvia. *A Short History of the Worshipful Company of Bakers of London* (n.p. info, 1933).

Thrush, Andrew. 'Naval Finance and the Origins and Development of Ship Money' in Mark Charles Fissel, ed. *War and Government in Britain, 1598–1650* (Manchester, 1991), 133–62.

Thrush, Andrew and John P. Ferris, eds. *The House of Commons 1604–1629*, 6 vols. (Cambridge, 2010).

Thrush, Coll. 'The Iceberg and the Cathedral: Encounter, Entanglement, and Isuma in Inuit London' *Journal of British Studies* 53:1 (2014), 59–79.

Thurley, Simon. *The Whitehall Palace Plan of 1670* (London, 1998).

———— 'A Country Seat Fit for a King' in Eveline Cruickshanks, ed. *The Stuart Courts* (Thrupp, 2000), 214–39.

Totman, Conrad. *The Green Archipelago: Forestry in Pre-Industrial Japan* (Berkeley, 1989).

———— *Early Modern Japan* (Berkeley, 1993).

———— *Japan: An Environmental History* (2014).

Travis, Toby. '"Belching Forth Their Sooty Jaws": John Evelyn's Vision of a 'Volcanic' City' *London Journal* 39 (March 2014), 1–20.

Tudor-Craig, Pamela. '*Old St Paul's*' *The Society of Antiquaries Diptych, 1616* (2004).

Tyacke, Nicholas. *Anti-Calvinists: The Rise of English Arminianism* (Oxford, 1987).

Uekotter, Frank. *The Age of Smoke: Environmental Policy in Germany and the United States, 1880–1970* (Pittsburgh, 2009).

Underdown, David. *Fire From Heaven: Life in an English Town in the Seventeenth Century* (New Haven, 1985).

Unger, Richard W. 'Energy Sources for the Dutch Golden Age: Peat, Wind, and Coal' *Research in Economic History* 9 (1984), 221–253.

Walmsley, Jonathan. 'John Locke on Respiration' *Medical History* 51 (2007), 453–76.

Walter, John. 'The Social Economy of Dearth in Early Modern England' in John Walter and Roger Schofield, eds. *Famine, Disease and the Social Order in Early Modern Society* (Cambridge, 1989), 75–128.

Warde, Paul. *Energy Consumption in England and Wales, 1560–2000* (Naples, 2007).

────── 'The Environmental History of Pre-Industrial Agriculture in Europe' in Paul Warde and Sverker Sörlin, eds. *Nature's End: History and the Environment* (Houndmills, UK: Palgrave Macmillan, 2009), 70–92.

────── 'The Idea of Improvement, c. 1520–1700' in Richard W. Hoyle, ed. *Custom, Improvement, and the Landscape in Early Modern Britain* (Farnham, Surrey, 2011), 127–48.

────── 'The Invention of Sustainability' *Modern Intellectual History* 8 (2011), 153–70.

────── 'Global Crisis of Global Coincidence?' *Past and Present* 228 (2015), 287–301.

Wear, Andrew. 'Place, Health, and Disease: The *Airs, Waters, Places* Tradition in Early Modern England and North America' *Journal of Medieval and Early Modern Studies* 38:3 (Fall 2008), 443–65.

Webster, Charles. *The Great Instauration: Science, Medicine, and Reform 1620–1660* (New York, 1975).

Wennerlind, Carl. *Casualties of Credit: The English Financial Revolution, 1620–1720* (Cambridge, MA, 2011).

Wheeler, Jo. 'Stench in Sixteenth-Century Venice' in Alexander Cowan and Jill Steward, eds. *The City and the Senses: Urban Culture since 1500* (Aldershot, 2007), 25–38.

White, Jerry. *London in the Eighteenth Century: A Great and Monstrous Thing* (2013).

White, Sam. *The Climate of Rebellion in the Early Modern Ottoman Empire* (Cambridge, 2011).

Whitehead, Mark. *State, Science and the Skies: Governmentalities of the British Atmosphere* (Oxford, 2012).

Willan, T. S. *The English Coasting Trade 1600–1750* (Manchester, 1967).

Williams, Laura. '"To Recreate and Refresh their Dulled Spirites in the Sweet and Wholesome Ayre": Green Space and the Growth of the City' in Julia Merritt, ed. *Imagining Early Modern London: Perceptions and Portrayals of the City from Stow to Strype, 1598–1720* (Cambridge, 2001), 185–213.

Williams, Raymond. *The Country and the City* (1973).

Wilson, Kathleen. *The Sense of the People: Politics, Culture, and Imperialism in England, 1715–1785* (Cambridge, 1995).

Winfield, P. H. 'Nuisance as a Tort' *Cambridge Law Journal* 4 (July, 1931) 189–206.

Wing, John T. *Roots of Empire: Forests and State Power in Early Modern Spain, c.1500–1750* (Leiden, 2015).

Winn, James Anderson. *Queen Anne: Patroness of the Arts* (Oxford, 2014).

Wolhcke, Anne. *The 'Perpetual Fair': Gender, Disorder, and Urban Amusement in Eighteenth-Century London* (Manchester, 2014).

Wood, Andy. *The Memory of the People: Custom and Popular Senses of the Past in Early Modern England* (Cambridge, 2013).

Wood, Diana. *Medieval Economic Thought* (Cambridge, 2002).

Woodward, Donald. 'Straw, Bracken and the Wicklow Whale: The Exploitation of Natural Resources in England since 1500' *Past and Present* 159 (May 1998), 43–76.

Wrightson, Keith. *English Society 1580–1680* (New Brunswick, NJ, 2000).

────── *Earthly Necessities: Economic Lives in Early Modern Britain* (New Haven, 2000).

────── 'Mutualities and Obligations: Changing social relationships in early modern England' *Proceedings of the British Academy* 139 (2007), 157–94.

Wrigley, E. A. 'A Simple Model of London's Importance in Changing English Society and Economy 1650–1750' *Past and Present* 37 (1967), 44–70.

—— *Energy and the English Industrial Revolution* (Cambridge, 2010).

Zahedieh, Nuala. *The Capital and the Colonies: London and the Atlantic Colonies* (Cambridge, 2010).

DIGITAL SOURCES

The Adams Papers Digital Edition, (University of Virginia Press, 2008), http://rotunda.upress.virginia.edu/founders/ADMS

Biomass Energy Centre, Calorific Values of Fuels, www.biomassenergycentre.org.uk/portal/page?_pageid=75,20041&_dad=portal&_schema=PORTAL

British History Online, www.british-history.ac.uk

City of London, *2013 Air Quality Progress Report*, www.cityoflondon.gov.uk/business/environmental-health/environmental-protection/air-quality/Pages/air-quality-reports.aspx

Clean Air Asia, CitiesACT Database, http://citiesact.org/data/search/aq-data

The Hartlib Papers 2nd ed. (Sheffield, HROnline, 2002).

The History of Parliament. The House of Commons, 1715–54, www.historyofparliamentonline.org/research/parliaments/parliaments-1715–1754

The House of Commons, 1754–90, www.historyofparliamentonline.org/research/parliaments/parliaments-1754–1790

Indoor Air Pollution, Cookstoves, www.epa.gov/iaq/cookstoves/index.html

National Ambient Air Quality Standards, www.epa.gov/ttn/naaqs/standards/so2/s_so2_history.html

Oxford Dictionary of National Biography, www.oxforddnb.com

Palmer, Robert. 'The level of litigation in 1607: Exchequer, King's Bench, Common Pleas' http://aalt.law.uh.edu/Litigiousness/Litigation.html

The Papers of Benjamin Franklin, franklinpapers.org

People's Republic of China Ambient Air Quality Standards, Clean Air Asia, http://cleanairinitiative.org/portal/node/8163

United Kingdom Department for Food, Environment, and Rural Affairs National Air Quality Objectives, http://uk-air.defra.gov.uk/assets/documents/National_air_quality_objectives.pdf

United States Environmental Protection Agency, Air Quality Information: Six Common Pollutants, www.epa.gov/airquality

World Health Organization, Children's Environmental Health, http://www.who.int/ceh/risks/cehair/en. Indoor Air Pollution; www.who.int/ceh/risks/cehair/en. Indoor Air Pollution; www.who.int/entity/indoorair/en

Index

CPSIA information can be obtained
at www.ICGtesting.com
Printed in the USA
LVOW13s1300031117

554894LV00017B/375/P